OPTIMUM VITAMIN NUTRITION

for More Sustainable Aquaculture

OPTIMUM VITAMIN NUTRITION

NUTRITION

for More Sustainable Aquaculture

A. Liu, A. Dumas, E. Santigosa, G. Litta and J.M. Hernandez

5m Books

First published 2024

Copyright © dsm-firmenich 2024

Published by
5m Books Ltd,
Lings, Great Easton,
Essex CM6 2HH, UK,
Tel: +44 (0)330 1333 580
www.5mbooks.com

Follow us on
Twitter @5m_Books
Instagram 5m_books
Facebook @5mBooks
LinkedIn @5mbooks

A Catalogue record for this book is available from the British Library

ISBN 9781789183467
eISBN 9781789183511
DOI 10.52517/9781789183511

Book layout by Cheshire Typesetting Ltd, Cuddington, Cheshire
Printed by Bell & Bain Ltd, Glasgow
Photos by the authors unless otherwise indicated

CONTENTS

Authors and acknowledgements

Angela Liu[1,2]
[1] The Center for Aquaculture Technologies Canada
[2] School of Biological, Earth and Environmental Sciences

André Dumas[1,2]
[1] The Center for Aquaculture Technologies Canada
[2] AD Aquaculture Nutrition Services Inc.

Ester Santigosa, Gilberto Litta and José Maria Hernandez
dsm-firmenich, Animal Nutrition and Health

The authors are grateful to Louise Buttle and Sebastien Rider from dsm-firmenich Animal Nutrition and Health for their contributions and advice.

Abbreviations

AChE	acetylcholinesterase
ACP	acyl carrier protein
ADP	adenosine diphosphate
AGPAT	acylglycerophosphate acyltransferase
ALT	alanine aminotransferase
ANOVA	analysis of variance
ARA	arachidonic acid
AST	aspartate aminotransferase
ATP	adenosine triphosphate
BA	betaine aldehyde
BADH	betaine aldehyde dehydrogenase
BCAA	branched-chain amino acids
BHMT	betaine homocysteine methyltransferase
BW	body weight
CAT	catalase
CC	choline chloride
CDP	cytidine diphosphocholine
ChAT	choline acetyltransferase
ChDH	choline dehydrogenase
CK	choline kinase
CMP	cytidine monophosphate
CNS	central nervous system
CoA	constituent of acetyl-coenzyme
CPT	choline phosphotransferase
CRALBP	cellular retinaldehyde binding protein
CS	choline-sufficient
CTP	cytidine triphosphate
DAG	diacylglycerol
DBP	d binding protein
DHA	docosahexaenoic
DMG	dimethylglycine
ELISA	enzyme-linked immunosorbent assay
EPA	protected eicosanoic acid
ETKA	erythrocyte transketolase activity
FAD	flavin adenine dinucleotide
FBP	folate-binding proteins
FCR	feed conversion ratio
FMN	flavin mononucleotide
GGCX	γ-carboxyglutamyl carboxylase
GHG	greenhouse gas

GIFT	genetically improved farmed tilapia
GLUT	glucose transporters
GOT	glutamate–oxaloacetate transaminase
GPD	glycerophosphodiesterase
GPT	glutamic pyruvic transferase
GPX	glutathione peroxidase
GSR	glutathione reductase
GST	glutathione-S-transferase
GWT	gutted weight tonnes
HC	haptocorrin
HDL	high-density lipoproteins
HPLC	high-performance liquid chromatography
HSP	heat shock protein
IgA	immunoglobulin A
IgG	immunoglobulin G
IgM	immunoglobulin M
IHNV	infectious hematopoietic necrosis virus
IU	international unit
LC	liquid chromatography
LDL	low-density lipoproteins
LF	low fishmeal
LMS	lipid malabsorption syndrome
LT	lethal time
MAT	methionine adenosyltransferase
MDA	malondialdehyde
MNB	menadione nicotinamide bisulfite
MS	mass spectrometry
mTHF	methyl tetrahydrofolate
NAD	nicotinamide adenine dinucleotide
NADP	nicotinamide adenine dinucleotide phosphate
NADPH	nicotinamide adenine dinucleotide phosphate
NLR	non-linear regression
NRC	National Research Council
PABA	para-aminobenzoic acid
PAF	platelet-activating factor
PAH	polycyclic aromatic hydrocarbons
PC	pyruvate carboxylase
PCT	phosphocholine cytidylyltransferase
PE	phosphatidylethanolamine
PEMT	phosphatidylethanolamine-N-methyltransferase
PER	protein efficiency ratio
PK	phylloquinones
PL	post-larvae
PLA	phospholipase A
PLD	phospholipase D
PLP	pyridoxal-5'-phosphate
PM	pectoralis major
PMP	pyridoxamine-5'-phosphate

PN	pyridoxol or pyridoxine PNP
PNS	peripheral nervous system
PPI	inorganic pyrophosphate
PTH	parathyroid hormone
PUFA	polyunsaturated fatty acids
RA	retinyl acetate
RAR	retinoic acid receptors
RBP	retinol-binding protein
RE	retinol equivalent
RIA	radioimmunoassay
RNA	ribonucleic acid
ROS	reactive oxygen species
RXR	retinoid X receptor
SAH	S-adenosylhomocysteine
SAM	S-adenosylmethionine
SD	standard deviation
SDG	Sustainable Development Goals
SM	sphingomyelin
SMase	sphingomyelinase
SOD	superoxide dismutase
SRS	salmonid rickettsial syndrome
TAN	total ammonia nitrogen
TBARS	thiobarbituric acid-reactive substance
TC	transcobalamin
TCA	tricarboxylic acid
THC	thiamine hydrochloride
THF	tetrahydrofolate
TMP	thiamine monophosphate
TPP	thiamine pyrophosphate
TTP	thiamine triphosphate
UA	unionized ammonia
USP	United States Pharmacopeia
UV	ultraviolet
VKD	vitamin K-dependent
VLDL	very-low-density-lipoprotein

Chapter 1

Contribution of vitamin nutrition to a more sustainable aquaculture farming

ADDRESSING THE CHALLENGES OF TODAY AND TOMORROW

Today, well into the 21st century, the crucial issues relating to food production are changing. Key concepts such as productivity and efficiency continue to be of vital importance. Increasingly, the emphasis is on the significance of strategies related to sustainability, animal health and welfare, food quality, and decreased food waste.

Everything indicates that continuous development in the field of animal nutrition is becoming essential to meet current and future challenges, such as:

- unpredictable production conditions, e.g., higher sea temperatures, increased frequency of marine heatwaves and algal bloom, increase in stocking densities, increase in fish handling
- higher incidence of more aggressive animal diseases and parasites, e.g., increased challenge from sea lice
- the call to reduce and replace antibiotics
- an industry trend in increasing robustness
- increased animal welfare and environmental protection requirements (social and regulatory challenges)
- the impact of new raw materials in feed
- the demand for improving product nutritional value for end consumers
- the goal to reduce food waste significantly
- a growing focus on a more sustainable farming in which our industry has a critical role to play in shaping a better world, in line with the Sustainable Development Goals (SDGs) agreed by the United Nations.

Aquafeed contributes 60–80% of the animal footprint. Therefore, it is important for the feed industry to lead by example, constantly seeking to reduce the carbon and environmental footprint of products and processes in order to reduce its impact. That means closely managing absolute greenhouse gas (GHG) emissions and energy efficiency in all phases of the production, starting with the feed footprint. A growing number of companies are setting long-term goals, validated by the Science Based Targets initiative (SBTi) aligned with the 2015 Paris agreement on climate change to reach net-zero emissions before 2050. Ambitious and long-term challenges in which the optimal use of vitamins in animal nutrition can be part of the solution.

COMMITMENT TO SUSTAINABILITY

Providing the right levels of high-quality and sustainably produced vitamins to feed mills, integrators, and farmers help them improve animal health, well-being, and performance while also protecting the environment, succeeding in a dynamic and ever-changing global market.

Optimizing the performance and improving the sustainability of feed additives and pre-mixes plays an important role in reducing the environmental challenges of animal protein production. Excipients such as rice hulls and calcium carbonate can make up to 50% of a premix composition, but little can be done to reduce their carbon footprint further. The critical products with the potential to contribute to carbon footprint and environmental impact reductions in premixes are the nutritional supplements like vitamins.

Reducing the impact of vitamins and other feed additives operations might enable feed mills and farmers to become more sustainable, reduce their risk profile, and potentially benefit from the value created from future carbon tax savings (Figure 1.1).

Carbon pricing – whether implemented as a tax or cap and trade system – seeks to reduce GHG emissions by putting a direct financial liability on industries and activities that are large GHG emitters. It is a policy intervention to encourage the reduction of harmful activities. The sustainability of mainstream animal production is increasingly questioned as demand rises. International bodies agree that animal protein production is one of the activities that need to reduce carbon emissions if we want to solve the climate crisis. Reductions in the impact of agricultural and animal production processes can be supported by greater use of sustainably produced nutritional solutions.

Additionally, governments may seek to impose low-carbon product standards, further environmental regulation, or tax schemes on animal protein production or products as an incentive to reduce emissions and steer consumer consumption. The groundwork for these interventions is already being laid.

The agricultural and animal protein products supply chain prepares to minimize the risk posed by these changes or face severe financial penalties. In consequence, the industry wants to significantly reduce animal production's impact on the climate and the environment.

This transition to a lower-carbon future for animal proteins can be facilitated by supplying nutritional ingredients such as vitamins together with industry-leading solutions to improve

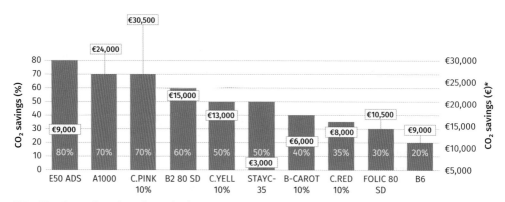

*Value of CO_2 saving assuming a carbon credit or tax of €50/ton CO_2

Figure 1.1 Carbon dioxide savings and potential value (carbon tax) per 10 t of feed additive: dsm-firmenich product *vs.* the main alternative (source: dsm-firmenich Animal Nutrition and Health, unpublished)

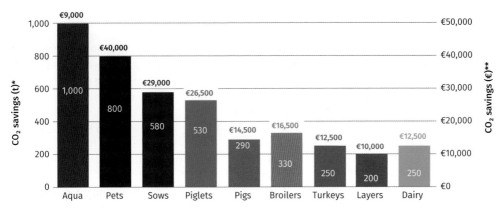

* Examples of CO₂ savings based on average vitamin levels used in 100k t feed
** Value of CO₂ saving assuming a carbon credit or tax of €50/ton CO₂

Figure 1.2 Carbon dioxide savings and potential value (carbon tax) of a more sustainable vitamin source for 100,000 t of feed produced (source: dsm-firmenich Animal Nutrition and Health, unpublished)

the efficiency of animal production systems while reducing the carbon footprint of these feed additives (Figure 1.2), as vitamin nutrition stands as an essential component for more sustainable aquaculture. Vitamins can contribute to at least 3 United Nations SDGs: 2 (Zero hunger), 3 (Good health and well-being), and 13 (Climate action).

UPDATING THE NUTRITIONAL STANDARDS OF VITAMINS IN A CONSTANTLY EVOLVING WORLD

Vitamins play a decisive role in both human and animal nutrition. As organic catalysts present in small quantities in most foods, they are essential for the normal functioning of metabolic and physiological processes such as growth, development, health, and reproduction. The requirements for vitamins in animals are dynamic: they vary according to physiological status, new genotypes, levels of yield, and production systems. Vitamin functions and requirements are becoming increasingly studied.

The concept of OVN Optimum Vitamin Nutrition® for animals is essential today. Its objective is to develop a new standard for vitamin supplementation in feed in order to improve the animal health status and resilience to diseases and environmental stress, which will translate into better animal productivity. Moreover, the quality of food produced by those animals can be enhanced, improving human health and reducing food waste. The latter is critical in a global society in which, unfortunately, many people still do not have access to the correct quantity and quality of food.

When we talk about optimum vitamin supplementation in the diet of animals, we refer to vitamin levels both over and above the established minimum requirements for avoiding deficiencies and adapted to the specific conditions of each animal species for achieving the objectives mentioned above.

Historically, the objective of the vitamin recommendations provided by international scientific organs – such as the National Research Council, USA (NRC, 2011) – was preventing nutritional shortages or deficiencies. Some of the studies on which they are based on are over 40 years old. We all know that today's aquaculture industry has massively evolved compared to the industry we had a few decades ago (Figure 1.3): the type of raw ingredient

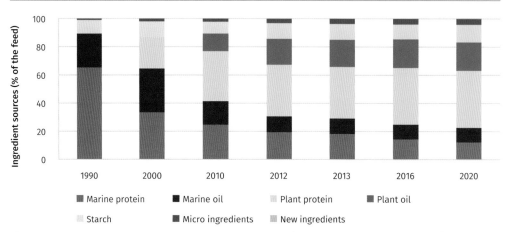

Figure 1.3 Changes in salmonid feed formulation, 1990–2022 (source: adapted Aas *et al.*, 2022)

sources has drastically changed, bringing changes in the vitamin contribution of new raw materials to the final feeds (Figure 1.4) and economic feed conversion ratios (eFCR) (Figure 1.5) have improved due to genetic and feed formulation efforts. Since 2010, the improvement in FCR for aqua species has slowed down and shows a lower rate of change than earlier. Nevertheless, in the last 10 years an FCR improvement of around 25% (2–3% per year) applies to some aqua species (Glencross *et al.*, 2023), highlighting the need to review vitamin requirements under todays commercial farming conditions more than in the land-based animal industry.

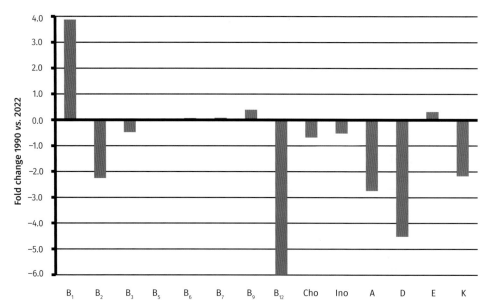

Figure 1.4 Increase or decrease in vitamin content in salmon feed in 2022 compared to feeds from 1990. Changes are related to the use of different raw materials, notably the reduction of marine ingredients, replaced by an increase of vegetable proteins and oils (source: Dumas, 2022)

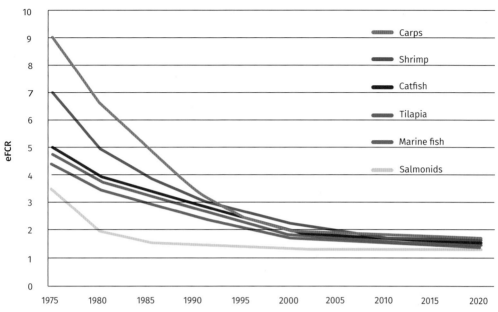

Figure 1.5 Changes in economic feed conversion ratios (eFCR) of major aquaculture species groups over the past 45 years (source: Glencross *et al.*, 2023)

Therefore, it is logical to infer that nutrition programs for farmed animals must be adjusted, including vitamin supplementation, consistently with improved animal management techniques and genetic make-up.

The case of pantothenic acid illustrates how an improved salmon FCR requires adjusted vitamin levels. With 40 mg pantothenic acid/kg diet and an FCR of 1.3 fish are receiving 52 mg/kg body weight. With an FCR improvement to 1.0 and the same amount of pantothenic acid in the diet, salmon receive 40 mg/kg body weight. This means that pantothenic acid level in feed should increase by around 20% to get the same vitamin intake with such an improvement in FCR. Both changes call for vitamin supplementation levels in aquaculture diets to be reviewed.

Likewise, in recent times there have been important legislative changes around the world that limit or ban the use of compounds such as antibiotics, substances that were regular additives to animals' diets as well as of the diet of the animal trials that vitamins requirements were based on. At the same time, many countries are developing new rules on animal welfare which, in short to medium term, will entail less "intensiveness" in the aquaculture industry, aiming to improve the animal health and well-being. In such a context, recent data show that adequate vitamin levels in the diet of fish and shrimp contribute not only to maximized growth but also to overall increased robustness, reduced oxidative stress, better wound healing, and at the same time increase fillet quality and reduce food loss and waste.

From the nutritional point of view, in these fast-changing circumstances, so different from those we have become accustomed to over the last decades, it is essential to re-evaluate the vitamin requirements of animals with the aim of safely and efficiently producing healthy and nourishing food that meets consumer expectations, always under sustainable farming practices.

VITAMINS: ESSENTIAL MICRONUTRIENTS IN THE ANIMAL ORGANISM

Vitamins are unique and crucial nutrients in the diet of people and animals. They are important elements in the organism's vital functions: maintenance, growth, development, health, and reproduction. They also combine 2 characteristics.

- The daily requirement for each of the vitamins is very small (mg or µg) an aspect in which they differ from macronutrients, such as carbohydrates, fats, and proteins.
- Vitamins are organic compounds, unlike other essential nutrients such as minerals (iron, iodine, zinc, etc.).

The discovery of vitamins and their function in preventing classical deficiency diseases are milestones that stand among the most important achievements of the last century. Vitamins are particularly important because they allow optimum metabolism of other nutrients in the animal diet. In general, humans and animals need to derive them from their diet as they cannot produce the appropriate quantities by themselves. Vitamins are present in many reactions of the cellular metabolism and play, particularly in combination, a critical role in biochemical pathways such as the Krebs or citric acid cycle (Figure 1.6).

Vitamins are present in trace amounts in the feed, but they are present in 100% of metabolic functions. This fact gives them the status of micronutrients of macro importance. Vitamins are found in minimal quantities in most feedstuffs. Their absence from the diet gives rise to specific deficiency diseases because of their significance for normal metabolism functioning.

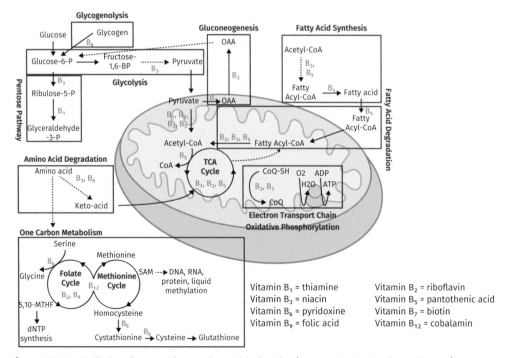

Figure 1.6 Metabolic functions and interactions of B vitamins (source: Godoy-Parejo *et al.*, 2020)

While the need to provide additional vitamins in feed is unquestioned, the levels of supplementation needed to achieve an optimum economic return in field conditions are open to debate. As a general rule, the optimum economic supplementation level is the one which achieves the best index of growth, feed conversion, health status – including immunocompetence – and which, in addition, provides the reserves appropriate for the organism. Nutrition is optimal when an animal efficiently utilizes the nutrients provided in the feed for survival, health, growth, and reproduction. Although all the nutrients, including proteins, fats, carbohydrates, minerals, and water, are essential for carrying out these vital functions, vitamins play a key role in basic functions such as an appropriate immune response in animals.

As mentioned already, several factors – e.g., increased productivity, intensive farming, and higher handling and susceptibility to diseases – give rise to a growing vulnerability to vitamin deficiencies and a yield below the maximal potential. A great majority of nutritionists and investigators recognize that the minimum vitamin requirements needed to prevent clinical deficiency symptoms may not be sufficient to achieve an optimum state of health and yield. In Chapter 4 we will review the multiple metabolic functions specific to each of the vitamins in greater depth.

VITAMIN LEVELS IN ANIMAL DIET: THE NUTRITIONIST'S GREAT UNKNOWN QUANTITY

Establishing the vitamin supplementation level should be an area of focus for all nutritionalists. Economic cost and benefit must be a fundamental reason for revising and determining vitamin supplements in feed. The cost of supplementing feed with essential vitamins must be assessed considering the risk of suffering losses from deficiency symptoms and productive yields below the optimum.

Unlike other species – where the great challenge for nutritionists is choosing a particular level from the numerous recommended tables – vitamin requirements for aquaculture are still underdeveloped and lack cohesion (Figure 1.7). The NRC in the USA periodically publishes nutritional recommendations for different species. Latest NRC recommendations (NRC, 2011) for salmonids show that not all vitamins had an established requirement. Since then, some of those gaps have been filled thanks to the EU-funded research project Arraina.

However, the aquaculture agree that the knowledge is fragmented and generally based on scientific research that uses a "reductionist" approach as recommendations are generally based on one single life stage, using healthy fish in optimal culture conditions. In addition, normally only 1 or 2 response criteria are analyzed with a single vitamin at graded levels. Ideally, challenged fish under suboptimal conditions should be studied using different nutrient matrixes in the investigation design and choosing multiple response criteria (e.g., robustness, gut health, immunity, in 2 or more life stages).

Nevertheless, NRC recommendations generally constitute reference sources of limited value from the viewpoint of commercial feed formulation as they are based on establishing vitamin levels necessary to prevent clinical deficiency symptoms in optimal trial conditions. We know, however, that external stressors can have an impact on the performance and health of the animals. Naderi et al. (2017a) demonstrated that the use of high levels of vitamin E improves the performance and health of chronically stressed fish after an acute stress event (Figure 1.8).

Salmon exposed to salmonid rickettsial syndrome (SRS) challenge experienced in high mortality rates. The use of high vitamin E inclusion in the feeds improved fish survival. This result was repeated when vitamin E levels were increased from 0.17 to 0.75 g/kg in diets containing 1.05, 1.73, and 3.73 g/kg arachidonic acid (Figure 1.9).

Vitamin	Atlantic salmon	Rainbow trout
Vitamin A	NT	0.75
Vitamin D$_3$*	NT	40
Vitamin E	60	50
Vitamin K	<10	R
Vitamin B$_1$	NT	1
Vitamin B$_2$	NT	4
Vitamin B$_6$	5	3
Pantothenic acid	NT	20
Niacin	NT	10
Biotin	NT	0.15
Vitamin B$_{12}$	NT	R
Folic acid	NT	1
Choline	NT	800
Vitamin C	20	20

Figure 1.7 Vitamin recommendations for Atlantic salmon and rainbow trout in seawater show that gaps are still to be closed for fish vitamin nutrition (NT: not tested; R: required but not determined) (source: NRC, 2011)

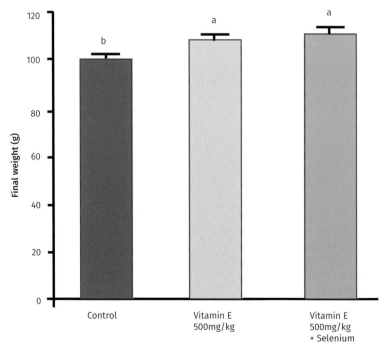

Figure 1.8 Supplementation of vitamin E can exert positive effects on the welfare of chronically stressed rainbow trout subjected to an additional acute stressor (source: Naderi et al., 2017a)

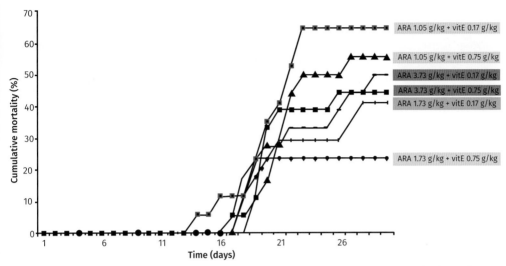

Figure 1.9 Benefits of increased levels of vitamin E on survival in fish exposed to salmonid rickettsial syndrome pathogen (source: Dantagnan *et al.*, 2017)

To make more efficient use of the NRC's vitamin recommendations, it is advisable to take into account the following considerations.

- The indicated levels were established to prevent deficiencies in the animal.
- They do not include any kind of safety margin to prevent loss of vitamin activity stemming from usual feed processing or feed storage conditions. In other words, the recommended NRC levels must be those present at feed intake and when it is eating it.
- They do not include safety margins for the eventuality that the animals are subjected to some sort of stress or subclinical disease.
- They do not consider possible adverse environmental conditions, such as high temperatures or increased handling, which may reduce the animal's food consumption.
- In most cases, they are not specific to the new animal genotypes and improved farming practices.
- NRC recommendations are in many cases based in diet formulations that no longer reflect today's formulation practices because the vitamin contribution from the raw materials has massively changed in the last decades (Figure 1.4).

There is a great disparity between the levels of supplementation prescribed by the industry and those indicated by NRC (2011). While the industry continues to adjust vitamin supplementation in feed to achieve an optimum yield and state of health in the animal, the NRC (2011) has introduced only a few minor changes for most animal species in the last few decades. Logically, the vitamin needs established decades ago do not apply to today's animals. Most nutritionists agree on this aspect, and, in general, commercial inclusion of many vitamins is delivered at levels 3 or 10 times higher than those recommended by the NRC (2011).

BIOAVAILABILITY OF VITAMINS IN ANIMALS

Many feed materials used in animal nutrition contain variable quantities of vitamins. Hence, depending on the ingredients used to create the diet, the vitamin levels in the diets can vary considerably. In general, the overall content in the final diet is low, and their presence in the feed does not guarantee their bioavailability or that the animal will indeed benefit from them.

Other causes of variability in the vitamin levels in feed materials are the geographical origin, time of harvesting, and climatic conditions at each harvest. Moreover, long storage periods and the use of preservatives, fungicides, etc., negatively affect the vitamin content. A non-exhaustive list of factors that will most adversely affect the level of vitamins in feed raw materials are:

- origin of the harvest
- use of fertilizers
- genetic modifications which increased productive yield
- climate
- agricultural practices such as crop rotation
- harvesting conditions
- storage conditions and the use of preservatives
- vitamin form and consequently vitamin stability and bioavailability.

The content of vitamins in feed is determined by complex chemical and microbiological analysis methods in authorized laboratories, which provide the real value at a given time for a certain sample or batch of the respective feed. But given the great number of factors that affect the stability of vitamins (temperature, humidity, light, and moisture), it would be necessary to undertake costly systematic analyses of the principal feed materials to be able to use those values reliably in the formulation – at minimum cost – of the feed, with the constant adaptation of the values to avoid possible variations of the desired level.

In many cases, vitamins in the feedstuffs are present in compound forms that are not bioavailable, hence not available to be absorbed and participate in the animal's different physiological and metabolic processes.

In commercial products, vitamins are formulated within a protective matrix with the aim to prevent losses from feed production and from aggressive environmental agents during storage. Therefore, it is essential to consider the bioavailability of these substances when determining the vitamin content of any feed ingredient. Figure 1.10 shows the commercial form of vitamin A – A 1000 prod. Z – is not bioavailable as the other 2 preparations in a commercial broiler feed.

In contrast, in nature, both in vegetable and animal origin feed materials, substances can effectively destroy the vitamin activity or limit their bioavailability. These antinutritional agents can also be released by certain types of bacteria or fungi – e.g., mycotoxins – as by-products of their metabolic activity, as well as being present in the normal environment of the production facilities. Their most frequent mechanism of action consists of deactivating the free form of the vitamin or preventing its absorption.

The addition of fats and oils as an energy source is common practice in the manufacture of feed. Attention should be paid to the total content of unsaturated fatty acids since they increase the likelihood that the oils and fats will become rancid and impact the absorption of fat-soluble vitamins. In fact, when adding unsaturated fats, it is advisable to increase the supplementation level of vitamin E to avoid oxidation and rancidity.

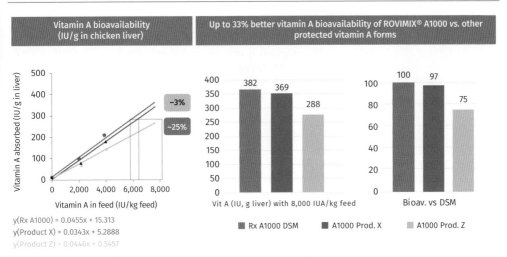

y(Rx A1000) = 0.0455x + 15.313
y(Product X) = 0.0343x + 5.2888
y(Product Z) = 0.0446x + 0.5457

Figure 1.10 Impact of different commercial forms of ethoxyquin-free vitamin A on vitamin A bioavailability in broilers (source: Ludwig-Maximilian University, Munich, unpublished)

STABILITY OF VITAMINS IN ANIMAL FEED

Diverse factors can affect the stability of substances as unstable as vitamins, whether in their pure commercial form, in vitamin-mineral premixes, or after the manufacture of compound feed and its subsequent storage. Some of these factors relate to the catalytic activity of the molecules themselves, the handling of commercial forms and their premixes, the characteristics of the blend, the presence of various antagonistic substances, and the storage conditions.

The vitamins present in feed materials are very susceptible to the adverse conditions mentioned above, and a heavy reduction of vitamin activity is a common occurrence in these macro ingredients. In contrast, the highest-quality commercial forms of vitamins are produced from industrial processes, which stabilize and protect the active molecules of active substance during manufacture and storage, both premixes and the feed.

Feed manufacturing exposes vitamins to high temperatures during grinding, pre-conditioning, extrusion, pelleting, post-conditioning, and drying. Heating is often by steam, and the combination of high temperature and high humidity can be particularly damaging to unprotected vitamins. Sometimes organic acids are added to feeds, and several B vitamins can be damaged if standard vitamin forms are combined with acids. Marchetti *et al.* (1999) compared the stability of fat-coated and un-coated crystalline vitamins during both pelleting and extrusion of fish feed (Figure 1.11). Vitamins most vulnerable to damage were vitamin C, vitamin K_3 (menadione), folic acid, and pyridoxine.

However, it must again be emphasized that the stabilization of the vitamins must not compromise their bioavailability in the animal. Different methods are used for stabilizing vitamins.

- **Use of antioxidants** – Antioxidants are included in the formulation of commercial vitamin products to prevent fat-soluble vitamin oxidation and prolong compound shelf life. In general, those commercial forms with an appropriate quantity of antioxidant substances have a longer effective shelf life. This period, during which vitamin content is guaranteed, will depend to a great extent on storage conditions. The ban of ethoxyquin in the EU and some

Vitamin	Before (%)	After pelleting (%)	After extrusion (%)
Ascorbic acid	100[a]	52[b]	20[c]
Biotin	100	95	94
Cobalamin	100	88	85
Folic acid	100[a]	64[b]	48[b]
Menadione	100[a]	51[b]	34[c]
Nicotinamide	100	96	92
Pantothenic acid	100	89	86
Pyridoxine	100[a]	76[b]	66[b]
Riboflavin	100	87	86
Thiamine	100	82	88

Note: Values within a row with different superscript letters are significantly different ($p < 0.05$).

Figure 1.11 Comparison of the content of un-coated crystalline vitamins in feeds before and after extrusion and pelleting (source: Marchetti *et al.*, 1999)

Additive	Temperature	Oxygen	Humidity	Light
Vitamin A	++	++	+	++
Vitamin D	+	++	+	+
25OHD$_3$ (calcifediol)	++	++	+	+
Vitamin E	0	+	0	+
Vitamin K$_3$	+	+	++	0
Vitamin B$_1$	+	+	+	0
Vitamin B$_2$	0	0	+	+
Vitamin B$_6$	++	0	+	+
Pantothenic acid	+	0	+	0
Nicotinates	0	0	0	0
Biotin	+	0	0	0
Folic acid	++	0	+	++
Vitamin C	++	++	++	0

++ Marked effect
+ Moderate effect
0 No effect

Figure 1.12 External factors influencing the stability of non-formulated vitamins (source: dsm-firmenich Animal Nutrition, dsm-firmenich Product Forms: Quality feed additives for more sustainable farming, 2022a)

other countries has put additional pressure on vitamin, premix, and feed manufacturers to adapt their product formulation technologies and sources to guarantee the right content of the active substance in the products they market.

- **Mechanical methods** – The process, in this case, covers the active substance with a stabilizing coat. This covering protects the vitamin molecule inside it from the adverse effect of aggressive external agents such as the presence of oxygen, ultraviolet radiation, sunlight, humidity, different temperatures, etc. Figure 1.12 shows that all of the fat-soluble vitamins (A, D, E, K), vitamin B$_1$, and vitamin C are sensitive to oxidizing agents such as inorganic

Physical characteristics	ROVIMIX® Biotin 2% SD	ROVIMIX® Biotin HP 10% SD	Biotin 2% Triturate A
Soluble in water	yes	yes	no
Flowability (sec-100g)	medium	medium	low flow, tapping required
Average practical size (µm)	66	73	296
Mixability in feed, CV %	6%	5.7%	10.8%
Total particles per g product	>21 mio	>20 mio	>10 mio
Active particles per g product	100%	100%	2% (98% carrier)
Active Biotin particles per animal/day @ 0.2 mg biotin/Kg feed @ 10 g feed/day/chick	2000	400	20

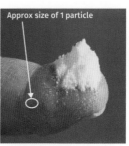

Approx size of 1 particle

Biotin *consumed by* **1 sow** *(or 20 birds, or 1 dog or 4 salmon) per year*

Figure 1.13 Confrontation of physical characteristics of biotin spray-dried form against biotin triturate (source: dsm-firmenich Animal Nutrition and Health, unpublished)

minerals. On a practical level, this method has proved highly effective in protecting these substances and, depending on their characteristics, can be combined with a process of spray drying, which provides a large number of active particles (all with the active form of the vitamin) which facilitates a subsequent homogenous mixture in the animal's food. This is particularly relevant for vitamins like biotin or vitamin B_{12} added into feed at very low levels (Figure 1.13).

OPTIMUM VITAMIN NUTRITION® IN PRACTICE

The objective of OVN™ is to supplement the diet of animals with the amounts of each vitamin considered most appropriate to optimize the state of health and the productivity of farmed fish or crustacean species while guaranteeing the efficiency (desired effect at minimum cost) of the recommended levels.

As already outlined, the levels of supplementation required for OVN™ are generally higher than those necessary to prevent clinical deficiency symptoms (Liu *et al.*, 2022). These optimum supplementation levels should likewise compensate for the stress factors affecting the animal and its diet, thus guaranteeing they do not limit the animal's productivity and health.

OVN™ describes the concept of a cost-effective window for vitamin supplementation. This level must satisfy but not exceed the aim of achieving a state of optimum health and productivity. Below are some definitions of terms applicable to the OVN™ concept (Figure 1.14).

1 **Average animal response** refers to productivity results – FCR, growth rate, reproductive status, level of immunity, the animal's health status, etc. – as a consequence of the ingestion of vitamins.

2 **Total vitamin intake** describes the total level of vitamins in the diet, feed raw material bioavailable quantity, and supplementation.

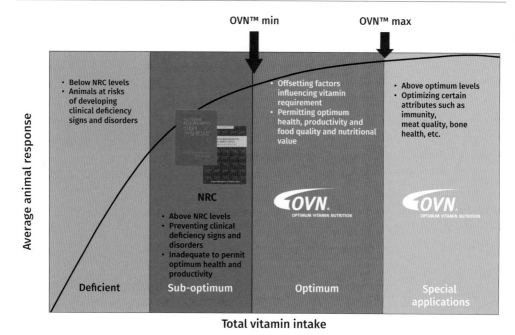

Figure 1.14 The Optimum Vitamin Nutrition® concept (Source: dsm-firmenich Animal Nutrition and Health (2022b)

3   **Deficient or minimum vitamin intake** refers to the vitamin level that puts the animal in danger of showing clinical deficiency symptoms and metabolic disorders and in which the level of vitamins falls short of the NRC supplementation level.

4   **A suboptimum intake** prevents the appearance of clinical deficiency symptoms. Its supplementation levels comply with or exceed the NRC's guidelines but are inadequate to permit an optimum state of health and productivity.

5   **An optimum intake** compensates for the negative factors which influence an animal's yield and therefore contributes to achieving an optimum state of well-being, health, and productivity.

6   **Special applications** are above optimum intake levels of vitamin supplementation for optimizing certain attributes such as immunity, meat quality, bone health, etc., or are directly used to produce vitamin-enriched animal origin foods. There is a growing demand for food with a greater added value, such as vitamin-rich eggs, meat, or milk, the occasional consumption of which contributes to a balanced human diet. The production of which necessitates a higher vitamin content in the feed. Concerning the safety of vitamins, only very large quantities (between 10 and 100 times the levels used in practice) of some vitamins such as A and D_3 in feed might occasionally cause some sort of disorder in animals.

In summary, we can say that implementing a nutritional program with the most appropriate levels of all the vitamins in an animal's diet aims to offer the following benefits to the food chain.

1   **Optimum health and welfare of animals** are a prerequisite for producing safe and healthy flesh.

2   **Optimum productivity**, given better sanitary conditions and greater efficiency in animal farming within parameters such as FCR, final weight, weight gain, mortality, etc.

3   **Reduced food waste and achieve optimum food quality** offered to consumers provides them with whole food with balanced nutrient content.

1. Optimizing an animal's health and welfare

It is common knowledge that there is a close relationship between nutrition, health, and welfare. Improving the robustness of an animal constitutes a crucial aspect in the production of any food type of animal origin, so one essential objective regarding nutrition and management programs would be limiting the incidence of diseases and their debilitating effect on animals. Supplementing an animal's diet with optimum quantities of vitamins at times of greatest vulnerability to infection reduces the risk of contracting a disease.

Health and immunity

Vitamins, in sufficient quantities, play a fundamental role in the capacity of an animal's organism to respond to a disease. Since the onset of an illness cannot be predicted, the immune system must be prepared to act before the disease attacks the organism.

Various studies have demonstrated a close relationship between low levels of vitamin E in the tissues and a decrease in immunocompetence (the immune system's response) long before any clinical signs of disease appear. Vitamins E and C are powerful antioxidants that protect cells from free radicals and other types of by-products harmful to an animal's metabolism.

Research shows that high levels of vitamin E have several benefits in different aquatic species:

- in salmonids:
  - improvement in the immune response by increasing lymphocyte counts (Rahimi *et al.*, 2015) and delaying apoptosis and increasing phagocytosis (DSM internal research)
  - improvement in the survival after infection (Dantagnan *et al.*, 2017)
  - improvement in stress resistance (Naderi *et al.*, 2017a)
- in tilapia:
  - improvement in inflammatory responses (Qiang *et al.*, 2019)
  - increased resistance to bacterial challenges (Lim, C., *et al.*, 2010b)
- in shrimp:
  - adequate dietary vitamin E levels promote phosphatase activity (Lee and Shiau, 2004; Li, Y., *et al.*, 2019). However, excessive vitamin E levels may inhibit its activity
  - vitamin E might be involved indirectly in modulating shrimp growth by its effect on immunological parameters (Amlashi et al., 2011).

Welfare

Animal well-being is very important for farmers. Moreover,in the food sector, there are many retailers, wholesale distributors, and even fast-food chains that have incorporated certain animal welfare standards in their codes of best practices, with the aim of giving animals a healthy life that will contribute to guaranteeing more nutritious food to their consumers. Optimum vitamin supplementation in an animal's diet will contribute to improving its welfare because:

- optimum vitamin levels contribute to improving the metabolism, nutrition, and well-being of the aquatic species
- improving immunity increases resistance to diseases

- some vitamins, such as 25OHD$_3$, a more active vitamin D$_3$ metabolite, can promote cell regeneration *in vitro* supporting wound and gill healing.

2. Optimizing productivity

In the aquaculture industry, advances have been achieved over decades through strain selection, improved feeds, and new feeding technologies (Glencross *et al.*, 2023).

Recent data have demonstrated that animals grow faster under experimental and commercial conditions when the supplementation of vitamins in the diet is increased according to their requirements. From the economic viewpoint, the improvement in productivity after optimizing the supply of vitamins in the feed gave rise to a significantly better cost-to-benefit ratio, in turn recompensing the farmer.

Different examples are available in the literature (see reviews published by Mai *et al.*, 2022; Liu *et al.*, 2022). Recent data on shrimp (Liu *et al.*, 2024) showed that the increase of vitamin levels in the diet of shrimp can significantly improve the final body weight by 30% and the survival by 10% (Figure 1.15).

In commercial conditions, stress represents a serious threat to achieving optimum yield. It reduces or stops, in some cases, feed intake, increasing the vitamin concentration required to satisfy an animal's needs. Moreover, stress alters the animal's metabolic needs, turning a nutritionally balanced meal into a diet with possible nutritional deficiencies. It has been shown that increased levels of vitamin C and/or E improve resistance to diseases in different aquaculture species, such as tilapia (Ibrahem *et al.*, 2010; Ibrahim *et al.*, 2020; Lim *et al.*, 2010b), salmon (Dantagnan *et al.*, 2017), or crustaceans (Chen Y.L. *et al.*, 2018). Likewise, increased levels of B vitamins across aquatic species have been shown to increase animal robustness in response to with external stressors.

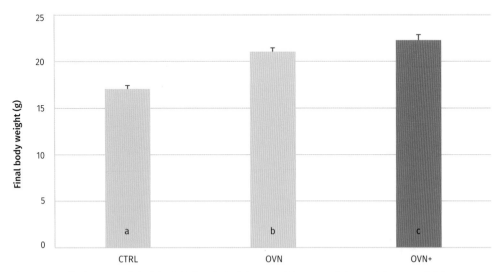

Figure 1.15 Final body weight (g) of shrimp fed with control (commercial premix), OVN™ (high inclusion level of vitamins), or OVN™+ (high inclusion diet of vitamins plus HyD®) diet for 58 days (source: dsm-firmenich internal research, 2023, unpublished)

3. Increasing shelf life and optimizing the quality of food for the consumer

Fish is a high-quality source of protein in the human diet. Likewise, global consumption of aquatic foods has increased at an average annual rate of 3.0% since 1961 – from 9.9 kg/per capita in the 1960s to 20.2 kg/per capita in 2020 – compared with a population growth rate of 1.6% (FAO, 2022).

On the other hand, fish diets contain high levels of fat. Following the replacement of fish oil by vegetable oils over the last 2 decades, fish diets are now rich in unsaturated vegetable fats, that can in consequence considerably increase levels of mono-unsaturated fatty acids in the flesh. The increase levels of mono-unsaturated fatty acids can significantly reduce shelf-life. Fortunately, high levels of vitamin E can counteract this effect and so improve the final quality of the meat, as has been shown in salmon (Faizan *et al.*, 2013; Rodríguez et al., 2015) and trout (Frigg *et al.*, 1990; Kamireddy *et al.*, 2011) and in tilapia (Wu F. *et al.*, 2017), where not only does oxidation decreases but also fillet hardness increases.

Considering the two factors above, and to increase the shelf-life to the maximum to better respond to the increase in the demand, an adequate vitamin nutritional strategy should be developed in animal diets to minimize lipid oxidation, which constitutes a problem for the conservation of meat.

Also, the level of dietary vitamins can modify the quantity of vitamins contained in the flesh of the animals. Faizan *et al.* (2013) and Verlhac *et al.* (unpublished) showed a direct relationship between the dietary amount of vitamin E and the levels in the flesh, which had in consequence better nutritional value for the final consumers (Figure 1.16).

OPTIMUM VITAMIN NUTRITION: A DYNAMIC PROCESS IN CONSTANT EVOLUTION

Optimum vitamin supplementation in an animal's diet, over and above the established minimum needs and adapted to the specific conditions of each animal species, will permit an improvement in the state of health and welfare of the animal, thus optimizing its productive potential at the same time as facilitating the production of high-quality food that is nutritionally balanced.

These optimum levels are based on many studies carried out in university and industrial centers, on the requirements published by different associations and vitamin manufacturers,

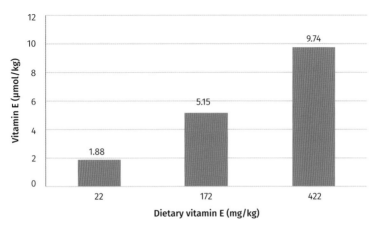

Figure 1.16 Vitamin E content in the flesh of Atlantic salmon after 14 weeks of feeding a diet with 22, 172, or 422 mg/kg Vitamin E (Faizan *et al.*, 2013)

and on the continuous experience of the worldwide aquaculture industry. These optimum levels guarantee farmers minimum impact of negative nutritional factors, such as variability in the natural content of feed ingredients, the existence of antinutritional factors, and different levels of stress.

Although the vitamin recommendations for feed also attempts to compensate for the majority of the factors mentioned above, which influence an animal's vitamin needs, in extreme conditions where the processing of premix or feed is very aggressive (e.g., the inclusion of trace minerals or choline chloride (CC), the use of feed expanders or extruded feed) supplementary quantities of some vitamins may be necessary. The negative effect on the stability of vitamins can be reduced by using high-quality commercial vitamin forms where their formulation and the bioavailability of the active substance they contain are key elements to be considered. And all this is considering the environmental impact of how vitamins are manufactured.

Given that animal farming is a dynamic process the levels of vitamin supplementation need to be reassessed more frequently – this is a change demanded, in the majority of cases, by society, for economic and environmental reasons related to the productivity of animals and by farming systems.

The concept of OVN™ always considers the costs of vitamin supplementation in an animal's diet (in many cases, less than 1% of the cost of feed) against the risk of suffering losses through vitamin deficiencies and through working with yield indices below the optimum.

Nutritionists who follow the recommended guidelines based on the OVN™ concept ensure that vitamins enable the development of an animal's genetic potential and contribute to a more sustainable animal farming.

Moreover, in addition to our vitamin recommendations, nutrient interactions need to be accounted for. Not only do vitamin-to-vitamin interactions need to be understood and further studied, but also interaction between vitamins and minerals and/or other micronutrients. These combinations can be additive or synergistic (as described by Hamre *et al.* (1997) for vitamins C and E; or vitamin C and iron interactions already described by Hilton *et al.* (1978)). However, in some cases we find antagonistic effects or a lack of research on how 2 different compounds interact in the animal (e.g., vitamins K and D (Sivagurunathan *et al.*, 2022)).

OPTIMUM VITAMIN NUTRITION WITH CONSIDERATIONS FOR VITAMIN QUALITY

There is strong evidence that supplementing vitamins in feeds at NRC (2011) recommended levels realistically provides only suboptimum nutrition under average aquaculture conditions, especially in recent years where the inclusion of marine ingredients has decreased substantially in feed formulas (Ytrestøyl *et al.*, 2015; Hamre *et al.*, 2016; Hemre *et al.*, 2016; Prabhu *et al.*, 2019; Vera *et al.*, 2020; Hamre *et al.*, 2022). These alterations in feed formulation have affected the vitamin content of salmon feeds over time (Table 1.1). The concept of optimum vitamin nutrition (OVN™), developed by dsm-firmenich, is based on scientific evidence and consists in feeding optimum levels to compensate for the negative factors that reduce animal performance and weaken animal health (Figure 1.14). It is a dynamic concept in which vitamin levels are optimized to account for the metabolic needs of (1) genetically improved fish strains for maximized growth or other phenotypes and (2) immune system functions that need to be reinforced during exposure to pathogens and under stressful conditions. Classical signs of true vitamin deficiency are rarely observed today. However, this is not to say that fish welfare needs and optimum performance objectives are consistently met under commercial conditions.

Table 1.1 Comparison of vitamin content of salmon diets not supplemented with vitamin premix between 1990 and 2022 (vitamin C excluded) and dietary vitamins supplied by raw materials of typical salmon feeds formulated between 1990 and 2022 (vitamin C excluded) (source: Liu, A., *et al.*, 2022)

Composition	1990	2022
Typical salmon feed formula (as-fed)	(%)	(%)
Fishmeal, anchovy	30.0	5.0
Poultry by-product meal	20.0	0.0
Corn gluten meal	7.5	10.0
Soy protein concentrate	0.0	30.0
Guar meal	0.0	3.5
Wheat gluten meal	0.0	10.0
Wheat flour	15.8	10.2
Fish oil, anchovy	21.5	8.1
Canola/Rapeseed oil	0.0	16.8
Supplement[1]	5.2	6.4
Proximate composition (as-fed)	(%)	(%)
Dry matter	94	95
Crude protein	43	41
Crude lipid	28	28
Ash	8	6
Vitamin/vitamin-like nutrients	(mg/kg)	(mg/kg)
Thiamine (B_1)	1.077	4.180
Riboflavin (B_2)	4.294	1.912
Niacin (B_3)	46.451	31.625
Pantothenic acid (B_5)	10.533	10.430
Pyridoxine (B_6)	3.616	3.885
Biotin (B_7)	0.160	0.174
Folate (B_9)	0.291	0.407
Cobalamin (B_{12})	105.608	17.605
Choline	2303.524	1379.468
Myo-inositol	316.589	209.248
Vitamin A	89.761	32.772
Vitamin D	1.059	0.013
Vitamin E	60.722	80.251
Vitamin K	1.190	0.549

Note: [1] Supplement: crystalline amino acids, lecithin, vitamin C (0.3% inclusion in both diets), choline chloride 60% (0.4% inclusion in both formulas), monocalcium phosphate, astaxanthin.

Until recently, vitamin nutrition research in aquaculture has followed a reductionist approach, attempting to avoid deficiencies, and focusing on individual vitamins in a limited number of aquatic species at one life stage, often at the larval/juvenile stage, with little consideration for sustainability (NRC, 2011; Acevedo-Rocha *et al.*, 2019; Wang, Y., *et al.*, 2021). Recent advances in the understanding of immune functions of vitamins, limited understanding of interactions with other nutrients (e.g., minerals, lipids), and a tight supply chain are driving further research to optimize vitamin nutrition in aquaculture.

Furthermore, innovations in feed formulation (e.g., decreasing inclusion of marine raw materials, usage of ingredients susceptible to oxidation such as blood meal), regulatory changes (e.g., suspension of ethoxyquin in Europe), and pressure to reduce the use of antibiotics demand that dietary vitamin levels be reviewed regularly (Cabello, 2006; Tocher, 2015; Shepherd *et al.*, 2017; Saito *et al.*, 2021).

Chapters 5, 6, and 7 review the impact of vitamins in aquaculture nutrition as reported in the international research literature. These studies have endeavored to emphasize the beneficial effects that optimum vitamin levels have on an animal, both on the level of health and welfare and concerning productivity. The chapters also identify aspects on which there is currently insufficient information available, intending to address these gaps in future research and editions of this book.

Chapter 2

A brief history of vitamins

Vitamins were mostly discovered in the 20th century and were once regarded as "unknown growth factors" (Eggersdorfer *et al.*, 2012). The first phase of developing the concept of vitamins began many centuries ago and gradually led to the recognition that night blindness, xerophthalmia, scurvy, beriberi, and rickets are dietary diseases. These diseases had long plagued humankind and were mentioned in the earliest written records. Records of medical science from antiquity attest that researchers had already linked certain foods and diseases or infirmities, postulating that food constituents played a causal or a preventive role. These are considered the nebulous beginnings of essential nutrients (Eggersdorfer *et al.*, 2012).

Beriberi is probably the earliest documented deficiency disorder being recognized in China as early as 2697 BC. By 1500 BC, scurvy, night blindness, and xerophthalmia were described in Egyptian writings. Two books of the Bible contain accounts that point to vitamin A deficiency (McDowell, 2006). Jeremiah 14:6 states: "and the asses did stand in high places, their eyes did fail because there was no grass." In addition, the Bible mentions that fish bile was used to cure a blind man named Tobias.

In 400 BC, the Greek physician Hippocrates, known as the Father of Medicine, reported using raw ox liver dipped in honey to prevent night blindness. He also described soldiers afflicted with scurvy. Scurvy took a heavy toll on the Crusades of the Middles Ages because the soldiers traveled far from home, and their diet was deficient in vitamin C. During the long sea voyages between 1492 and 1600, scurvy posed a serious threat to the health of sailors and undermined world exploration. For example, while sailing worldwide, Magellan lost 80% of his crew to the disease. Vasco de Gama, another great explorer, lost 60% of his 160-man crew while mapping the coast of Africa. In 1536, during Jacques Cartier's expedition to Canada, 107 out of 110 men became sick with scurvy. However, the journey was saved when the Indians shared their knowledge of the curative value of pine needles and bark. In 1593, British Admiral Richard Hawkins wrote: "I have seen some 10,000 seamen die of scurvy; some sailors tried treating themselves by trimming the rotting, putrid black flesh from their gums and washing their teeth in urine."

In 1747, James Lind, a British naval surgeon, carried out the first controlled clinical experiment aboard a ship to find a cure for scurvy. Twelve patients with scurvy were divided into 6 treatment groups. Two sailors received a dietary supplement of oranges and lemons, while the other treatment groups were given nutmeg, garlic, vinegar, cider, and seawater, respectively. The 2 men who had received the citrus fruit were cured of scurvy. Where did Lind get the idea that scurvy was related to nutrition? He had been told a story of an English sailor with scurvy who was left to die on a lonely island with no food. Feeling hungry, the man nibbled a few blades of beach grass. The next day, he felt stronger and ate some more grass. After a few weeks on this "diet," he was completely well.

In the second half of the 19th century, there was another disease that killed thousands of sailors in the Japanese navy. In 1880, the Japanese navy recorded almost 5,000 deaths from beriberi in 3 years. Patients with beriberi became weak and eventually partially paralyzed, lost weight, and died. Doctors tried to find the germ that was causing beriberi. Finally, they listened to Japanese naval surgeon Kamekiro Takaki, who believed the sailors' diet was causing beriberi. Takaki noted a 60% incidence of beriberi on a ship returning from a 1-year voyage during which the sailors' diet had been mostly polished rice and some fish. He sent out a second ship under the same conditions but substituted barley, meat, milk, and fresh vegetables for some of the rice. The dietary change eliminated beriberi, but Takaki incorrectly concluded that the additional protein prevented the beriberi. Regardless, the Japanese knew they could avoid beriberi by not relying on polished rice as the only dietary staple.

Before the beginning of the 20th century, there was a growing body of evidence that nutritional factors, later known as vitamins, were implicated in certain diseases. Louis Pasteur was the chief opponent of the "vitamin theory," which held that certain illnesses resulted from a shortage of specific nutrients in foods. Pasteur believed there were only 3 classes of organic nutrients: carbohydrates, fats, and proteins. His research showed that microorganisms caused disease and made scientists with medical training reluctant to believe the vitamin theory. It has been said that the immensely successful "germ theory" of disease, coupled with toxin theory and the successful use of antisepsis and vaccination, convinced scientists of the day that only a positive agent could cause disease (Guggenheim, 1995). Until the mid-1930s, most US doctors still believed that pellagra was an infectious disease (McDowell, 2006).

VITAMIN THEORY TAKES SHAPE

Beginning in the mid-1850s, German scientists were recognized as leaders in the field of nutrition. In the late 1800s, Professor C. von Bunge, who worked at the German university in Dorpat, Estonia, and then at Basel, Switzerland, had some graduate students conduct experiments with purified diets for small animals (Wolf and Carpenter, 1997). In 1881, N. Lunin, a Russian student studying in von Bunge's laboratory, observed that some mice died after 16 to 36 days when fed a diet composed solely of purified fat, protein, carbohydrate, salts, and water. Lunin suggested that natural foods such as milk contain small quantities of "unknown substances essential to life."

Many great scientific advances have come about due to chance observations made by men and women of inspiration. In 1896, Dutch physician and bacteriologist Christiaan Eijkman made a historic finding concerning a cure for beriberi. Eijkman was researching in Indonesia to identify the causal pathogen of beriberi. He astutely observed that a polyneuritis condition in chickens produced clinical signs similar to those in humans with beriberi. This chance discovery was made when a new head cook at the hospital discontinued the supply of "military" rice (polished rice), and the chickens fed the wholegrain "civilian" rice recovered from the polyneuritis. After extensive experimentation, Eijkman proved that both polyneuritis and beriberi were caused by eating polished white rice. Both afflictions could be prevented or cured when the outer portions of the rice grain (e.g., rice bran) were consumed. Thus, Eijkman became the first to produce a vitamin deficiency disease in an experimental animal. He also noted that prisoners with beriberi eating polished rice tended to get well when fed a less milled product. In 1901, Grijns, one of Eijkman's colleagues in Indonesia, was the first to come up with a correct interpretation of the connection between the excessive consumption of polished rice and the etiology of beriberi. He concluded that rice contained "an essential nutrient" found in the grain's outer layers.

In 1902, a Norwegian scientist named Holst conducted some experiments on "ship-beriberi" (scurvy) using poultry, but the experiments failed. In 1907, Holst and Frolich produced experimental scurvy in guinea pigs. Later it was learned that poultry could synthesize vitamin C while guinea pigs could not.

In 1906, Frederick Hopkins, working with rats in England, reported that "no animal can live upon a mixture of pure protein, fat, and carbohydrate and even when the necessary inorganic material is supplied, the animal cannot flourish." Hopkins found that small amounts of milk added to purified diets allowed rats to live and thrive. He suggested that unknown nutrients were essential for animal life, calling them "accessory food factors." Hopkins' experiments were like those of Lunin; however, they were more in depth. He played an important role by recording his views in memorable terms that received wide recognition (McCollum, 1957). Hopkins also expressed that various disorders were caused by diets deficient in unidentified nutrients (e.g., scurvy and rickets). He was responsible for opening a new field of discovery that largely depended on experimental rats.

In 1907, Elmer McCollum (Figure 2.1) arrived in Wisconsin to work on a project to determine why cows fed wheat or oats (versus yellow corn) gave birth to blind or dead calves. The answer was found to be that wheat and oats lacked the vitamin A precursor carotene. Between 1913 and 1915, McCollum and Davis discovered 2 growth factors for rats, "fat-soluble A" and "water-soluble B." By 1922, McCollum had identified vitamin D as a substance independent of vitamin A. He bubbled oxygen through cod liver oil to destroy its vitamin A; the treated oil remained effective against rickets but not against xeropthalmia. Thus, "fat-soluble vitamin A" had to be 2 vitamins, not just one (DeLuca, 2014).

Figure 2.1 Elmer McCollum (source: Roche Historical Archive)

In 1912, Casimir Funk (Figure 2.2), a Polish biochemist working at the Lister Institute in London, proposed the "vitamin theory" (Funk and Dubin, 1922). He had reviewed the literature and made the important conclusion that beriberi could be prevented or cured by a protective factor present in natural food, which he successfully isolated from rice by-products. What he had isolated was named "beriberi vitamin" in 1912. This term "vitamin" denoted that the substance was vital to life and chemically an amine (vital + amine). In 1912, Funk proposed the theory that other "deficiency diseases" in addition to beriberi were caused by a lack of these essential substances, namely scurvy, rickets, sprue, and pellagra. He was the first to suggest that pellagra was a nutrient deficiency disease.

In 1923, Evans and Bishop discovered that vitamin E deficiency caused reproductive failure in rats. Steenbock (1924) showed that irradiation of foods as well as animals with ultraviolet light produced vitamin D. In 1928, Szent-Györgyi isolated hexuronic acid (later renamed "ascorbic acid") from foods such as orange juice. One year later, Moore proved that the animal body converts carotene to vitamin A. This experiment involved feeding 1 group of rats carotene and finding higher levels of vitamin A in their livers compared to controls. By 1928, Joseph Goldberger and Conrad Elvehjem had shown that vitamin B was more than one substance. After the "vitamin" was heated, it was no longer effective in preventing beriberi (B_1), but it was still good for rat growth (B_2). The 1930s and 1940s were the golden age of vitamin research.

Figure 2.2 Casimir Funk (source: Roche Historical Archive)

During this period, the traditional approach was to (1) study the effects of a deficient diet, (2) find a food source that prevents the deficiency, and (3) gradually concentrate the nutrient (vitamin) in a food and test potency. Laboratory animals were used in these procedures.

Henrick Dam of Denmark discovered vitamin K in 1929 when he noted hemorrhages in chicks fed a fat-free diet. Ironically 1 year earlier, Herman Almquist, working in the United States, had discovered both forms of the vitamin (K_1 and K_2) in studies with chicks. Unfortunately, university administrators delayed the review of his paper, and when it was finally submitted to the journal *Science*, it was rejected. Therefore, only Henrick Dam received a Nobel prize for discovering vitamin K.

Vitamin B_{12} was the last traditional vitamin to be identified, in 1948. Shortly after that, it was discovered that cobalt was an essential component of the vitamin. Simple monogastric animals were found to require the vitamin, whereas ruminants and other species with large microbial populations (e.g., horses) require dietary cobalt rather than vitamin B_{12}.

Compared with the situation for night blindness, xeropthalmia, beriberi, scurvy, and rickets, there were no records from the ancient past of the disease of pellagra. The disease was caused by niacin deficiency in humans, a problem prevalent mainly in cultures where corn (maize) was a key dietary staple (Harris, 1919). Columbus took corn to Spain from America. Pellagra was not recognized until 1735, when Gaspar Casal, physician to King Philip V of Spain, identified it among peasants in northern Spain. The local people called it "mal de la rosa," and Casal associated the disease with poverty and spoiled corn. The popularity of corn spread eastward from Spain to southern France, Italy, Russia, and Egypt, and so did pellagra. James Woods Babcock of Columbia, South Carolina, who identified pellagra in the United States by establishing a link with the disease in Italy, studied the case records of the South Carolina State Hospital and concluded that the disease condition had occurred there as early as 1828. Most cases occurred in low-income groups, whose diet was limited to inexpensive foodstuffs. Diets characteristically associated with the disease were the 3 Ms, specifically meal (corn), meat (backfat), and molasses.

The word pellagra means rough skin, which relates to dermatitis. Other descriptive names for the condition were "mal de sol" (illness of the sun) and "corn bread fever." In the early 1900s in the United States, particularly in the South, it was common for 20,000 deaths to occur annually from pellagra. It was estimated that there were at least 35 cases of the disease for every death due to pellagra. Even as late as 1941, 5 years after the cause of pellagra was known, 2,000 deaths were still attributed to the disease. The clinical signs and mortality associated with pellagra are the 4 Ds: dermatitis (of areas exposed to the sun), diarrhea, dementia (mental problems), and death. Several mental institutions in the United States, Europe, and Egypt were primarily devoted to caring for pellagra sufferers or pellagrins.

In 1914, Joseph Goldberger, a bacteriologist with the US Public Health Service, was assigned to identify the cause of pellagra. His studies observed that the disease was associated with poor diet and poverty and that well-fed persons did not contract the disease (Carpenter, 1981). The therapeutic value of good diets was demonstrated in orphanages, prisons, and mental institutions in South Carolina, Georgia, and Mississippi. Goldberger, his wife, and 14 volunteers constituted a "filth squad" who ingested and injected various biological materials and excreta from pellagrins to prove that pellagra was not an infectious disease. These extreme measures did not result in pellagra, thus demonstrating the non-infectious nature of the disease. At the time, researchers and physicians did not want to believe that pellagra resulted from poor nutrition. They sought to link it to an infection in keeping with the popular "germ theory" of diseases (McDowell, 2006). An important step toward isolating the preventive factor for pellagra involved the discovery of a suitable laboratory animal for testing its potency in various concentrated preparations. It was found that a pellagra-like disease (black tongue) could be produced in dogs. Elvehjem and his colleagues (1974) isolated nicotinamide from the liver and identified it as the factor that could cure black tongue in dogs. Reports of niacin's dramatic therapeutic effects in human pellagra cases quickly followed from several clinics.

In 1824, James Scarth Combe first discovered fatal anemia (pernicious anemia) and suggested it was linked to a digestive disorder. George R. Minot and William Murphy reported in 1926 that large amounts of the raw liver would alleviate the symptoms of pernicious anemia. In 1948, E. L. Rickes and his colleagues in the United States and E. Lester Smith in England isolated vitamin B_{12} and identified it as the anti-pernicious anemia factor (McDowell, 2006). Much earlier, in 1929, W. B. Castle had shown that pernicious anemia resulted from the interaction between a dietary factor (extrinsic) and a mucoprotein substance produced by the stomach (intrinsic factor). Castle used an unusual but effective method to relieve the symptoms of pernicious anemia patients. He ate some beef, and after allowing enough time for the meat to mix with gastric juices, he regurgitated the food and mixed his vomit with the patients' food. With this treatment, the patients recovered because they received both the extrinsic (vitamin B_{12}) and intrinsic (a mucoprotein) factors from Castle's incompletely digested beef meal.

The importance of vitamins was well accepted in the first 3 decades of the 20th century. Table 2.1 provides an overview of the chronological evolution of the discovery, isolation, and assignment of the chemical structure and first production of the individual vitamins. The development of synthetic production of vitamins started in 1933 with ascorbic acid/vitamin C from Merck (Cebion®), which was isolated from plant leaves. However, the first industrial-scale chemical production of vitamin C was achieved by F. Hoffmann-La Roche in 1934 based on a combined fermentation and chemical process developed by Tadeus Reichstein. These scientific innovations were recognized with 12 Nobel Prizes and 20 laureates (Table 2.2). A complete description of the history of discovery, first syntheses, and current industrial processes used for producing each vitamin was described by Eggersdorfer et al. (2012) and McDowell (2013).

Table 2.1 Discovery, isolation, structural elucidation, and synthesis of vitamins (source: Eggersdorfer et al., 2012)

Vitamin	Discovery	Isolation	Structural elucidation	First synthesis
Vitamin A	1916	1931	1931	1947
Vitamin D	1918	1932	1936	1959
Vitamin E	1922	1936	1938	1938
Vitamin B_1	1912	1926	1936	1936
Vitamin B_2	1920	1933	1935	1935
Niacin	1936	1936	1937	1837/1940*
Pantothenic acid	1931	1938	1940	1940
Vitamin B_6	1934	1938	1938	1939
Biotin	1931	1935	1942	1943
Folic acid	1941	1941	1946	1946
Vitamin B_{12}	1926	1948	1956	1972
Vitamin C	1912	1928	1933	1933

Note: * 1837: synthesis of niacin used in photography before discovering its nutritional function; 1940: nicotinamide.

Table 2.2 Nobel prizes for vitamin research (source: Eggersdorfer *et al.*, 2012)

Year	Recipient	Field	Citation
1928	Adolf Windaus	Chemistry	Research into the constitution of steroids and connection with vitamins
1929	Christiaan Eijkman	Medicine, Physiology	Discovery of antineuritic vitamins
1929	Sir Frederick G. Hopkins	Medicine, Physiology	Discovery of growth-stimulating vitamin
1934	George R. Minot, William P. Murphy, George H. Whipple	Medicine, Physiology	Discoveries concerning liver therapy against anemias
1937	Sir Walter N. Haworth	Chemistry	Research into the constitution of carbohydrates and vitamin C
1937	Paul Karrer	Chemistry	Research into the constitution of carotenoids, flavins, and vitamins A and B_2
1937	Albert Szent-Györgyi	Medicine, Physiology	Discoveries in connection with biological combustion processes, with special reference to vitamin C and catalysis of fumaric acid
1938	Richard Kuhn	Chemistry	Work on carotenoids and vitamins
1943	Carl Peter Henrik Dam	Medicine, Physiology	Discovery of vitamin K
1943	Edward A. Doisy	Medicine, Physiology	Discovery of chemical nature of vitamin K
1953	Fritz A. Lipmann	Medicine, Physiology	Discovery of coenzyme A and its importance for intermediary metabolism
1964	Konrad E. Bloch, Feodor Lynen	Medicine, Physiology	Discoveries concerning mechanism and regulation of cholesterol and fatty acid metabolism
1964	Dorothy C. Hodgkin	Chemistry	Structural determination of vitamin B_{12}
1967	Ragnar A. Granit	Medicine, Physiology	Research which illuminated electrical properties of vision by studying wavelength discrimination by eye
1967	Halden K. Hartine	Medicine, Physiology	Research on mechanisms of sight
1967	George Wald	Medicine, Physiology	Research on chemical processes that allow pigments in the eye retina to convert light into vision

Vitamins became available in the following years through chemical synthesis, fermentation, or extraction from natural materials (Table 2.1). From 1930 to 1950, there was mainly small-scale production in several countries to reach local markets, but, as demand grew, larger plants became more common from 1950 to 1970. Still, it was not until 1987 that all the vitamins were accessible by industrial processes. Nowadays, chemical synthesis is still the dominant method of industrial production.

The large companies Hoffmann-La Roche (Figure 2.3) and BASF were market leaders, but numerous European and Japanese pharmaceutical companies produced and sold vitamins.

Figure 2.3 Early production of vitamin A at Roche Nutley, USA (source: Roche Historical Archive).

Between 1970 and 1990, production plants became even bigger and with global reach, and since 2000 China has become a larger producer. Fermentation technology started to gain importance, especially for vitamin B_{12} and B_2. New technologies have been emerging in the past 10 years, such as the overexpression of vitamins in plants by either using traditional breeding or genetically modified plants (Eggersdorfer *et al.*, 2012).

Chapter 3

Introduction to vitamins

VITAMIN DEFINITION AND CLASSIFICATION

A vitamin is an organic substance that is:

- a component of a natural compound but distinct from other nutrients such as carbohydrates, fats, proteins, minerals, and water
- present in most foods in a minute amount
- essential for normal metabolism in physiological functions such as growth, development, maintenance, and reproduction
- a cause of a specific deficiency disease or associated with a syndrome if absent from the diet or if improperly absorbed or utilized
- not (with very few exceptions) synthesized by the host in sufficient amounts to meet physiological demands and therefore must be obtained from the diet.

Vitamins are differentiated from trace elements, also present in the diet in small quantities, by their organic nature. Some vitamins deviate from the preceding definition in that they do not always need to be constituents of food (McDowell, 2000; Combs and McClung, 2022). For example, companion animals and farm livestock can synthesize vitamin C (ascorbic acid) but not all fish (Drouin et al., 2011; Moreau and Dabrowski, 2000; Cho et al., 2007). It has now been established that all cartilaginous and non-teleost bony fish species are able to synthesize vitamin C and that no teleost fish species can do so. Nevertheless, deficiencies have been reported in some species that synthesize vitamin C, and supplementation with this vitamin has been shown to have value for particular diseases or stress conditions (Rahimi et al., 2015; Lim, C., et al., 2010b; Mustafa et al., 2013; Vieira et al., 2018; Feng, T., et al., 2018), toxicoses (Mirvaghefi et al., 2015; Su et al., 2018), and to restore productivity or maximize performance (Falahatkar et al., 2011; Ibrahim et al., 2020; Solyanik et al., 2021).

Likewise, for most species, niacin can be synthesized from the amino acid tryptophan (but not by the cat or fish species studied to date) and choline from the amino acid methionine. Nevertheless, dietary supplementation with niacin and choline is necessary for animal farming. Finally, vitamin D can also be synthesized in the skin under UV stimulation. Still, the diets of all farm animals are supplemented with this vitamin to provide the required quantity primarily for proper bone development.

The quantities of vitamins required are tiny, but they are essential for tissue integrity, normal development or physiological functions, and health maintenance. Their physiological and metabolic roles vary and are of great importance. They are involved in many biochemical reactions and participate in nutrient metabolism derived from the digestion of carbohydrates, lipids, and proteins.

A single vitamin may have several different functions, and many interactions between them are known. Classically, vitamins have been divided into 2 groups based on their solubilities in

Table 3.1 Main functions of vitamins and symptoms of deficiency in aquaculture species

Vitamin	Basic function(s)	Deficiency disorders/diseases
Vitamin A	• Essential for growth, health (immunity), reproduction (steroid synthesis), vision, development and integrity of skin, epithelia and mucosa	• Blindness or night-blindness (xeropthalmia) • Loss of appetite, poor absorption of nutrients, impaired growth and, in severe cases, death • Reduced immune response and increased risk of infections (respiratory and intestinal) • Reproduction defects such as failure of spermatogenesis in the male and fetal resorption or death in the female swine • Dry and scaly skin • Keratinization of epithelial tissues
Vitamin D$_3$	• Homeostasis of calcium and phosphorus (intestine, bones and kidney) • Regulation of bones calcification • Modulation of the immune system • Muscular cell growth	• Rickets, osteomalacia • Bone disorders (e.g., soft bones) and lameness • Stiff and hesitant gait • Reduced growth rate • Muscular weakness • Insufficient production of plasma 25OHD$_3$
Vitamin E	• Most powerful fat-soluble antioxidant • Immune system modulation • Tissue protection • Fertility • Meat quality	• Muscular dystrophy and myopathy • Mulberry heart disease • Reduced immune response • Reduced fertility and Mastitis, Metritis and Agalactia (MMA) in sows • Meat quality defects: drip-loss, off-flavours
Vitamin K$_3$	• Blood clotting and coagulation • Coenzyme in metabolic process related to bone mineralization (Ca binding proteins) and protein formation	• Increased clotting time • Haemorrhages diseases, anemia and weakness • Bone disorders • Hematomas or blood swelling in the ears
Vitamin B$_1$	• Coenzyme in several enzymatic reactions • Carbohydrate metabolism (conversion of glucose into energy) • Involved in ATP, DNA and RNA production • Synthesis of acetylcholine, essential in transmission of nervous impulses	• Loss of appetite up to anorexia and vomiting • Reduced growth rate • Neuropathies (polioencephalomalacia-PEM), general muscle weakness, poor leg coordination • Fatty degeneration and necrosis of heart fibers (cardiac failure) • Mucosal inflammation with gastrointestinal malfunction
Vitamin B$_2$	• Fat and protein metabolism • Flavin coenzyme (FMN and FAD) synthesis, essentials for energy production (respiratory chain) • Involved in synthesis of steroids, red blood cells and glycogen • Integrity of mucosa membranes and antioxidant system within cells	• Reduced feed intake and growth • Reduced absorption of zinc, iron and calcium • Inflammation of the mucous membranes of the digestive tract • Scours and ulcerative colitis • Fertility impairments
Vitamin B$_6$	• Aminoacids, fats and carbohydrate metabolism • Essential for DNA and RNA synthesis • Involved in the synthesis of niacin from tryptophan	• Growth retardation, lesser feed intake and protein retention • Dermatitis, rough hair coat, scaly skin • Disorders of blood parameters • Brown exudate of the eyes • Anemia and ascites • Muscular convulsions, incoordination of movements and paralysis

Vitamin	Basic functions	Deficiency symptoms
(continued)	• involved in methionine metabolism • Coenzyme in nucleic acids (DNA and RNA) and protein metabolism • Metabolism of fats and carbohydrates	• growth retardation and lower feed conversion • Leg weakness • Embryo mortality, reduced piglet survival
Niacin or Vitamin B$_3$	• Coenzyme (active forms NAD and NADP) in amino acids, fats and carbohydrates metabolism • Required for optimum tissue integrity, particularly for the skin, the gastrointestinal tract and the nervous system	• Nervous system disorders • Inflammation and ulcers of mucous membranes • Reduced growth and feed efficiency • Dermatitis (pellagra), hair loss • Ulcerative necrotic lesions of the large intestine, diarrhoea • Reduced reproductive performance
Biotin or Vitamin B$_7$	• Coenzyme in protein, fat and carbohydrates metabolism • Normal blood glucose level • Synthesis of fatty acids, nucleic acids (DNA and RNA) and proteins (keratin)	• Reduced appetite, retarded growth • Fertility disorders • Skin ulcers, alopecia, hair loss and dermatitis • Inflammation of the hooves and hoof-sole lesions • Diarrhoea, eye inflammation and changes in oral mucosa
d-Pantothenic acid or Vitamin B$_5$	• Present in Coenzyme A (CoA) and Acyl Carrier Protein (ACP) involved in carbohydrate, fat and protein metabolism • Biosynthesis of long-chain fatty acids, phospholipids and steroid hormones	• Functional disorders of nervous system • Locomotive disorders • Scaly skin, dermatitis • Fatty degeneration of the liver • Reduced antibody formation • Reduced appetite, poor feed utilization and growth depression
Folic acid or Vitamin B$_9$	• Coenzyme in the synthesis of nucleic acids (DNA and RNA) and proteins (methyl groups) • Stimulates hematopoietic system • With vitamin B$_{12}$ it converts homocysteine into methionine	• Megaloblastic (macrocytic) anemia • Skin damages and hair loss • Retarded growth and reduced appetite • Compromised reproduction in sows • Embryonic mortality and smaller litter size
Vitamin C	• Intracellular (water-soluble) antioxidant • Immune system modulation: stimulation of phagocytosis • Collagen biosynthesis • Formation of connective tissues, cartilage and bones • Synthesis of corticosteroids and steroid metabolism • Conversion of vitamin D$_3$ to its active form 1,25(OH)$_2$D$_3$	• Weakness, fatigue and dyspnea • Bone pain • Haemorrhages of the skin, muscle and certain organs • Reduced fertility in both males (reduced sperm quality) and females (termination of corpus luteum)
Choline	• Membrane structural component (phosphatidylcholine) • Fat transport and metabolism in the liver • Support nervous system function (acetylcholine) • Source of methyl donors for methionine regeneration from homocysteine	• Fatty liver • Growth retardation
β-carotene	• Source of vitamin A • Stimulation of progesterone synthesis • Reproductive system function	• Poor reproductive performance e.g., prolonged estrus, retarded follicle maturation and ovulation • Increased susceptibility of young animals to infectious diseases

Note. *25OHD$_3$ Basic functions: Precursor metabolite for production of active vitamin D hormone in plasma and tissues. More efficacious than D$_3$ for raising plasma 25OHD$_3$. Plasma and muscle 25OHD$_3$ correlates with body weight.

fat solvents or water. Thus, fat-soluble vitamins include A, D, E, and K, while B complex vitamins, vitamin C, and choline are water-soluble. An overview of metabolic functions, symptoms of deficiency, and sensitivity of vitamins in for salmonids, warm water fish, and shrimp are listed respectively later in Tables 5.2, 6.1, and 7.1.

Fat-soluble vitamins are found in feedstuffs in association with lipids. The fat-soluble vitamins are absorbed along with dietary fats, apparently by mechanisms like those involved in fat absorption. Conditions favorable to fat absorption, such as adequate bile flow and good micelle formation, also favor the absorption of fat-soluble vitamins (La Frano et al., 2018; Harrison, 2012; Gonçalves et al., 2015; Maurya and Aggarwal, 2017). Water-soluble vitamins are not associated with fats, and alterations in fat absorption do not affect their absorption. The fat-soluble vitamins A, D, and, to a lesser extent, E are generally stored in appreciable amounts in the animal body (Guillou et al., 1989; Gesto et al., 2012; Shiau and Chen, 2000). Water-soluble vitamins are not stored, and excesses are rapidly excreted, except for vitamin B_{12} and perhaps biotin. Table 3.1 lists the solubility characteristics of vitamins classified as either fat- or water-soluble.

Vitamins can seldom be regarded as nutrients in isolation because they display various interactions with each other and other nutrients (Calderón-Ospina and Nava-Mesa, 2020). For example, the fat-soluble vitamins compete for intestinal absorption, so an excess of one may cause deficiencies in the others (West et al., 1992; Stacchiotti et al., 2021). The vitamins of the B group are regulators of intermediary metabolism. Some metabolic processes are interdependent: for example, choline, B_{12}, and folic acid interact in the methyl groups' metabolism, so a lack of one of them increases the requirement for the others (Stabler, 2006; Bailey et al., 2013). The same happens between B_{12} and pantothenic acid (Rucker and Bauerly, 2013). It may also occur that an excess of one vitamin induces a deficiency of others. Thus, biotin status deteriorates if the diet is supplemented with high levels of choline and other vitamins of the B group (e.g., in pigs (Kopinski and Leibholz, 1989). High choline levels may similarly affect other vitamins during feed storage.

Vitamins are also known to interact in diverse ways with other nutrients, such as amino acids. However, the biosynthesis of niacin from tryptophan appeared limited or absent in fish and shrimp (Shiau and Suen, 1994; NRC, 2011; Xia et al., 2015). Both methionine and choline can be a source of methyl groups, which are needed to synthesize both, and this relationship is of commercial importance because supplementation entails economic cost. Biotin, folic acid, and B_6 play a part in metabolic interconversions of amino acids, so their requirement increases if protein levels are high (Hardy et al., 1979; Mock, 2013). The same applies to those vitamins involved in the metabolism of carbohydrates (biotin, B_1), the requirements for which are higher with low-fat diets (Camporeale and Zempleni, 2006; Mock, 2013). Finally, there are also interactions between minerals and vitamins. The best-documented example is selenium and vitamin E (Poston et al., 1976) as well as the interactions among vitamin E, vitamin C, and folic acid (Mirvaghefi et al., 2015; Naderi et al., 2019; Hamre et al., 2022).

These interactions make it somewhat difficult to estimate the requirements for each vitamin precisely, and it is probably more appropriate to focus on the problem generally. Chapters 5, 6, and 7 provide more detailed explanation of the situation in the different species. The classic evaluation of the dose–response curve, so widely used to estimate the requirements of other nutrients, is not an appropriate technique for vitamins, as their cost is generally low compared to the value of the response and the potential negative consequences of inadequate levels. For these reasons, the usual practice is to define vitamin requirements by considering the maximum response obtained with the chosen evaluation criteria, traditionally weight gain or growth and FCR. However, growth is not a specific response: it may be affected by other factors associated with the feed (palatability, particle size, levels of other nutrients, etc.). This issue may explain the variability in response levels between studies on a particular vitamin, even if they are almost concurrent. According to the extensive literature presented in this book, it would not be possible to establish precise mathematical relationships regarding vitamin requirements until all their interactions are known in detail, taking at the same time into account many factors like various diet types and changes in physical composition.

VITAMIN CONVERSION FACTORS

The recommended vitamin supplementation level is given in vitamin activity. Commercial products indicate the amount of vitamin activity: e.g., for vitamin A 1,000,000 international units (IU) per gram or for vitamin B_6 99% pyridoxine hydrochloride. In the latter case an additional correction must be applied: when supplementing 1 mg vitamin B_6 (pyridoxine) it is advised 1.215 mg pyridoxine hydrochloride is required. Table 3.2 provides the conversion factors, the product forms, and their content. In some countries regulatory authorities may have stated different rules for vitamin declaration in premixes and feeds. It could be, for example, the case for vitamin E: usually the declaration is in milligrams whereas in some countries the declaration could be required in IUs. However, as indicated in the table, the international standard is that 1 mg all-rac-α-tocopheryl acetate is equal to 1 IU.

VITAMINS IN FEEDSTUFFS

The vitamin content of feedstuffs is highly variable, and current values have not been completely evaluated recently. More details are provided in Chapter 4, when discussing individual vitamins, with specific reference to feedstuffs commonly used in fish nutrition. There are severe limitations in relying on average tabular values of vitamins in feedstuffs. As an example, the vitamin E content of 42 varieties of corn varied from 11.1 to 36.4 mg/kg, a 3.3-fold difference (McDowell and Ward, 2008; Combs and McClung, 2022). As a reference, the average vitamin levels in some common feed ingredients are presented in Table 3.3 (Chen, Y.F., et al., 2019). Some sources of information (Combs and McClung, 2022; IAFFD – International Aquaculture Feed Formulation Database) still rely on data published in the 1980s or before.

Generally, these average values are based on a limited number of assays and were not adjusted for bioavailability and variations of vitamin levels within ingredients. Therefore, they may not reflect the changes in genetic characteristics, handling and storage of crops, cropping practices, and processing of feedstuffs over the years. Heat treatments in feed processing, like pelleting and extrusion, improve nutrient digestibility, reduce antinutritional factors, and eventually control *Salmonella* and other pathogens, resulting in more significant vitamin destruction (Gadient, 1986; Svihus and Zimonja, 2011; Spasevski et al., 2015). In addition, values for some vitamins were not determined by current, more precise assay procedures. Additional information on the limitations of using average values of vitamins in feedstuffs when formulating animal rations have been reported (see, for example, Yang, P., et al. (2021) discussing the case for piglets' feed). Diets based on few ingredients often exclude or contain lower amounts of the more costly vitamin-rich ingredients. The vitamin fortification levels in these simpler diets should be increased to "fill in the gaps" resulting from the reduced amounts of vitamins supplied by feedstuffs. Finally, it is very important to consider recent innovations in feed formulation in fish nutrition, e.g., decreasing inclusion of marine raw materials, usage of ingredients susceptible to oxidation such as blood meal, etc. Since ingredient changes are frequent and unpredictable in computerized least-cost diet formulation, the low levels of vitamins

Table 3.2 Vitamin conversion factors (source: dsm-firmenich Animal Nutrition and Health, unpublished)

Vitamin (active substance)	Unit	Conversion factor active substance form to vitamin form	Product form
Vitamin A (retinol)	IU	1 IU Vitamin A = 0.344 µg Vitamin A acetate (retinyl acetate)	ROVIMIX® A 1000
			ROVIMIX® A 500 WS
			ROVIMIX® A Palmitate 1.6
			ROVIMIX® AD3 1000/200
Vitamin D_3 (cholecalciferol)	IU	1 IU Vitamin D_3 = 0.025 µg Vitamin D_3	ROVIMIX® D_3-500
			ROVIMIX® AD3 1000/200
25OHD$_3$ (25 hydroxy-cholecalciferol)	mg	1 µg 25OHD$_3$ = 40 IU Vitamin D_3	ROVIMIX® Hy–D™ 1.25%
Vitamin E (tocopherol)	mg	1 mg Vitamin E = 1 IU Vitamin E = 1 mg all-rac-α-tocopheryl acetate	ROVIMIX® E-50 Adsorbat
			ROVIMIX® E 50 SD
Vitamin K_3 (menadione)	mg	1 mg of Vitamin K_3 = 2 mg of Menadione Sodium Bisulfite (MSB)	K_3 MSB
		1 mg of Vitamin K_3 = 2.3 mg of Menadione Nicotinamide Bisulfite (MNB)	ROVIMIX® K_3 MNB
Vitamin B_1 (thiamine)	mg	1 mg of Vitamin B_1 = 1.233 mg of Thiamine mononitrate	ROVIMIX® B_1
Vitamin B_2 (riboflavin)	mg		ROVIMIX® B2 80-SD
Vitamin B_6 (pyridoxine)	mg	1 mg Vitamin B_6 = 1.215 mg Pyridoxine hydrochloride	ROVIMIX® B_6
Vitamin B_{12} (cobalamin)	mg		Vitamin B_{12} 1% Feed Grac
			ROVIMIX® B_{12} 1% Feed Gr
Vitamin B_3 (Niacin; nicotinic acid and nicotinamide)	mg	1 mg Nicotinic acid = 1 mg Niacin	ROVIMIX® Niacin
		1 mg Nicotinamide (or Niacinamide) = 1 mg Niacin	ROVIMIX® Niacinamide
Vitamin B_7 (d-Biotin)	mg	1 mg of Biotin = 1 mg D-Biotin	ROVIMIX® Biotin ROVIMIX® Biotin HP
Vitamin B_5 (d-Pantothenic acid)	mg	1 mg d-Pantothenic acid = 1.087 mg Calcium d-pantothenate or 2.174 mg Calcium dl-pantothenate	ROVIMIX® Calpan
Vitamin B_9 (Folic acid)	mg		ROVIMIX® Folic 80 SD
Vitamin C	mg	1 mg Vitamin C = 1 mg L-Ascorbic acid	STAY-C® 35
			STAY-C® 50
			ROVIMIX® C-EC
			Ascorbic acid
β-Carotene	mg		ROVIMIX® β-Carotene 10
			ROVIMIX® β-Carotene 10

* M: Mash; P: Pellet; EXP: Expansion; EXT: Extrusion; W: Water
For more information about further dsm-firmenich products and product forms please ask your local dsm-firmenich representative

t (min.)	Formulation technology	Application*
00 IU/g	Beadlet	M, P, EXP, EXT
0 IU/g	Spray-dried powder, water dispersible	W/MR
00 IU/g	Oily liquid, may crystalize on storage	Oily solution
n A 1,000,000 IU/g n D$_3$ 200,000 IU/g	Beadlet	M, P, EXP, EXT
0 IU/g	Spray-dried powder, water dispersible	M, P, EXP, EXT, W/MR
n A 1,000,000 IU/g n D$_3$ 200,000 IU/g	Beadlet	M, P, EXP, EXT
25OHD$_3$ (12.5 g/kg)	Spray-dried powder, water dispersible	M, P, EXP, EXT, W/MR
00 g/kg)	Adsorbate on silicic acid	M, P, EXP, EXT
00 g/kg)	Spray-dried powder, water dispersible	M, P, EXP, EXT, W/MR
ione: 51.5% (515 g/kg)	Fine crystalline powder	M, P, EXP, EXT, W/MR
ione: 43% (430 g/kg) namide: 30.5% (305 g/kg)	Fine crystalline powder	M, P, EXP, EXT
80 g/kg)	Fine crystalline powder	M, P, EXP, EXT
00 g/kg)	Spray-dried powder	M, P, EXP, EXT, W/MR
90 g/kg)	Fine crystalline powder	M, P, EXP, EXT, W/MR
g/kg)	Fine powder	M, P, EXP, EXT
g/kg)	Spray-dried powder	M, P, EXP, EXT
(995 g/kg)	Fine crystalline powder	M, P, EXP, EXT
(995 g/kg)	Fine crystalline powder	M, P, EXP, EXT, W/MR
g/kg) 00 g/kg)	Spray-dried powder, water dispersible	M, P, EXP, EXT, W/MR
alcium d-pantothenate (980 g/kg) m 8.2 – 8.6% (82 – 86 g/kg)	Spray-dried powder, water dispersible	M, P, EXP, EXT, W/MR
00 g/kg)	Spray-dried powder, water dispersible	M, P, EXP, EXT, W/MR
f total phosphorylated ascorbic acid y (350 g/kg)	Spray-dried powder	M, P, EXP, EXT
f total phosphorylated sodium salt ascorbic ctivity (500 g/kg)	Spray-dried powder	M, P, EXP, EXT, W/MR
(975 g/kg)	Ethyl-cellulose coated powder	M, P, W/MR
00% (990 – 1000 g/kg)	Crystalline powder	W/MR
00 g/kg)	Encapsulated beadlet	M, P, EXP, EXT
00 g/kg)	Cross linked beadlet	M, P, EXP, EXT

Table 3.3 Vitamin concentrations of feedstuffs, mean (SD) (source: adapted from Chen, Y.F., et al., 2019)

Feed ingredients	Vitamins, mg/kg – mean (standard deviation)									
	β-carotene	Vitamin E	Vitamin B$_6$	Folic acid	Niacin	Pantothenic acid	Vitamin B$_2$ (riboflavin)	Vitamin B$_1$ (thiamine)	Choline
Corn	1.62 (0.44)	18.67 (9.48)	3.05 (1.19)	0.08 (0.07)	3.68 (1.73)	3.98 (1.20)	1.30 (0.20)	0.61 (0.17)	292.82 (80.55)
Corn DDGS	2.16 (0.58)	39.24 (9.97)	0.99 (0.47)	0.57 (0.31)	23.23 (13.13)	15.55 (3.02)	3.73 (0.66)	2.14 (0.54)	337.36 (173.38)
Corn germ meal	0.36 (0.09)	19.31 (5.89)	4.44 (1.80)	0.72 (0.49)	29.42 (5.73)	5.35 (2.98)	3.95 (2.28)	4.14 (2.34)	818.00 (513.98)
Corn gluten meal	11.10 (4.05)	7.00 (1.51)	2.13 (0.93)	0.02 (0.02)	36.46 (5.90)	3.93 (2.06)	0.72 (0.36)	0.37 (0.13)	80.04 (13.16)
Corn gluten feed	0.56 (0.11)	9.89 (2.62)	4.15 (1.01)	1.30 (0.46)	52.65 (11.94)	12.56 (3.14)	1.89 (0.80)	5.11 (1.04)	1525.18 (242.75)
Wheat	–	7.27 (1.73)	2.23 (1.33)	0.40 (0.18)	32.86 (8.63)	7.36 (3.60)	0.57 (0.19)	2.12 (1.09)	234.29 (81.65)
Wheat bran	–	19.79 (5.86)	6.52 (1.19)	0.40 (0.18)	102.53 (14.81)	12.35 (4.02)	0.96 (0.26)	2.52 (0.58)	931.59 (121.89)
Wheat shorts	–	9.03 (4.28)	3.93 (1.23)	0.62 (0.27)	4.54 (1.31)	8.18 (0.96)	1.18 (0.12)	3.69 (1.26)	830.48 (226.78)
Soybean meal	–	2.42 (0.85)	6.43 (1.07)	0.51 (0.11)	62.03 (6.21)	3.63 (0.85)	1.77 (0.37)	9.13 (1.51)	1686.54 (144.07)
Rapeseed meal	–	8.66 (2.40)	3.46 (1.80)	0.65 (0.47)	72.82 (10.53)	6.73 (2.75)	1.53 (0.41)	2.14 (0.90)	2276.32 (193.22)
Peanut meal	–	0.67 (0.27)	5.44 (1.11)	104.96 (13.40)	104.96 (14.07)	18.27 (3.46)	1.65 (0.41)	4.22 (1.47)	1527.60 (130.54)
Cottonseed meal	–	10.57 (7.83)	6.30 (1.64)	9.15 (1.87)	9.15 (1.87)	6.75 (1.92)	2.19 (0.32)	1.63 (0.48)	1546.24 (272.49)
Sunflower seed meal	–	1.04 (0.30)	7.57 (0.49)	183.91 (14.60)	183.91 (14.60)	27.25 (1.52)	3.65 (0.16)	1.78 (0.34)	399.96 (69.15)

likely to be supplied by feedstuffs should be disregarded, and adequate dietary vitamin fortification provided.

Vitamins, as pure substances, are almost all sensitive to various physical stress factors. Table 3.4 provides a simple qualitative overview of the sensitivity of each vitamin to these factors, which explains the importance, for industrial application, of properly formulating each vitamin to make it more stable and ensuring that the calculated amount per kilogram of feed reaches the animal.

FACTORS AFFECTING VITAMIN REQUIREMENTS AND UTILIZATION

1. Physiological make-up, genetics, and production function

Vitamin requirements of animals and humans depend significantly on their physiological make-up related to the traits given by decades of genetic selection, life stage, health, and nutritional status. In aquatic species, vitamin requirements can vary between species and be affected by body weight, rearing environment (e.g., salinity; intensive versus extensive conditions), dietary factors (e.g., levels of polyunsaturated fatty acids), physiological changes (e.g., flatfish metamorphosis, smoltification), disease and stress (Fernández and Gisbert, 2011; NRC, 2011; Mai *et al.*, 2022). The following chapters will describe further these endogenous and exogenous factors along with their impact on vitamin requirements.

Selection for a faster growth and improvement of feed conversion allow fish and crustaceans to reach much higher weights at much younger ages with less feed consumption. In fish

Table 3.4 Sensitivity of vitamins to physical stress factors (source: dsm-firmenich Animal Nutrition and Health, unpublished)

Additive	Temperature	Oxygen	Humidity	Light
Vitamin A	Marked effect	Marked effect	Marked effect	Marked effect
Vitamin D	Marked effect	Marked effect	Marked effect	Marked effect
25OHD$_3$ (calcifediol)	Marked effect	Marked effect	Marked effect	Marked effect
Vitamin E	No effect	Marked effect	No effect	Marked effect
Vitamin K$_3$	Marked effect	Marked effect	Marked effect	No effect
Vitamin B$_1$	Marked effect	Marked effect	Marked effect	No effect
Vitamin B$_2$	No effect	No effect	Marked effect	Marked effect
Vitamin B$_6$	Marked effect	No effect	Marked effect	Marked effect
Vitamin B$_{12}$	Marked effect	Marked effect	Marked effect	Marked effect
Pantothenic acid	Marked effect	No effect	Marked effect	No effect
Nicotinates	No effect	No effect	No effect	No effect
Biotin	Marked effect	No effect	No effect	No effect
Folic acid	Marked effect	No effect	Marked effect	Marked effect
Vitamin C	Marked effect	Marked effect	Marked effect	No effect
Carotenoids	Marked effect	Marked effect	Marked effect	Marked effect

Marked effect	Moderate effect	No effect

and shrimp nutrition, vitamin requirements were often estimated several years, if not decades, ago when growth and feed conversion were inferior to current genotypes and production practices. Applying past requirements to high performing strains can lead to suboptimal vitamin nutrition which, in turn, compromise growth potential and welfare. There is thus a need to update and adapt recommendations for dietary vitamin levels.

Finally, the effects of sex, ploidy status (diploid versus triploid) and genotype on vitamin requirements are yet to be described for most aquatic species of interest in aquaculture.

2. Antioxidants and role of vitamins on stress and immune system modulation

Stress and disease conditions influence the antioxidant vitamins E, C, and β-carotene (Bacou *et al.*, 2021; Combs and McClung, 2022). Antioxidant vitamins play essential roles in animal health by inactivating harmful free radicals produced through regular cellular activity, and more intensely under various stressful conditions (e.g., low oxygen event, disease outbreak). Free radicals can damage biological systems. Free radicals, including hydroxy, hypochlorite, peroxyl, alkoxy, superoxide, hydrogen peroxide, and singlet oxygen, are generated by autoxidation, radiation, or some oxidases, dehydrogenases, and peroxidases. Also, phagocytic granulocytes undergo respiratory bursts to produce oxygen radicals to destroy pathogens. However, these oxidative products can, in turn, damage healthy cells if they are not eliminated. Antioxidants quench these highly reactive free radicals, maintaining cells' structural and functional integrity (McCay, 1985; Mai *et al.*, 2022).

Tissue defense mechanisms against free-radical damage generally involve vitamins C, E, and carotenoids as significant vitamin antioxidant sources. In addition, antioxidant vitamins can protect several metalloenzymes, which include glutathione peroxidase (selenium), catalase (iron), superoxide dismutase (copper, zinc, and manganese), and even pyridoxine (Dalto and Matte, 2017) that are also critical in preventing oxidative damages in internal cellular constituents (Combs and McClung, 2022).

A compromised immune system will reduce animal production efficiency through increased susceptibility to disease, leading to animal morbidity and mortality. Both *in vitro* and *in vivo* studies show that antioxidant vitamins enhance cellular and noncellular as well as innate and adaptive immunity (Mai *et al.*, 2022). The antioxidant function of these vitamins could, at least in part, enhance immunity by maintaining the function and structural integrity of critical immune cells.

In the following chapters, several examples are presented where antioxidant vitamins enhanced stress resistance and immunocompetence of salmonids, warm water fish species, and crustaceans.

3. Vitamin antagonists

Vitamin antagonists (anti-metabolites) interfere with the activity of various vitamins. Oldfield (1987) summarized the action of antagonists, which:

- could cleave the vitamin molecule and render it inactive, as occurs with thiaminase, found in raw fish and some feedstuffs, and thiamine (pyrithiamine is another thiamine antagonist)
- could bind with the metabolite, with similar results, as happens between avidin, found in raw egg white, and streptavidin, from *Streptomyces* mold and biotin
- could, because of structural similarity, occupy reaction sites and deny them to the vitamin, as with dicumarol, found in certain plants or sulphonamides (antibiotics used in aquaculture) and vitamin K
- inactivate, through rancid fats, biotin and destroy vitamins A, D, E, and possibly others.

These effects were also reviewed by Wolley (2012). The presence of vitamin antagonists in animal and human diets should be considered when adjusting vitamin allowances, as most vitamins have antagonists that reduce their utilization. For instance, sulphonamides could interfere with folic acid synthesis by the gut microflora and may require supplemental dietary folate to avoid clinical signs of deficiency when aquatic species are exposed to such chemo-therapeutants (Ovung and Bhattacharyya, 2021; Mai et al., 2022) Mycotoxins are antagonists in the feed that can substantially decrease antioxidant nutrient assimilation and increase their requirements to prevent the damaging effects of free radicals and toxic products. It is now increasingly recognized that at least 25% of the world's grains are contaminated with myco-toxins (Surai and Dvorska, 2005). Mycotoxins cause digestive disturbances such as vomiting and diarrhea, and internal bleeding and interfere with the absorption of dietary vitamins A, D, E, and K (Surai and Dvorska, 2005; McDowell, 2006). Vitamin C can reduce the toxicity of certain mycotoxins (Su et al., 2018). Mycotoxins are antagonists in the feed that can substan-tially decrease antioxidant nutrient assimilation and increase their requirements to prevent the damaging effects of free radicals and toxic products (Royes and Yanong, 2010; Oginni et al., 2020). According to a quite recent scientific paper (Eskola et al., 2020) the majority (60–80%) of the world's grains are contaminated with detectable levels of mycotoxins (Eskola et al., 2020). Mycotoxins cause digestive disturbances and interfere with the absorption of dietary vitamins A and vitamins D, E, and K (Surai and Dvorska, 2005; McDowell, 2006). Vitamin C can reduce the toxicity of certain mycotoxins (Su et al., 2018; Shalaby, 2004) Toxic minerals may be antagonists and will likewise increase vitamin requirements. Vitamin E protects against the toxicity of certain heavy metals (e.g., cadmium, mercury, and lead), which increases the need for the vitamin (McDowell, 2000). Rothe et al. (1994) reported that dietary vitamin C (1,000 mg) reduced by 35–40% the elevated levels of cadmium in pig kidneys and liver induced by high supplementation of copper (175 mg Cu/kg feed).

Specific vitamins can likewise be antagonistic to other vitamins. Excess vitamin A (50,000 IU/kg) can affect the metabolism (e.g., absorption) of other fat-soluble vitamins (Grisdale-Helland et al., 1991).

4. Use of antimicrobial drugs

Some antimicrobial drugs will increase the vitamin needs of animals by altering intestinal microflora and inhibiting the synthesis of specific vitamins. Certain sulfonamides may increase the requirements for biotin, folic acid, vitamin K, and possibly others when intestinal synthe-sis is reduced. These gut health issues may be insignificant except when antagonistic drugs toward a particular vitamin are added in excess, i.e., sulfaquinoxaline versus vitamin K and sulfonamide potentiators versus folic acid (Perry, 1978).

5. Levels of other nutrients in the diet

The fat level in the diet may affect the absorption of the fat-soluble vitamins A, D, E, and K and the requirements for vitamin E and possibly other vitamins. Fat-soluble vitamins may fail to be absorbed if fat digestion is impaired by liver damage or when the enterohepatic recirculation of bile acids is interrupted. Type (e.g., animal fats, vegetable oils, and blends) and quality (e.g., cis versus trans, saturated versus polyunsaturated fatty acids (PUFAs), and oxidized sources) of fats can influence individual vitamin allowances (Ellis and Madsen, 1944; Thode Jensen et al., 1983; Astrup and Langebrekke, 1985; Cerolini et al., 2000). For example, a precise vitamin E : PUFA ratio may not apply to all diet and health status types. Therefore, there has been no consensus on the exact vitamin E : PUFA ratio to determine the vitamin requirement. However, the published human data for a diet with an average concentration of PUFA and containing mainly linoleic acid indicates that the additional vitamin E requirement ranges from 0.4 to 0.6 mg RRR-α-tocopherol/g of PUFA in the diet. A ratio of 0.5 mg RRR-α-tocopherol/g of linoleic acid was used in the diet, and the degree of unsaturation of the dietary fatty acids was also considered to evaluate the required vitamin E. Thus, using the proposed equation, humans' estimated requirement for vitamin E varied from 12 to 20 mg/day for a typical range of dietary PUFA intake (Raederstorf *et al.*, 2015). Calculations in animal diets indicate that high dietary PUFA increases the vitamin E requirement by 3 mg/g PUFA (Bieber-Wlaschny, 1988). This information in humans and various animal species calls for further investigation in aquaculture.

Many interrelationships of vitamins with other nutrients exist and affect requirements. For example, prominent interrelationships exist for vitamin E with selenium, vitamin D with calcium and phosphorus, choline with methionine, and niacin with tryptophan.

6. Body vitamin reserves

The fat-soluble vitamins A, D, and E but not vitamin K, are more inclined to remain in the body. This is especially true of vitamin A and carotene, which may be stored by an animal in its liver and fatty tissue in sufficient quantities to meet its requirements for varying periods. Body storage of B group vitamins, except for vitamin B_{12}, is irrelevant. Overall, a daily supplementation at the proper levels typical of each species and growth stage is generally recommended in animal husbandry in industrial conditions. In Chapter 4, when discussing individual vitamins, we have summarized in specific tables the average concentrations of each vitamin in tissues of different farmed species (e.g. Table 4.1 provides the average concentration of vitamin D in tissues of different farmed species).

Vitamin description

FAT-SOLUBLE VITAMINS

Vitamin A
Chemical structure and properties

Vitamin A is used as a generic term for all the non-carotenoid β-ionone derivatives possessing the biological activity of all-trans-retinol (Combs and McClung, 2022). Retinol ($C_{20}H_{30}O$) is the alcohol form of vitamin A_1. Replacement of the alcohol group (-OH) by an aldehyde group (-CHO) yields retinal, and replacement by an acid group (-COOH) gives retinoic acid (Ross and Harrison, 2013). Unlike mammals, birds, and marine species, most fish in freshwater contain vitamin A_2 ($C_{20}H_{26}O$), also referred to as dehydroretinol and 3,4-didehydroretinol, that frequently predominates over vitamin A_1 in tissues like liver, pyloric caeca, intestine, and gonads (Guillou et al., 1989; NRC, 2011; Gesto et al., 2012; La Frano et al., 2018). Vitamin A_2 can be derived from vitamin A_1 by dehydrogenation or synthesized de novo directly from certain carotenoids such as astaxanthin (Lambertsen and Braekkan, 1969; Schiedt et al., 1985; Al-Khalifa and Simpson, 1988). Vitamin A_1 and A_2 differ in their chemical structure only by the presence of an additional desaturation between carbons 3 and 4 of the β-ionone ring in the latter vitamin (Figure 4.1). The presence of vitamin A_2 has not been detected in crustaceans (Fisher and Kon, 1959; Goldsmith and Cronin, 1993). Vitamin A products for feed use include retinyl acetate, propionate, and palmitate esters (Figure 4.2).

Vitamin A alcohol (retinol) is a nearly colorless, fat-soluble, long-chain, unsaturated compound with 5 bonds. Since it contains double bonds, vitamin A can exist in different isomeric forms. Vitamin A and its precursors, carotenoids, are rapidly destroyed by oxygen, heat, light, and acids. Moisture and trace minerals reduce feed vitamin A activity (Olson, 1984). The number

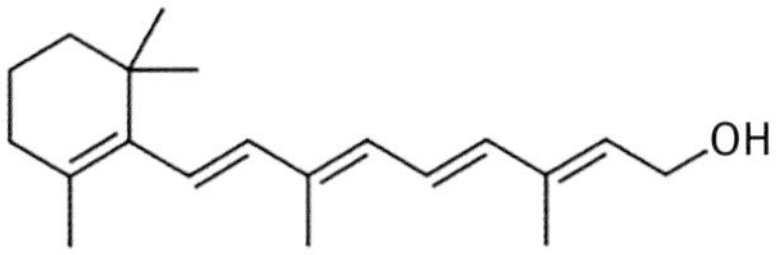

Vitamin A1

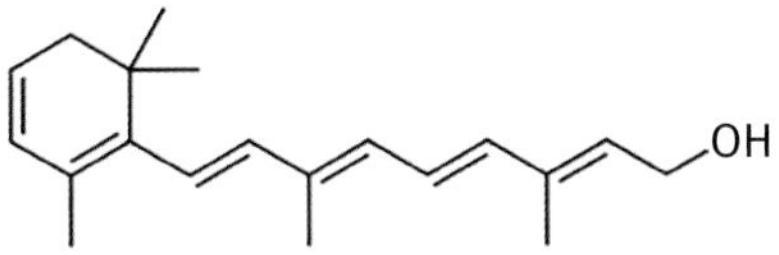

Vitamin A2

Figure 4.1 Chemical structure of vitamins A_1 (retinol) and A_2 (dehydroretinol) (source: https://link.springer.com/chapter/10.1007/978–981–10–8022–7_102)

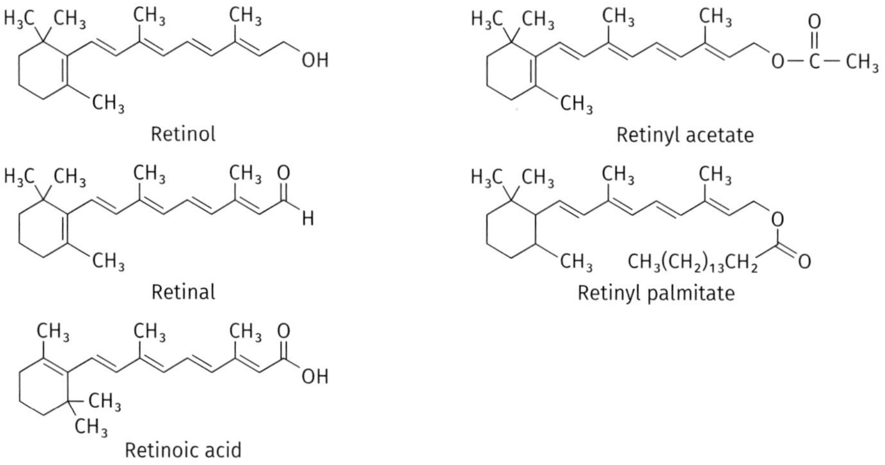

Figure 4.2 Vitamin A chemical structure of some natural and synthetic forms (source: Oviedo-Rondón et al., 2023b)

of carotenoids that serve as provitamin A is estimated to be more than 50 (Simpson and Chichester, 1981).

The vitamin A activity from retinoids and provitamin A carotenoids are standardized and expressed as international units (IU) and retinol equivalent (RE) (Combs and McClung, 2022):

- 1 IU = 0.3 µg all-trans-retinol = 0.344 µg all-trans-retinyl acetate = 0.55 µg all-trans-retinyl palmitate.
- 1 RE = 1 µg all-trans-retinol = 3.3 IU all-trans-retinol = 2.0 µg all-trans-β-carotene in dietary supplements = 12.0 µg all-trans-β-carotene in foods.

Some fish nutrition studies do not report the conversion factor or apply the 0.3 conversion factor despite supplementing vitamin A exhibiting various activities that can lead to miscalculation of feed formulation.

The conversion of carotenoids into vitamin A has been well described in terrestrial and avian species (e.g., Olson and Hayaishi, 1965; Olson, 1989; Surai et al., 2001) compared to fish and crustaceans (Green and Fascetti, 2016).

Astaxanthin is of particular importance for aquaculture species since it represents the main pigment stored in tissues of salmonids and shrimp and is responsible for the pigmentation of the flesh and exoskeleton (reviewed by Tan et al., 2022). It can also be considered a strong antioxidant, immunostimulant and precursor of vitamin A (reviewed by Liu et al., 2022 and Mai et al., 2022).

Current knowledge in fish shows that the provitamin A carotenoids and their conversion efficiency vary between aquatic species, carotenoids, vitamin A status, and environmental conditions. Nearly a century ago, Morton and Creed (1939) were the first to demonstrate that certain carotenoids, such as β-carotene, served as precursors of vitamins A_1 and A_2 in freshwater fish using European perch (*Perca fluviatilis*) as model. Their findings were confirmed in subsequent studies with various species (e.g., Gross and Budowski, 1966). However, β-carotene is not always a precursor of vitamin A_2 in freshwater fish (Czeczuga and Czepak, 1976). For instance, β-carotene was not a precursor of vitamin A in brook trout (*Salvelinus fontinalis*) according to Poston (1968). Results from Del Tito Jr. (1983) indicated that lutein, but not β-carotene, can be converted into vitamin A_2 in goldfish (*Carassius auratus*). In other species, lutein was not a precursor of any of the vitamin A (Gross and Budowski, 1966). The role of canthaxanthin as provitamin A_1 and astaxanthin as direct precursor of vitamin A_2 has been well demonstrated in Nile tilapia (*Oreochromis niloticus*) and rainbow trout (Katsuyama and Matsuno, 1988; Guillou *et al.*, 1989). Evidence from studies conducted in rainbow trout and salmon in freshwater suggests that water temperature is a driving factor of bioconversion of carotenoids into vitamin A (Guillou *et al.*, 1989; Ørnsrud *et al.*, 2004a), which contrasts with endotherms based on a study conducted with rats at 5, 25 and 35°C (Smith and Borchers, 1972). Possible effect of genetics (within species) on carotenoid conversion into vitamin A has not been studied in fish or shrimp yet.

Quantitative description of carotenoid conversion efficacy is scarce in the fish and shrimp literature. Nevertheless, past studies have exposed factors affecting the conversion rate of targeted carotenoids into vitamin A. One of these factors refers to the molecular form of pigments as already discussed. Furthermore, lutein can be converted into vitamin A_1, albeit at a slower rate than β-carotene in goldfish (Del Tito Jr., 1983).

The vitamin A status of fish is another factor affecting the conversion rate of carotenoids. The conversion of astaxanthin into vitamins A_1 and A_2 reached between 11 and 17% in vitamin A-deficient salmonids but was nil in vitamin A-repleted fish (Schiedt *et al.*, 1985; Al-Khalifa and Simpson, 1988). These results indicated that (1) astaxanthin has potential to replace dietary vitamin A in case of shortage, and (2) vitamin A status of fish may affect astaxanthin retention which, in turn, may have implications for salmon flesh pigmentation. The role of astaxanthin as provitamin A in shrimp has been suspected, but not demonstrated yet (Pangantihon-Kühlmann *et al.*, 1998; Fawzy *et al.*, 2022).

Lastly, the conversion rate of carotenoids also depends on the ontogenetic developmental stage. Ørnsrud *et al.* (2004a) observed the conversion of carotenoids into vitamin A at the fry stage in salmon, after the liver became functional, but not at the egg stage. Levels of vitamin A in livers of rainbow trout <50 g were lower than in trout >200 g (Schiedt *et al.*, 1985). The low conversion ratio of β-carotene to vitamin A of 19 : 1 observed in hybrid tilapia may be explained by the small size of experimental fish that had an initial body weight of 1.6 g (Hu *et al.*, 2006). It is recommended to describe further the conversion rates of carotenoids at various life stages to match dietary vitamin A supplementation to metabolic scope of species that are commercially relevant.

Natural sources

In its active form, vitamin A is scarce in nature, as it is found predominantly as an ester only in fish oil, lard, and kelp meal (NRC, 2011). Vitamin A of animal origin is found mainly in the liver and adipose tissues. Green plants contain provitamin A carotenoids such as β-carotene, which content varies greatly according to the species, state of maturity, and preservation. Maize and its derivatives contain significant quantities of pigmenting carotenoids that can serve as provitamin A such as zeaxanthin in fish (Katsuyama and Matsuno, 1988). There is still a need to determine which carotenoids have provitamin A activities in fish and shrimp species of economic importance.

There is evidence that yellow corn may lose carotene rapidly during storage. For instance, a hybrid corn high in carotene lost one-half of its carotene in 8 months of storage at 25°C (77°F) and about three-quarters in 3 years. Less carotene was lost during storage at 7°C (45°F) (Quackenbush, 1963). The bioavailability of natural β-carotene was less than chemically synthesized forms (White *et al.*, 1993). Aside from yellow corn and its by-products, practically all concentrates used in feeding animals are devoid of vitamin A value, or nearly so.

Commercial forms

The vitamin A activity contained in ingredients and aquaculture feeds is difficult to predict. Therefore, the provision of vitamin A in fish and shrimp diets is achieved mainly through synthetic forms. The most convenient and effective means of providing vitamin A to fed aquaculture species is inclusion in premixes added to feed.

The major sources of supplemental vitamin A used in animal diets are trans-retinyl acetate and trans-retinyl palmitate. The propionate ester is much less common (McGinnis, 1988). These are available in gelatin or glycerin beadlet product forms for protection against oxidative destruction in premixes, mash, and pelleted and extruded feeds. Carbohydrates, gelatin, and antioxidants are generally included inside the beadlets to stabilize vitamin A to provide physical and chemical protection against factors either normally present in the feed or due to feeding treatment and storage that are destructive to vitamin A.

The vitamin A acetate products most frequently used in aquaculture feeds contain 1,000,000 IUs or United States Pharmacopeia Units (USP) per gram of product (United States Pharmacopeia, 1980). The values of 1 IU or 1 USP are the same and equal the activity of the 0.3 µg of all-trans-retinol or 0.344 µg of all-trans-retinyl acetate, or 0.6 µg of β-carotene. In other words, 1 mg of β-carotene is equivalent to 1,667 IU of vitamin A.

The stability of vitamin A in extruded feeds after 3 months of storage at room temperature can vary between 40 and 80% (NRC, 2011). Loss of vitamin A is affected by several factors such as interactions with other feed components and conditions during processing and storage. Trace minerals in animal feeds and supplements, particularly inorganic copper, and zinc, can be detrimental to vitamin A stability (e.g., Yang, P., et al., 2021), but this negative effect has not been reported in premixes for fish or crustacean feeds. Vitamin A (and carotene) destruction also occurs from processing feed with steam and pressure. Pelleting effects of vitamin A in the feed are determined by die thickness and hole size, which produce frictional heat and a shearing effect that can break supplemental vitamin A beadlets and expose the vitamin. In addition, steam application exposes feed to heat and moisture. Running fines back through the pellet mill or extruder exposes vitamin A to the same factors a second time. Vitamin A supplements should not be stored for prolonged periods (~90 days) before feedings, especially if temperature and relative humidity are above 22°C and 28%, respectively (Saensukjaroenphon et al., 2020). Chen J. (1990) measured the stability of 3 commercial cross-linked vitamin A beadlets on the market in trace mineral premixes and feeds. After 3 months of storage at high temperature and humidity, vitamin A retention varied from 30 to 80%, depending on the antioxidant present in the beadlet.

It is therefore important to carefully assess the quality of the commercial product. The gelatin beadlet, in which the vitamin A ester is emulsified into a gelatin-antioxidant viscous liquid formulation and spray dried into discrete dry particles, results in products with good chemical stability, physical stability, and excellent biologic availability (Shields et al., 1982). The reaction between gelatin and sugar makes the beadlet insoluble in water and gives it a more resistant coating that can sustain higher pressure, friction, temperature, and humidity (Frye, 1994).

Metabolism

Absorption and transport

Most studies on vitamin A metabolism have been conducted in terrestrial species. Therefore, it is assumed in this chapter that absorption and transport mechanisms are the same for fish and crustaceans, unless reported otherwise in the aquaculture literature.

Evidence suggests that the absorption and transport of retinol (vitamin A_1) and dehydroretinol (vitamin A_2) follow similar mechanisms in fish than in terrestrial species (La Frano *et al.*, 2018). For the sake of conciseness, vitamin A will be used in this section and applied to both vitamins A_1 and A_2. When absorption and transport of retinol are described, it is assumed that these mechanisms also apply to dehydroretinol.

Dietary retinyl esters (e.g., acetate) are hydrolyzed to retinol in the intestine by pancreatic retinyl ester hydrolase, absorbed as free alcohol retinol, and then re-esterified in the mucosa, mostly to palmitate. Vitamin A and β-carotene become dispersed in micelles before absorption from the intestine. These micelles are composed of mixtures of bile salts, monoglycerides, and long-chain fatty acids, together with vitamins D, E, and K, all of which influence the transfer of vitamin A and β-carotene to the intestinal cell. Here, most of the β-carotene is converted to vitamin A, which is converted to various esters (Ross and Harrison, 2013).

The retinyl esters are transported mainly in association with lymph chylomicrons to the liver. Hydrolysis of the ester storage form mobilizes vitamin A from the liver as free retinol or dehydroretinol. Retinol and dehydroretinol are released from the hepatocyte as a complex with retinol-binding protein (RBP) and transported to peripheral tissues. Retinol, in association with RBP circulates in peripheral tissues complexed to a thyroxine-binding protein, transthyretin (Blomhoff *et al.*, 1991; Ross, 1993; Mai *et al.*, 2022). The retinol-transthyretin complex is transported to target tissues, where the complex binds to a cell-surface receptor. The receptor was found in all tissues known to require retinol for their function, particularly the pigment epithelium of the eye (Wolf, 2007). Once the retinoids are transferred into the cell, they are quickly bound by specific binding proteins in the cell cytosol (Figure 4.3).

An enzyme converts β-carotene in feed to retinal in the intestinal mucosa. The retinal is then reduced to retinol (vitamin A_1). However, extensive evidence exists also for random (eccentric) cleavage, resulting in retinoic acid and retinal, with a preponderance of apocarotenals formed as intermediates (Wolf, 1995). The main site of vitamin A and carotenoid absorption is the mucosa of the proximal jejunum. Although carotenoids are normally converted to retinol and

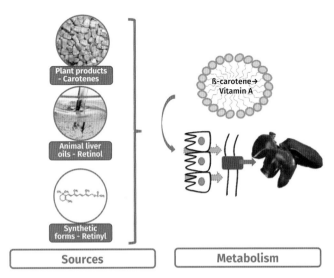

Figure 4.3 Main sources and types of the precursors of vitamin A and schematic illustration of absorption (source: Oviedo-Rondón *et al.*, 2023b)

dehydroretinol in the intestinal mucosa, they may also be converted in the liver and other organs (Lambertsen and Braekkan, 1969; McGinnis, 1988; La Frano *et al.*, 2018).

Intestinal absorption of vitamin A is calculated to be between 40 and 80%. Many factors may modify it, either positively, such as the inclusion of fats in the diet, the addition of antioxidants, and the use of moderate levels of vitamin E (Abawi *et al.*, 1985; Noel and Brinkhaus, 1998) or negatively, such as high levels of vitamin E, the presence of aflatoxins or enteric infections (West *et al.*, 1992). Thus, enteritis may reduce vitamin A levels both in plasma and in the hepatic reserves, hence increasing the requirements of vitamin A because of poor absorption and oxidation induced by the cellular immune response (Augustine and Ruff, 1983; Allen, 1988, 1997; Allen *et al.*, 1996).

Several factors affect the absorption of carotenoids. Cis-trans-isomerism of the carotenoids is important in determining their absorbability, with the transforms being more efficiently absorbed (Stahl *et al.*, 1995). Dietary fat is important for carotenoids absorption (Fichter and Mitchell, 1997). Cholesterol improved astaxanthin absorption in Atlantic salmon (Chimsung *et al.*, 2014). Dietary antioxidants (e.g., vitamin E) also affect carotenoids' utilization and perhaps absorption. It is uncertain whether the antioxidants contribute directly to the efficient absorption or whether they protect both carotene and vitamin A from oxidative breakdown.

The intracellular retinoid-binding proteins bind retinol, dehydroretinol, retinal, and retinoic acid to protect against decomposition, solubilize them in an aqueous medium, render them nontoxic, and transport them within cells to their site of action (Ross and Harrison, 2013; La Frano *et al.*, 2018). These binding proteins also function by presenting the retinoids to the appropriate enzymes for metabolism (Wolf, 1991). Some of the principal forms of intracellular (cytoplasmic) retinoid-binding proteins are cellular retinol-binding proteins (CRBP, I and II), cellular retinoic acid-binding proteins (CRABP, I and II), cellular retinaldehyde binding protein (CRALBP), and 6 nuclear retinoic acid receptors (RAR and retinoid X receptor [RXR], with alpha, beta, and gamma forms). There are 2 classes of nuclear receptors with all-trans-retinoic acid, the ligand for RAR; 9-cis-retinoic acid is the ligand for RXR (Kasner *et al.*, 1994; Kliewer *et al.*, 1992). Retinol and dehydroretinol are readily transferred to the egg in fish (Gesto *et al.*, 2012).

The color of salmonids and shrimp products is a valuable attribute for the consumer. Pigmenting carotenoids are mainly astaxanthin and canthaxanthin. Their absorption and transport have been recently reviewed by Turchini *et al.* (2022). Briefly, carotenoid pigments are hydrolyzed into mixed micelles before they are delivered to the glycocalyx and microvilli on the apical surface of enterocyte. Carotenoids are then catabolized (e.g., converted into vitamin A) or transported by lipoproteins to be deposited mostly in the muscle, integument, and gonads depending on species and life stage. Digestibility of carotenoids is highly variable. For instance, apparent digestibility coefficients of astaxanthin varied between 20 to 60% in post-smolt Atlantic salmon (Ytrestøyl *et al.*, 2005).

Storage and excretion

Unlike mammals and avian species, the preferred tissue for storage of vitamin A is not always the liver in fish and crustaceans and varies depending on species and sex. The liver of rainbow trout contains between 44 and 95% of the total body vitamin A (Guillou *et al.*, 1989; Gesto *et al.*, 2012). The pyloric caeca are another important tissue for vitamin A storage in mature rainbow trout. Guillou *et al.* (1989) reported that 41 and 27% of total vitamin A was accumulated in the pyloric caeca of mature female and male rainbow trout, respectively. The intestine is the main tissue for vitamin A storage followed by the liver and kidney in lamprey (*Lampetra japonica*) (Wold *et al.*, 2004), whereas most of vitamin A is concentrated in the pyloric caeca of arrowtooth halibut (*Atheresthes evermanni*) followed by intestine and liver (Yoshikawa *et al.*, 2006). In crustaceans, vitamin A has been found mostly in the eye (Wolfe and Cornwell, 1965; Shiau and Chen, 2000). The relatively large storage capacity of vitamin A in fish and crustaceans must be considered in studies of vitamin A requirements to ensure that intakes that appear adequate for a given function are not being supplemented by reserves stored before the observation period.

The chemical form of vitamin A stored in aquaculture species depends on life stage. For instance, the primary form of vitamin A storage is aldehyde retinal in fish eggs and retinol in fish just before the onset of first feeding (Rønnestad *et al.*, 1998; Irie and Seki, 2002). The proportion of vitamin A_1 (retinol) increased in fish eggs with incubation temperature, most likely as the result of conversion of aldehyde retinal into retinol (Ørnsrud *et al.*, 2004a). Vitamin A_1 (retinol, $C_{20}H_{30}O$) is stored in fish tissues mostly as the biologically inactive form retynil esters.

Biochemical functions

Vitamin A is necessary to support growth, vision, cell differentiation, immune and antioxidant responses, reproduction, and embryonic and larval development in fish and shrimp. These biochemical functions of vitamin A have been reviewed recently by Hernandez and Hardy (2020) and Mai *et al.* (2022). Current knowledge on the biochemical functions of vitamin A in fish and crustaceans has remained limited relative to birds and mammals.

Vitamin A is essential for growth and life. Vitamin A and its derivatives, the retinoids, profoundly influence organ development, cell proliferation, and cell differentiation, and their deficiency originates or predisposes several disabilities (McDowell, 2000; Esteban-Pretel *et al.*, 2010). During embryogenesis, retinoic acid has been shown to influence processes governing the patterning of neural tissue and craniofacial, eye, and olfactory system development; retinoic acid affects the outcome, regeneration, and well-being of neurons in rodents (Asson-Batres *et al.*, 2009). In zebrafish, retinoic acid acts as a signaling molecule for vertebrae formation, fin regeneration, and osteogenic cells and osteoblasts differentiation (Draut *et al.*, 2019).

Recent discoveries have revealed that most, if not all, actions of vitamin A in development, differentiation, and metabolism are mediated by nuclear receptor proteins that bind retinoic acid, the active form of vitamin A (Iskakova *et al.*, 2015). A group of retinoic acid-binding proteins (receptors) function in the nucleus by attaching to promoter regions in several specific genes to stimulate their transcription and thus affect growth, development, and differentiation. RAR in cell nuclei are structurally homologous and functionally analogous to the known receptors for steroid hormones, thyroid hormone (triiodothyronine), and vitamin D $1,25(OH)_2D_3$. Thus, retinoic acid is now recognized as a hormone regulating more than 500 genes' transcription activity (Ross, 1993; Shin and McGrane, 1997; Draut *et al.*, 2019).

Dietary supplementation with vitamin A can alleviate negative effects of contaminants such as polycyclic aromatic hydrocarbons (PAHs) present mostly in plant oils included in aquaculture feeds. Total PAHs can be up to 10× higher in feed containing vegetable oils than fish oils and nearly 2× higher in fillet of fish fed plant oils than fish oil-based diets (Berntssen *et al.*, 2010). Supplementing retinyl acetate at 8,721 IU kg⁻¹ diet at least partially alleviated the harmful effects of PAHs on energy metabolism, growth, and storage of vitamin A in post-smolt Atlantic salmon (Berntssen *et al.*, 2016). More details on this positive effect of vitamin A on contaminants are provided in Chapter 5.

Vitamin A interacts with vitamins E, carotenoids and may disrupt the phosphorus-calcium homeostasis in aquatic animals.

Nutritional assessment

Retinyl palmitate, dehydroretinol, retinol and its metabolite retinoic acid are the compounds relevant for the determination of vitamin A status in fish and shrimp. Storage of these retinoids in the liver has been used as a general indicator of vitamin A status. Vitamin A level in serum is not reliable to assess vitamin A status because blood serves as a vehicle to transport retinoids, not a storage tissue (Dunning, 2018). The provitamin A carotenoids could be used as an indirect measurement of vitamin A status.

The levels of vitamin A in fish tissues can be highly variable with sometimes 10-fold difference observed between aquaculture-fed species and between studies (e.g., Guillou et al., 1989; Gesto et al., 2012; Stancheva and Dobreva, 2013). It is thus challenging to establish typical ranges as indicators of deficiency or excess in fish and shrimp.

To determine retinoid levels, lipids are first extracted from tissues using solvent (e.g., Yerlikaya et al., 2022). Thereafter, retinoids are separated using reversed-phase high-performance liquid chromatography (HPLC) and detected by ultraviolet spectrometry (Combs and McClung, 2022). This widespread method lacks specificity and has analytical limitations, especially since retinoids are often present in fish at the pmol g^{-1} level (Gesto et al., 2012). A more sensitive method to measure retinoids has been adopted recently and consists in pairing mass spectrometry (MS) with HPLC (HPLC-MS). This method has been proven reliable to measure retinoids in livers and has been applied to detect detrimental effects of contaminants on vitamin A status in fish (Dunning, 2018).

Deficiency signs

Signs of vitamin A deficiency are associated with various clinical conditions, including eye pathologies (Figure 4.4), bone deformities, fin and skin lesion, anorexia, and poor growth (NRC, 2011; Mai et al., 2022). Deficiency signs along with vitamin A requirements vary between species and are thus described more categorically in the following chapters.

Figure 4.4 Coho salmon with cataracts (source: courtesy R.W. Hardy)

Safety
Excess of vitamin A is physiologically possible in aquatic animals, but unlikely to happen in aquaculture-fed species due to the extremely high dietary levels needed to observe clinical signs of hypervitaminosis. Quantitative information on risks of vitamin A excess among different species is presented in the following chapters.

Vitamin D
Chemical structure and properties
The term vitamin D covers a group of closely related steroids, sometimes referred to as "sunshine vitamin" because it is synthesized by the action of ultraviolet (UV) radiation on the skin of most terrestrial vertebrates (Combs and McClung, 2022). In fish, cutaneous synthesis of vitamin D via UV radiation and blue light is possible, but likely insufficient to meet requirements (Barnett *et al.*, 1982a; Rao and Raghuramulu, 1997; Pierens and Fraser, 2015). Furthermore, biosynthesis of vitamin D from the provitamin D_3 or 7-dehydrocholesterol present in the skin appeared low in fish compared to other vertebrates and indicated vitamin D must be obtained exogenously (Blondin *et al.*, 1967; Lock *et al.*, 2010; Pierens and Fraser, 2015).

The most important forms of vitamin D are ergosterol (vitamin D_2, $C_{28}H_{44}O$) and cholecalciferol (vitamin D_3, $C_{27}H_{44}O$) (Figure 4.5). Ergocalciferol is characteristic of plants, whereas vitamin D_3 or cholecalciferol is found almost exclusively in animals. The biological activity of vitamin D_2 is generally lower than that of vitamin D_3 in aquatic animals (NRC, 2011).

Natural sources
Ergocalciferol is found mostly in plants and eukaryotes. Yeast, in particular *Saccharomyces cerevisiae*, and UV-irradiated mushrooms are significant sources of ergocalciferol and its provitamin form ergosterol with concentrations varying from 0.4 to 87.5 mg/g of dry weight (Rattray *et al.*, 1975; Janoušek *et al.*, 2022).

Cholecalciferol is found mostly in animals but is particularly abundant in oily fish like herring (15–27.5 µg/100 g fresh sample). Tuna liver oil and cod liver oil contain cholecalciferol at levels of 25,000–112,500 and 85–1,250 µg/ 100 g oil, respectively (Takeuchi *et al.*, 1984; Janoušek *et al.*, 2022).

Cholecalciferol is one of the vitamins most affected by replacement of fishmeal and fish oil in aquaculture feeds (Figure 4.6). It is estimated that current commercial salmon feeds contain

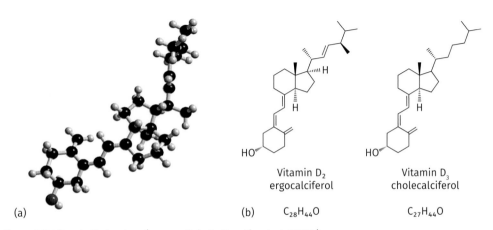

(a) (b) $C_{28}H_{44}O$

Vitamin D_2
ergocalciferol

Vitamin D_3
cholecalciferol

$C_{27}H_{44}O$

Figure 4.5 Vitamin D structure (source: Oviedo-Rondón *et al.*, 2023b)

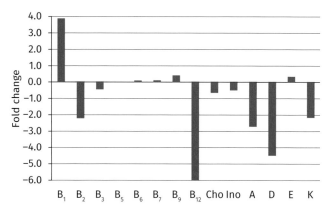

Figure 4.6 Changes in dietary vitamins supplied by ingredients (excluding vitamin premix) in salmon feeds between 1990 and 2022 (source: Dumas, A., unpublished)

less than 4.5× vitamin D_3 than feeds formulated in 1990 (see Chapter 5 for more details). It is therefore critical to adjust dietary vitamin D_3 supplementation consequently by increasing its content in vitamin premixes.

Commercial forms

The majority of vitamin D_3 used to fortify animal feeds is in the form of a spray-dried formulation containing 500,000 IU/g. In poultry, a combination of vitamin A and D_3, usually in a 5:1 ratio – i.e., vitamin A 1,000,000 IU/g and vitamin D_3 200,000 IU/g – is often used in feed fortification. Such recommendation is worth exploring with aquaculture species given possible interactions between vitamins A to D_3 (Lock *et al.*, 2010).

The first vitamin D_3 metabolite 25(OH)D_3, also referred to as calcidiol or calcifediol, has been shown to be efficacious and safe in rainbow trout (Rider *et al.*, 2023), but has not been largely used as a feed additive in aquaculture as it is in poultry and swine (Oviedo-Rondón *et al.*, 2023a, b). Comparative studies to assess the efficiency of 25OHD$_3$ *vs.* vitamin D_3 are needed in fish and crustaceans.

Calcifediol is commercially available under the tradename Rovimix® Hy-D® in spray-dried form. A form of calcitriol as 1,25(OH)$_2$D$_3$ – glycoside is also available as feed material but has not been tested in fish yet to our knowledge.

Metabolism
Absorption, conversion to active forms, and transport
Dietary vitamin D (D_2 and D_3) is absorbed from the small intestine in association with fats and bile salts, as are all the fat-soluble vitamins (Combs and McClung, 2022). Like the others, vitamin D is absorbed predominantly via chylomicron passively and perhaps actively through cholesterol transporters (Reboul *et al.*, 2011).

The efficiency of enteric absorption of vitamin D has not been described extensively in fish and shrimp, but it appeared to be only 50% and highly variable in avian and terrestrial species (e.g., Oviedo-Rondón *et al.*, 2006; Combs and McClung, 2022). Evidence suggests that dietary component from plants could impair vitamin D absorption efficiency in carnivorous fish (Ferreira *et al.*, 2020; Koskela *et al.*, 2021). Koskela *et al.*, (2021) fed rainbow trout with feeds containing between 54 and 62% plant-based ingredients with vitamin D_3 above known requirements (97–107 µg/kg diet) and reported levels of vitamin D_3 between 6.0 and

7.7 μg/kg muscle, which is at least 4× lower than previously reported data for salmonids (e.g., Horvli *et al.*, 1998). Ferreira *et al.* (2020) observed a negative correlation between vitamin D concentrations in rainbow trout fillets and inclusion level of kelp in experimental diets.

After entering the blood circulation, vitamin D is transported mostly via plasma vitamin D binding protein (DBP) to the liver where it is hydroxylated under the action of 25-hydroxylase into calcidiol or 25-hydroxy-D_3 [25(OH)D_3] and other metabolites. Calcidiol, which is the main vitamin D_3 metabolite circulating, is in turn hydroxylated predominantly in the liver, but also in kidney and other tissues in fish (Graff *et al.*, 1999; Lock *et al.*, 2010; Cheng, K., *et al.*, 2023). The resulting hydroxylation products are calcitriol or 1,25-dihydroxyvitamin D_3 [1,25(OH)$_2D_3$] and 24,25-dihydroxyvitamin D_3 [24,25(OH)$_2D_3$]. Calcitriol [1,25(OH)$_2D_3$] represents the most studied active form of vitamin D in fish and crustaceans (NRC, 2011; Mai *et al.*, 2022). However, more studies are needed to understand the role, transport, and utilization of 24,25-dihydroxyvitamin D_3 [24,25(OH)$_2D_3$], especially in anadromous fish where synthesis of this metabolite increased during adaptation to sea water (Lock *et al.*, 2010).

Finally, 1,25(OH)$_2D_3$ binds to vitamin D receptors and is transported to the intestine, bone, and other tissues to exert its functions (e.g., opening of calcium channels).

Storage and excretion
In contrast to avian and terrestrial species, oily fish can store an appreciable amount of vitamin D_3 and can prevent hypovitaminosis D in human populations at risk, which was estimated at more than one billion people (Combs and McClung, 2022; Janoušek *et al.*, 2022).

Vitamin D concentrations vary between species and among tissues (Table 4.1). Its content in fish is driven predominantly by dietary intake (Horvli *et al.*, 1998).

Vitamin D metabolites accumulate in bile acids in the gall bladder before flowing through the bile duct that opens in the proximal intestine (Pierens and Fraser, 2015; Small, 2022). The extent to which vitamin D metabolites are metabolized by the microbiota, reabsorbed in the distal intestine, or excreted has not been described yet.

Biochemical functions
Dietary vitamin D is required to regulate calcium and phosphorus metabolism, improve hepatic fatty acid metabolism, and support innate and humoral immunity. These biochemical functions of vitamin D have been reviewed extensively by Lock *et al.* (2010), Mai *et al.* (2022) and Cheng, K., *et al.* (2023) and are summarized below. Vitamin D also participates in glucose homeostasis of zebrafish (Shao, R., *et al.*, 2022) and maturation of intestinal enzymatic activity in European seabass at larval stage (Darias *et al.*, 2010). How these findings apply to other fish species and shrimp remain unknown. The biological functions of vitamin D metabolites are carried out at the intestine, gills, plasma, and bone levels in fish (Lock *et al.*, 2010).

The active form of vitamin D is 1,25(OH)$_2D_3$, which acts together with parathyroid hormone (PTH) and calcitonin as a multihormonal system (Combs and McClung, 2022). The production rate of 1,25(OH)$_2D_3$ is under physiological control. Its synthesis is upregulated by low levels of calcium and/or phosphorus and its breakdown is increased by high levels of its product (through feedback-inhibition) and phosphorus.

Current knowledge on the biochemical functions of vitamin D in crustaceans is rather limited relative to fish, birds, and mammals.

Table 4.1 Average concentrations of vitamin D_3 (µg/g) wet tissue) in tissues of different farmed species

Species	Tissue	Vitamin D_3 (µg/g wet tissue)	References
Atlantic salmon (*Salmo salar*)	Plasma	0.03–4.07	Horvli *et al.* (1998)
	Intestines	0.09–6.04	Horvli *et al.* (1998)
	Liver	0.08–0.57 0.13–0.17	Horvli et al. (1998) Graff et al. (2004)
	Posterior kidney	0.05–2.68	Horvli *et al.* (1998)
	Anterior kidney	0.04–2.73	Horvli *et al.* (1998)
	Spleen	0.06–2.21	Horvli *et al.* (1998)
	Gills	0.01–1.62	Horvli *et al.* (1998)
	Muscle	0.01–2.10 0.06 (farmed) 0.25 (wild)	Horvli *et al.* (1998) Lu *et al.* (2007) Lu *et al.* (2007)
	Skin	0.01–1.58 0.02	Horvli *et al.* (1998) Pierens and Fraser (2015)
Rainbow trout (*Oncorhynchus mykiss*)	Skin	0.30	Pierens and Fraser (2015)
	Muscle	0.10 (farmed) 0.14–0.15 0.073–0.078 (farmed)	Lu Z. *et al.* (2007) Byrdwell *et al.* (2013) Mattila *et al.* (1995)
Tilapia (species not mentioned)	Muscle	0.179–0.753	Bilodeau *et al.* (2011)
Yellowfin tuna (*Thunnus albacares*)	Skin	0.072	Pierens and Fraser (2015)
Common carp (*Cyprinus carpio*)	Plasma	0.036	Lock *et al.* (2010)
Lamprey (*Entosphenus japonicus*)	Plasma	0.023	Lock *et al.* (2010)
Atlantic cod (*Gadus morhua*)	Plasma	0.033	Lock *et al.* (2010)
	Muscle	0.026 0.069	Lu Z. *et al.* (2007) Mattila *et al.* (1995)
	Liver oil	0.085–12.500 2.100–2.500	Janoušek *et al.* (2022) O'Mahony *et al.* (2011)
Red seabream (*Pagrus major*)	Plasma	0.005	Lock *et al.* (2010)
Bastard halibut (*Paralichthys olivaceus*)	Plasma	0.132	Lock *et al.* (2010)
Japanese amberjack (*Seriola quinqueradiata*)	Plasma	0.210	Lock *et al.* (2010)
Shrimp (species not mentioned)	Muscle	0.000–0.005 <0.002	Byrdwell *et al.* (2013) Mattila *et al.* (1995)
Chinese mitten crab	Hepatopancreas	0.0004–0.0005	Liu S. *et al.* (2021)
Chinese mitten crab	Intestine	0.0002–0.0003	Liu S. *et al.* (2021)

Calcium and phosphorus metabolism

The classical function of vitamin D is to regulate the absorption, transport, deposition, and mobilization of calcium and phosphate by stimulating specific ion pump mechanisms in the gills, intestine, bone, and kidney (Mai *et al.*, 2022).

Calcitriol increases the expression of calcium (Ca^{2+}) membrane transporter (CaT1) in gills and small intestine (Lock *et al.*, 2010; Combs and McClung, 2022). This transporter allows the uptake of Ca^{2+}, which is thereafter transported across the mucosal cells before it is released into the blood circulation through the active Ca^{2+}-ATPase pump system. 24,25-dihydroxyvitamin D_3 [24,25(OH)$_2$D$_3$] downregulated Ca^{2+} uptake in the intestine *in vitro* of Atlantic cod (Larsson *et al.*, 1995).

Calcitriol contributes to the active absorption of phosphorus by increasing the active sodium-dependent transporter system before its release into the blood circulation. However, calcitriol does not appear to affect the passive paracellular absorption pathway of phosphorus.

Development and remodeling of bones and scales involve several hormones, among which is the calcitriol endocrine system (Lock *et al.*, 2010). Calcitriol contributes to the co-deposition of Ca^{2+} and phosphorus (bone mineralization) as well as their dissolution (bone demineralization).

Finally, calcitriol regulates the reabsorption and excretion of Ca^{2+} and phosphorus at the kidney level.

Fatty acid metabolism

Barnett *et al.* (1979) first observed an accumulation of lipids in tissues of rainbow trout and reported a relationship between dietary vitamin D intake and lipid metabolism in fish. This relationship was explained more than 40 years later by He *et al.* (2021). These authors described an inverse relationship between dietary vitamin D level and fatty acid synthase expression, while dietary vitamin D level and hepatic lipase were positively correlated in orange-spotted grouper. Lipid accumulation in the liver of grouper increased beyond 4,000 IU/kg. Dietary vitamin D also modulated expression of genes associated with lipolysis and lipogenesis in the shrimp *Litopenaeus vannamei* (Dai *et al.*, 2022a). Fatty acid synthesis and catabolism are thus regulated to some extent by vitamin D.

Immune system modulation

Vitamin D in its active form 1,25(OH)$_2$D$_3$ is involved in immune responses of fish and crustaceans. Several studies have shown that vitamin D modulates expression of antiinflammatory enzymes (e.g., catalase, superoxide dismutase) in several species including yellow catfish (Cheng, K., *et al.*, 2020a,b), Jian carp (Jiang, J., *et al.*, 2015), black carp (Wu, C., *et al.*, 2020), turbot (Liu, J., *et al.*, 2021), largemouth bass (Li, X., *et al.*, 2021) and shrimp (Dai *et al.*, 2021). Vitamin D also activates other innate immune responses associated with leukocytes, such as phagocytosis, and mitophagy in fish and crustaceans (Cerezuela *et al.*, 2009; Dioguardi *et al.*, 2017; Cheng, K., *et al.* 2020a,b; Dai *et al.*, 2022a). Dietary vitamin D (0.1 mg/kg) has been shown to improve survival of kuruma shrimp challenged with *Vibrio parahaemolyticus* by upregulating the expression of genes associated with antimicrobial peptides (Yang, M.C., *et al.*, 2021).

With regards to adaptive immunity – present in fish, but absent in crustaceans – vitamin D increased T cell transcription factor in intestine of turbots (Liu, J., *et al.*, 2021), expression of genes associated with antigen-presenting cells of the histocompatibility complex (Feng, 2017, cited by Cheng, K., *et al.*, 2023) and expression of immunoglobulin M mRNA in turbots (Liu, J., *et al.*, 2021).

Dietary levels of vitamin D were excessively high in some of these aforementioned studies (e.g., 16,600 IU/kg feed in Cheng, K., et al., 2020a,b). Therefore, there is a need to validate the immunological function of vitamin D at practical levels in some instances.

More specific description of the biochemical functions of vitamin D_3 are provided in the following chapters on salmonids, warmwater fish species and crustaceans.

Nutritional assessment

Vitamin D_3 content in liver or hepatopancreas, intestine and muscle can serve as a reliable marker of vitamin D status because aquatic species can store vitamin D in these tissues. This contrasts with humans, mammals, and poultry where $25OHD_3$ is recognized as the best status marker (Höller et al., 2018; Oviedo-Rondón et al., 2023a,b). Other potential markers of vitamin D status in fish and crustaceans are clinical signs of muscle tetany, hepatic steatosis, and poor growth.

Vitamin D and its metabolites are analyzed using immunochemical methods (e.g., chemiluminescence detection) and physical detection methods (e.g., HPLC). Competitive chemiluminescence immunoassays, HPLC and LC coupled with tandem mass spectrometry (LC-MS/MS) assays are used in clinical practice to analyze vitamin D and its metabolites. The advantages and disadvantages of these technologies have been reviewed in depth (Van den Ouweland, 2016; Janoušek et al., 2022). The enzyme-linked immunosorbent assay (ELISA) is another method that has been used for estimating vitamin D levels in feeds and animal tissues, but its negative biases are considerable (He, C S., et al., 2013). For this reason, ELISA is not recommended to determine vitamin D content.

The varying selectivity of the antibodies for $25OHD_3$ (and $25OHD_2$) and the potential for cross-reactivity with related metabolites such as 24,25-dihydroxy-vitamin D impact the repeatability between different immune-based assays. This aspect is extremely critical when using immune assays with plasma from ectotherm species like fish and crustaceans as most commercial immune kits are based on human antibodies, and few of them are "optimized" for use in aquatic species. Furthermore, most immunochemical methods are unable to differentiate between vitamin D_3 and D_2.

HPLC-MS/MS has been referred to as the gold standard, but, in common with several delicate analyses, the result can also be erroneous as this technique requires the skills of an experienced analyst (Atef, 2018).

Deficiency signs

The most common clinical signs of vitamin D deficiency are increased liver lipid (e.g., steatosis), muscle tetany, growth retardation, bone deformities and skin and fin erosion (NRC, 2011; Mai et al., 2022; Figure 4.7). Deficiency signs along with vitamin D requirements vary between species and are thus described more categorically in the following chapters.

Safety

Risk of hypervitaminosis D is low in aquatic species. Vitamin D levels up to 1.50 mg/kg diet (60,000 IU/kg diet), which correspond to more than 25× the dietary requirements for most species (NRC, 2011; Liu et al., 2022; Mai et al., 2022), are safe according to EFSA FEEDAP (2017). This recommendation by the EFSA FEEDAP is conservative since fish can tolerate a megadose of dietary vitamin D. For instance, no clinical sign associated with excess vitamin D intake nor growth retardation were observed at 25 mg/kg diet (1,000,000 IU/kg diet) in rainbow trout (Hilton and Ferguson, 1982) and 57 mg/kg diet (2,280,000 IU/kg diet) in Atlantic salmon fry (Graff et al., 2002b).

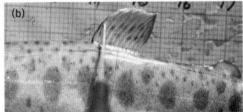

Figure 4.7 Dorsal fin erosion in rainbow trout (source: courtesy R. W. Hardy)

Although highly unlikely in aquaculture-fed species, excessive vitamin D intake can produce hypercalcemia, bone deformity, hypermelanosis, slow growth, and high mortality. The main pathological effect of ingestion of massive doses of vitamin D is widespread calcification of soft tissues. Pathological changes in these tissues are observed to be inflammation, cellular degeneration, and calcification.

The maximum tolerance level for using the 25(OH)D$_3$ metabolite is yet to be determined in fish and crustaceans.

Vitamin E
Chemical structure and properties
The term vitamin E includes all tocopherols, tocotrienols and tocomonoenols that qualitatively have α-tocopherol activity. This definition was given by the International Union of Pure and Applied Chemistry-International Union of Biochemistry (IUPAC-IUB, 1982) Commission on Biochemical Nomenclature. Tocopherols, tocotrienols and tocomonoenols consist of 2 aromatic rings (phenolic ring and heterocyclic ring) and one hydrophobic isoprenoid side chain (Figure 4.8). The hydroxyl group attached to C-6 of the phenolic ring is responsible for the biological (α-tocopherol) activity of vitamin E and confers antioxidant ability by donating its hydrogen atom (Combs and McClung, 2022). Tocopherols have a saturated side chain, whereas the tocotrienols and tocomonoenols have an unsaturated side chain containing 3 and one double bonds, respectively (Figure 4.8).

Tocopherols and tocotrienols exist in 4 different vitamers: α, β, γ, and δ (Figure 4.8). Tocomonoenols have 2 possible vitamers, namely α and γ. These vitamers are differentiated by the number of methyl (-CH$_3$) or H groups at positions 5, 7, or 8 of the phenolic rings, also referred to as chroman ring. α-Tocopherol, the most biologically active of these compounds, is the predominant vitamin E active compound in feedstuffs and the form used commercially to supplement animal diets. The biological activity of the other tocopherols is limited, but other functions have been found for non-α-tocopherol forms of vitamin E (Schaffer *et al.*, 2005; Freiser and Jiang, 2009; Traber, 2013).

The tocopherol molecule has 3 carbon atoms that are bonded to 4 different types or groups of atoms, -CH$_3$ in the case of α-tocopherol. These chiral or asymmetric carbons are located at positions C-4 and C-8 on the side chain and at C-2 on the heterocyclic ring (i.e., at the point where the side chain attached to the ring). These asymmetric carbon atoms generate 8 forms or stereoisomers following the R/S notational system: RRR-, RSR-, RRS-, RSS-, SRR-, SSR-, SRS-, and SSS-α-tocopherol. The RRR-form, also called d-form, of α-tocopherol is the unique stereoisomer occurring naturally and has all the -CH$_3$ groups facing one direction. It is the form found in plants. The all-rac (all racemic), or chemically synthesized form of α-tocopherol or dl-α-tocopherol, has an equal mixture of the R and S configurations at each of the 3 asymmetric carbons (Traber, 2013).

(a)

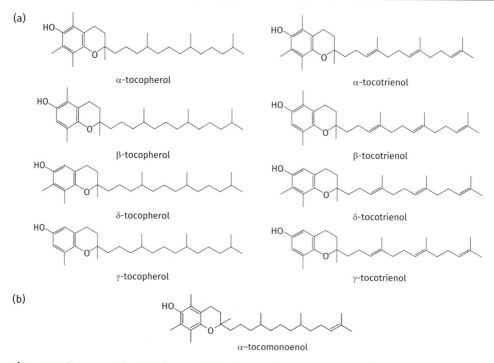

(b)

Figure 4.8 Vitamers of vitamin E (source: Reboul., 2017; Irías-Mata *et al.*, 2020)

α-Tocopherol is a yellow oil that is insoluble in water but soluble in organic solvents, resistant to heat but readily oxidized. The relative biological activity of vitamin E is expressed as a function of its α-tocopherol activity in international units. The factor used to convert international units into milligrams depends on the source of α-tocopherol:

- Natural vitamin E: 0.67 IU corresponds to the activity of 1 mg of d-α-tocopherol. Pure natural vitamin E is not commercially available.
- Nature-derived vitamin E: 1 IU corresponds to the activity of 1 mg of (all-rac)-α-tocopheryl acetate (dl-α-tocopherol).

Natural sources

Vitamin E is widespread in nature, with the richest sources being vegetable oils, cereal products containing these oils, eggs, liver, legumes, and, in general, green plants. It is abundant in whole cereal grains, particularly in germ, and thus in by-products containing the germ (McDowell, 2000; Traber, 2006, 2013). Naturally occurring vitamin E activity of feedstuffs cannot be accurately estimated from earlier published vitamin E or tocopherol values, and feed table averages are often of little value in predicting individual content of feedstuffs or bioavailability of vitamin E because of its variable concentrations. For instance, vitamin E content of 42 varieties of corn varied from 11.1 to 36.4 IU/kg, a 3.3-fold difference (McDowell and Ward, 2008).

α-Tocopherol is especially high in wheat germ oil and sunflower oil. Corn and soybean oil contain predominantly γ-tocopherol and some tocotrienols (McDowell, 2000; Traber, 2006, 2013). Cottonseed oil contains both α- and γ-tocopherols in equal proportions.

In practice, supplements are added in the more stable form of α-tocopherol acetate (dl-α-tocopherol).

The stability of all naturally occurring tocopherols is poor, and substantial losses of vitamin E activity occur in feedstuffs when processed and stored and in the manufacturing and storage of finished feeds (Gadient, 1986; Dove and Ewan, 1991; McDowell, 1996). Vitamin E sources in these ingredients are unstable under conditions that promote oxidation of feedstuffs, i.e., in the presence of heat, oxygen, moisture, oxidizing fats, and trace minerals. Vegetable oils that normally are excellent sources of vitamin E can be extremely low in the vitamin if oxidation has been promoted. Oxidized oil has little or no vitamin E, and it will destroy the vitamin E in other feed ingredients and deplete animal tissue stores of vitamin E.

Oxidation of vitamin E increases after grinding, mixing with minerals, adding fat, and pelleting or extruding for balanced feed. When feeds are pelleted or extruded, the destruction of vitamins E and A may occur if the diet does not contain sufficient antioxidants to prevent their accelerated oxidation under moisture and high-temperature conditions, and if iron salts (i.e., ferric chloride) is present.

Commercial forms

Vitamin E is usually incorporated into feeds as dl-α-tocopherol acetate. Other vitamin E esters are produced like propionate or succinate, but the acetate form has been shown to be the one providing the best bioavailability. All-rac-α-tocopheryl acetate, is manufactured by condensing trimethyl hydroquinone and isophytol and conducting ultra-vacuum molecular distillation, producing a highly purified form of α-tocopherol. This material may then be acetylated. All-rac-α-tocopherol is a mixture of α-tocopheryl acetate's 8 stereoisomers (4 enantiomeric pairs). The enantiomeric pairs, racemates, are present in equimolar amounts (Cohen et al., 1981; Scott et al., 1982). This finding indicates that the manufacturing processes lead to all-rac-α-tocopheryl acetate with similar proportions to all 8 stereoisomers (Weiser and Vecchi, 1982).

The vitamin E mostly used in animal feeding is an adsorbate that provides good storage stability and responds well to physical treatments applied in feed manufacturing, like pelleting or extrusion with the use of temperature and steam. Vitamin E acetate is also available in the spray-dried form, which is water-soluble. The spray-dried formulation is also indicated when stability may be critical, like aggressive premixes with very high pH or canned pet food.

Commercially there is not truly "natural" tocopherol product available since the d-form or RRR-form of α-tocopherol commercial products are obtained from the original raw material only after several chemical processing steps. Hence, it should be referred to as "naturally derived" and not natural.

Studies in several species comparing the naturally derived RRR to the synthetic all-rac-α-tocopheryl acetate have shown the former to have higher biopotency and be more effective in elevating plasma and tissue concentrations when administered on an equal IU basis (Jensen et al., 2006; Combs and McClung, 2022). Research in humans, poultry, sheep, pigs, guinea pigs, fish, and horses, in which the elevation of plasma concentrations was measured, indicated that the biopotency of RRR-α-tocopherol compared to all-rac-α-tocopherol can vary from the "official" figure of 1.36 : 1 up to closer to 2 : 1 (Traber, 2013), with differences among species. Considering the 1.36 : 1 ratio, 1 mg of all-rac-α-tocopheryl acetate can be replaced by 0.74 mg RRR-α-tocopheryl acetate.

Metabolism

Absorption, conversion to active forms, and transport

Tocopherol's absorption occurs predominantly in the small intestine in association with fats and is facilitated by bile acids, monoglycerides, free fatty acids, and pancreatic lipase necessary for the solubilization of vitamin E.

Figure 4.9 illustrates the absorption, conversion and transport of vitamin E. Before absorption, tocopherols are almost completely hydrolyzed in the intestinal lumen by duodenal pancreatic esterase, the enzyme which releases free fatty acids from dietary triacylglycerides. The resulting alcohol form in the case of dl-α-tocopherol acetate is α-tocopherol. Likewise, any ester form of vitamin E, i.e., tocopheryl acetate, succinate, or propionate, is converted to the alcohol form before its absorption. Thereafter, α-tocopherol is incorporated into mixed micelles prior to absorption (Sitrin *et al.*, 1987). The same applies to γ-tocopherol. At the site of absorption, the mixed micelles are broken down and tocopherols are absorbed predominantly via chylomicron passively, but active transport by intestinal brush border transmembrane proteins could be possible (Traber, 2006; Rigotti, 2007; Traber, 2013).

The efficiency of vitamin E absorption has not been described extensively in fish and shrimp. In poultry, vitamin E absorption is variable from 35 to 50%, lower than that of vitamin A (Oviedo-Rondón *et al.*, 2023a,b). Medium-chain triglycerides enhance vitamin E absorption, whereas PUFAs and phosphatidylcholine reduce it in avian species, rodents, swine and human (Reboul, 2017; Oviedo-Rondón *et al.*, 2023a,b). In aquatic species, vitamin E requirements increase as dietary lipid and PUFAs increase, but it is not clear if it is resulting from reduced vitamin E absorption or decreased α-tocopherol to PUFAs ratio in fish tissues (Hamre, 2011; NRC, 2011). Alpha-tocopherol is absorbed best, with γ-tocopherol absorption slightly less than α-forms but with a more rapid excretion. As mentioned earlier, most of the vitamin E activity within plasma and other animal tissues is α-tocopherol (e.g., Ullrey, 1981; Hamre, 2011).

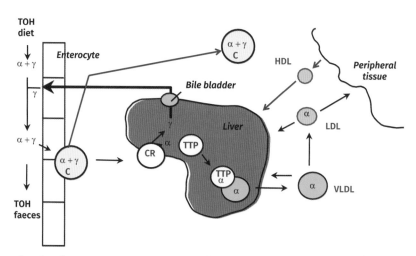

Figure 4.9 Vitamin E (α-tocopherol and γ-tocopherol) absorption, conversion and transport pathway (source: Hamre, 2011).

Notes: TOH: tocopherols; C: chylomicrons; CR: chylomicron remnants; TTP: tocopherol transfer protein; HDL: high-density lipoprotein; LDL: low-density lipoprotein; VLDL: very low-density lipoprotein.

Newly absorbed vitamin E is secreted in chylomicrons and transported in plasma to the liver. No plasma-specific vitamin E transport proteins have been described for vitamin E, unlike other fat-soluble vitamins. The vitamin E-containing chylomicrons are catabolized by lipoprotein lipase and tocopherols are taken up by the liver and, to a lesser extent, by peripheral tissues (Rigotti, 2007).

After hepatic uptake, the α-tocopherol form of vitamin E is preferentially returned into the circulation. The α-tocopherol transfer protein (α-TTP) is a critical regulator of vitamin E status and facilitates the secretion of tocopherol from hepatocytes. The liver releases the vitamin in combination with low-density and very-low-density lipoproteins (LDL and VLDL) (Traber, 2013).

Plasma vitamin E concentrations depend on α-tocopherol secretion from the liver (Kaempf-Rotzoli et al., 2003). Additionally, the newly absorbed vitamin E, rather than that returning from the periphery, appears to be preferentially secreted into the plasma from the liver (Traber et al., 1998). Thus, the liver, not the intestine, discriminates between tocopherols (Traber, 2013). The transport of α-tocopherol by high-density lipoproteins (HDLs) from peripheral tissues to liver has been demonstrated in Atlantic salmon during vitellogenesis, but little is known about the role of HDLs at other life stages and in other species (Lie et al., 1994; Hamre, 2011).

Negative interactions between vitamin E and other liposoluble vitamins such as vitamin A and provitamins A, polyphenols and phytosterols have not been studied comprehensively in aquatic species compared to poultry, rodents, and human (Reboul, 2017). With the increased inclusion of plant products in aquaculture feed formulas, there is a need to explore these possible interactions that could impact recommendations for dietary inclusion of vitamin E.

Storage and excretion

Vitamin E is stored throughout all body tissues, with the highest storage in the liver of fish or hepatopancreas of crustaceans (Table 4.2). Vitamin E content in tissues is driven by dietary intake (Bell et al., 1985; Hsu and Shiau, 1999a; Lee and Shiau, 2004; Linn et al., 2014). Vitamin E is stored predominantly in the form of α-tocopherol, but γ-tocopherol and δ-tocopherol accretion takes place particularly in adipose and red and white muscle tissues (Hamre, 2011).

Although ingested vitamin E can be stored in tissues, large amounts are excreted in the feces (Combs and McClung, 2022). Gamma-tocopherol and excess α-tocopherol are secreted in the bile and returned to the intestine to be excreted or reabsorbed by enterocytes. The extent to which γ-tocopherol is excreted or recirculated may depend on α-tocopherol status (Rigotti, 2007).

Biochemical functions

The most well-known biochemical functions of vitamin E in aquaculture relate to its antioxidant capacity, positive impact on reproductive performance, immunomodulatory effect, and enhancement of physiological detoxification (NRC, 2011; Mai et al., 2022). Vitamin E also contributes to flesh quality of fresh as well as frozen aquaculture products by maintaining organoleptic properties and preventing oxidation of highly unsaturated fatty acids (reviewed by El-Sayed and Izquierdo, 2022). These biochemical functions are reviewed in this chapter and supported with practical examples that apply to various species in Chapters 5, 6, and 7.

Classical α-tocopherol functions of vitamin E

Antioxidant

Under physiological conditions, cells maintain redox homeostasis by producing oxidants, i.e., reactive oxygen species (ROS) and other free radicals and eliminating them through

Table 4.2 Average concentrations of vitamin E (µg/g wet tissue) in tissues of different farmed species

Species	Tissue	α-tocopherol (µg/g wet tissue)	References
Rainbow trout (*Oncorhynchus mykiss*)	Whole blood	1.67–16.02	Bell *et al.* (1985)
	Plasma	6.5–80.6	Frigg *et al.* (1990)
	Liver	2.26–35.60 10.2–596.5	Bell *et al.* (1985) Frigg *et al.* (1990)
	Muscle	0.92–6.22 2.0–19.7 8.3–49.1 33–155 11.12	Bell *et al.* (1985) Frigg *et al.* (1990) Chaiyapechara *et al.* (2003) Kamireddy *et al.* (2011) Stancheva and Dobreva (2013)
Masu salmon (*Oncorhynchus masou*)	Egg	120.6	Yamamoto *et al.* (2001)
	Gonad	218.8	Yamamoto *et al.* (2001)
	Muscle	26.0	Yamamoto *et al.* (2001)
	Spleen	103.4	Yamamoto *et al.* (2001)
	Liver	447.9	Yamamoto *et al.* (2001)
Atlantic salmon (*Salmo salar*)	Liver	493–661	Hamre and Lie (1997)
	Plasma	38–56	Hamre and Lie (1997)
	Posterior kidney	87–107	Hamre and Lie (1997)
	Head kidney	123–158	Hamre and Lie (1997)
	Spleen	48–92	Hamre and Lie (1997)
	Gills	26–35	Hamre and Lie (1997)
	Adipose tissue	104–136	Hamre and Lie (1997)
	Red muscle	56–74	Hamre and Lie (1997)
	White muscle	22–27 5–13	Hamre and Lie (1997) Sigurgisladottir *et al.* (1994)
	Intestine	59–79	Hamre and Lie (1997)
	Gonads	74–126	Hamre and Lie (1997)
Common carp (*Cyprinus carpio*)	Serum	4–8	Wang K. *et al.* (2016)
	Muscle	6.18	Stancheva and Dobreva (2013)
Red seabream (*Pagrus major*)	Liver	20.4–238.8	Linn *et al.* (2014)
	Muscle	6.5–20.5	Linn *et al.* (2014)
Turbot (*Schophthalmus maximus*)	Muscle	5–40	Ruff *et al.* (2004)
Atlantic halibut (*Hippoglossus hippoglossus*)	Muscle	13–30	Ruff *et al.* (2004)
African catfish (*Clarias gariepinus*)	Muscle	6.00–13.87	Ng *et al.* (2004)
Grass shrimp (*Penaeus monodon*)	Hepatopancreas	68.63–269.28	Lee and Shiau (2004)
	Muscle	8.21–32.24	Lee and Shiau (2004)

an antioxidative system. When the balance favors oxidants, we have oxidative distress (Oviedo-Rondón *et al.*, 2023a,b).

Free-radical reactions are ubiquitous in biological systems and are associated with energy metabolism, biosynthetic reactions, natural defense mechanisms, detoxification, and intra- and intercellular signaling pathways. Redox homeostasis is an essential mechanism for aerobic organisms (bacteria, plants, animals, and humans). Highly ROS, such as the superoxide anion radical (O_2-), hydroxyl radical (OH), hydrogen peroxide (H_2O_2), and singlet oxygen (O_2-), are continuously produced during normal aerobic cellular metabolism. Additionally, phagocytic granulocytes undergo respiratory bursts to produce oxygen radicals to destroy intracellular pathogens. However, these oxidative products can, in turn, damage healthy cells if they are not eliminated. Antioxidants serve to stabilize these highly reactive free radicals, thereby maintaining the structural and functional integrity of cells (Chew, 1995).

Free radicals under oxidative stress conditions can be extremely damaging to biological systems by disrupting homeostasis and integrity of proteins, lipids, and DNA components (Lushchak and Bagnyukova, 2006). Oxidatively damaged proteins can inactivate enzymes and unsettle metabolic reactions. Free radicals can initiate lipid peroxidation – PUFA are particularly sensitive to oxidation – that releases toxic and mutagenic substances. ROS and other free radicals can also induce point mutation susceptible to alter gene expression and encoding of proteins.

The oxidation of PUFA takes place when free radicals attract one hydrogen atom, along with its electron, away from the chain structure, satisfying the electron needs of the original free radical, but leaving the PUFA short one electron. Thus, a fatty acid-free radical is formed that joins with molecular oxygen to form a peroxyl radical that steals a hydrogen-electron unit from yet another PUFA. This reaction can continue in a chain, destroying thousands of PUFA molecules (Gardner, 1989).

Vitamin E is a quenching agent for free-radical molecules with single, highly reactive electrons in the hydroxyl group of the phenolic ring. Vitamin E has a crucial role within the cellular defense system in the face of oxidation at both intracellular and extracellular levels. The bioactive α-tocopherol is integrated within the cellular membrane and protects lipids from oxidation, preventing them from being attacked by reactive oxygen and free radicals (Oviedo-Rondón *et al.*, 2023a,b). Tocopherols remove the peroxyl radical, donating a hydrogen atom and converting it to peroxide. Support for the antioxidant role of vitamin E *in vivo* also comes from observations that synthetic antioxidants can either prevent or alleviate certain clinical signs of vitamin E deficiency diseases. Therefore, antioxidants are very important to human and animal immune defense and health.

Vitamin E supplies become depleted when it acts as an antioxidant, which explains the frequent observation that the presence of dietary PUFA (susceptible to peroxidation) increases or induces a vitamin E deficiency. Consequently, vitamin E supplementation should increase in parallel to the amount of PUFA and the degree of oxidation of the fat added to the feed.

Vitamin E requirements also depend on the presence or absence of other compounds that intervene in the tissue oxidation defense system, such as selenium. It should be borne in mind that depending on the feed ingredients, and hence the content of tocopherols, carotenoids, and other antioxidants, there will be a variation in the oxidative state of the animal and, therefore, the requirements of antioxidants and specifically of vitamin E. Interruption of fat peroxidation by tocopherol explains the well-established observation that dietary tocopherols protect body supplies of oxidizable materials such as vitamin A, vitamin C, and the carotenes.

Relationship with selenium in tissue protection

Vitamin E and selenium spare each other and work synergistically to protect tissues against lipid oxidation (Bell *et al.*, 1985; Chen, Y. J., *et al.*, 2013; Le *et al.*, 2014a,b). These interactions have not been confirmed in crustaceans, but it is reasonable to assume they apply to these invertebrates until more studies are conducted.

Tissue breakdown occurs in most species receiving diets deficient in vitamin E and selenium, mainly through peroxidation. Selenium has been shown to act in aqueous cell media (cytosol and mitochondrial matrix) by destroying hydrogen peroxide and hydroperoxides via the enzyme glutathione peroxidase (GSH-Px) (Oviedo-Rondón *et al.*, 2023a,b). This capacity prevents the oxidation of PUFA materials within cells.

Reproduction

Vitamin E status affects reproductive performance of fish and crustaceans (NRC, 2011; Mai *et al.*, 2022). For instance, Lee K. J. and Dabrowski (2004) reported significantly higher testis-somatic index and survival of yellow perch (*Perca fluviatilis*) fed 160 mg vitamin E kg/diet compared to fish that received a non-supplemented diet. Vitamin E at 100 mg kg^{-1} diet significantly increased the number and size of pleopodal eggs of the freshwater crayfish *Astacus leptodactylus* (Harlıoğlu and Barım, 2004).

Although relationships between dietary levels of vitamin E and different reproductive responses have been well demonstrated, the underlying mechanisms driving these responses remain poorly described in aquaculture species.

Immune response

The immunomodulatory effect of vitamin E in aquaculture species has been known since the early 1980s (Blazer and Wolke, 1984). Since then, evidence never ceased to accumulate that indicates vitamin E is an essential nutrient for the normal function of the innate and adaptive immune system.

Like in the antioxidative response, vitamin E and selenium interact with regards to immunity by increasing, for example, serum lysosome and bactericidal activity in fish (Le *et al.*, 2014a,b).

There is a need to determine the extent to which vitamin E is exacerbated during bacterial and viral infections that cause an overload of free radicals and can lead to oxidative stress. During such infection, there is an increase in the production of glucocorticoids, epinephrine, eicosanoids, and of phagocytic activity. The protective effects of vitamin E on animal health may reduce glucocorticoids, which are known to be immunosuppressive (Golub and Gershwin, 1985). Eicosanoid and corticoid synthesis and phagocytic respiratory bursts are prominent producers of free radicals, which challenge the animal's antioxidant systems. These metabolic processes deplete vitamin E and likely require higher dietary vitamin E supplementation.

Physiological detoxification

Vitamin E can protect aquatic species from toxic effects of heavy metals. The mode of action has not been clearly described yet, but it has been suggested that vitamin E protection comes from its capacity to scavenge free radicals released during metabolism of heavy metals (Moniruzzaman *et al.*, 2017). Biochemical markers of liver damage (e.g., aspartate transaminase, alkaline phosphatase) were downregulated by vitamin E (100 mg /kg diet) in common carp (*Cyprinus carpio*) exposed to sub-lethal dose of waterborne cadmium (1.5 mg/L free ion) for 2 weeks (Mohiseni *et al.*, 2017). Vitamin E at 400 mg/kg diet protected Nile tilapia (*Oreochromis niloticus*) from genotoxicity and cytotoxicity associated with sub-lethal concentrations of waterborne cadmium, copper, lead and zinc (Harabawy and Mosleh, 2014). Vitamin E and selenium also interact synergistically to protect against the toxicity of various heavy metals (Whanger, 1981; Moniruzzaman *et al.*, 2017). For instance, dietary vitamin E supplementation at ≥100 mg/kg diet and selenium at 2–4 mg kg/diet depleted mercury concentration in muscle, liver, and kidney of olive flounder (*Paralichthys olivaceus*) (Moniruzzaman *et al.*, 2017).

Flesh quality
To improve oxidative stability and thus increase the shelf life of aquaculture products, antioxidants have been successfully added to feeds. Several compounds, such as carotenoids, vitamin E, vitamin C, and selenium, are known to have potent antioxidant effects on meat (Oviedo-Rondón *et al.*, 2023a,b).

Dietary vitamin E supplementation represents a safe and reliable approach to sustain flesh quality of aquaculture products over time and contribute to minimizing food waste by decreasing spoilage. The positive effect of vitamin E on the quality of aquaculture frozen products was demonstrated as early as 1978 and conducted with channel catfish (*Ictalurus punctatus*) (O'Keefe and Noble, 1978). In this study, dietary supplementation with ≥400 mg dl-α-tocopherol kg/diet prevented lipid oxidation of frozen fillets for 3 months. Similar findings were reported by Frigg *et al.* (1990) where vitamin E at with ≥200 mg dl-α-tocopherol kg/diet effectively stabilized lipids in fresh trout fillets. Compared to synthetic antioxidant, dietary vitamin E supplementation at 101 mg/kg diet prevented rancid and putrid odors and maintained fillet cohesiveness in coho salmon fillets frozen for 18 months (Rodríguez *et al.*, 2015). Vitamin E at ≥450 mg/kg diet blocked the development of lipid oxidation by-products and protected eicosanoic acid (EPA) from degradation in muscle of meager (*Argyrosomus regius*) (Rodríguez Lozano *et al.*, 2017). In a study with tilapia, Wu, F., *et al.* (2017) reported that vitamin E at ≥40 mg/kg diet improved the texture profile (i.e., cohesiveness, resilience and chewiness) of fresh fillet. Finally, vitamin E at ≥200 mg/kg diet hampered PUFA oxidation in rainbow trout fillets stored for 14 days at 2°C (Kamireddy *et al.*, 2011).

Non-α-tocopherol functions of vitamin E
Although α-tocopherol has been the most widely studied form of vitamin E, other tocopherols and tocotrienols have been shown to have biological significance (Qureshi *et al.*, 2001; Schaffer *et al.*, 2005; Nakagawa *et al.*, 2007; Combs and McClung, 2022). The greater emphasis on α-tocopherol undoubtedly arises from observations that γ-tocopherol and δ-tocopherol are only 10 and 1% as effective as α-tocopherol, respectively, in experimental animal models of vitamin E deficiency (Oviedo-Rondón *et al.*, 2023a,b).

Research with tocotrienols and non-alpha tocopherols has been conducted with laboratory animals and *in vitro* studies. Gamma-tocopherol has beneficial properties as an antiinflammatory agent (Wolf, 2006). Tocotrienols have been shown to regulate blood glucose and reduce insulin resistance in rodents (Combs and McClung, 2022). The metabolic functions of non-α-tocopherols deserve further attention in aquaculture species.

Other functions
Head *et al.* (2021) used zebrafish as model and reported additional functions of vitamin E that include a role in (1) glucose and energy metabolism, (2) gene transcription, and (3) intercellular signaling. These functions have not been studied extensively in fish species and crustaceans of importance in aquaculture.

Nutritional assessment

Vitamin E encompasses 8 naturally occurring vitamers – 4 tocopherols and 4 tocotrienols but only α-tocopherol is routinely measured and used for status determination since this form is preferably maintained in circulation. Tocopherols are commonly analyzed spectrophotometrically or spectrofluorometrically in animal tissues (Desai, 1984). Vitamin E is also analyzed using more sophisticated and sensitive detection methods by HPLC with fluorescence or UV detection (Oviedo-Rondón et al., 2023a,b). Using normal-phase HPLC, tocopherols and tocotrienols can also be separated. A recently introduced fast and sensitive reversed-phase HPLC method resolves the challenging separation of β- and γ-tocopherol, while the separation and quantification of the 8 stereoisomers of α-tocopherol are much more challenging (Höller et al., 2018).

Typical marker of vitamin E status in fish is erythrocyte fragility test (NRC, 2011; Mai et al., 2022). In vitro measurement of thiobarbituric acid-reactive substance (TBARS) in liver microsomes in which lipid peroxidation has been induced is another marker (Cowey et al., 1981). Erythrocyte peroxide hemolysis has also been suggested to assess vitamin E status in aquatic species, but this method has not been always sensitive enough (NRC, 2011; Mai et al., 2022). Vitamin E concentrations in different tissues (e.g., liver, muscle, plasma) could serve as indicators of vitamin E status since dietary and tissue levels are highly correlated. However, plasma tocopherol levels can be affected by blood lipid transport capacity.

Deficiency signs

Immunodeficiency decreased stress resistance, anemia, and myopathy (Figure 4.10) are among the most common deficiency signs associated with vitamin E (NRC, 2011; Mai et al., 2022). Several vitamin E deficiency symptoms flow from disorders of the cellular membrane due to the oxidative degradation of PUFAs and phospholipids (Chow, 1979) and critical sulfhydryl groups (Brownlee et al., 1977).

Safety

Vitamin E is relatively nontoxic but not entirely devoid of undesirable effects. Hypervitaminosis E studies in rats, chicks, and humans indicate maximum tolerable levels in the range of 1,000 to 2,000 IU kg/diet or 900 to 1,800 mg/kg diet (NRC, 1987). For fish and shrimp, no absolute value can be proposed across species because excess levels of vitamin E depend on multiple factors that have not been studied systematically at present. Excessive

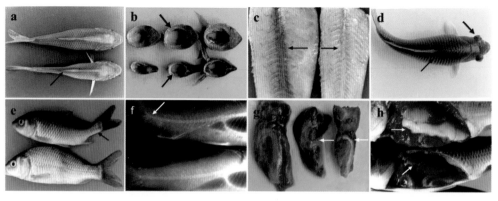

Figure 4.10 Muscle disorders caused by vitamin E deficiency in fish: (a) muscle atrophy (bottom fish); (b) cross section showing muscle atrophy; (c) white muscle disease where red muscle color faded (right fillet) (source: Wang K. et al., 2016)

levels vary depending on species, life stage, dietary components (e.g., PUFA, vitamin C, selenium), exposure to oxidative stress and infection by pathogens. Not surprisingly, dietary vitamin E levels that could be considered as excessive in aquaculture are incomplete and vary greatly between studies. For instance, growth reduction has been reported for juvenile grass carp (*Ctenopharyngodon idella*) fed vitamin E supplementation ≥180 mg/kg diet (Pan et al., 2017), but no such negative response was observed up to 1,000 mg/kg diet with rohu (*Labeo rohita*) at fingerling stage (Fatima et al., 2019). Typical signs of excess vitamin E are depressed growth rate, reduced hematocrit, and lipid peroxidation (reviewed by NRC, 2011; Mai et al., 2022).

Finally, excess supplementation of α-tocopherol could be detrimental to the other vitamin E forms. In humans, extra supplementation of diets with α-tocopherol reduced serum concentrations of gamma tocopherols (Wolf, 2006). The effects of high supplemental α-tocopherol levels on other forms of vitamin E are unknown for aquaculture species.

Vitamin K
Chemical structure and properties
The generic term vitamin K refers to different fat-soluble compounds of the quinone group that manifests the antihemorrhagic activity of phylloquinone (Combs and McClung, 2022). The basic molecule is a naphthoquinone (2-methyl-1,4-naphthoquinone), and the various vitamers differ in the nature and length of the side chain (Figure 4.11).

- Vitamin K_1 or phylloquinone is derived from plants (Figure 4.11a). Vitamin K_1 can be divided into subtypes indicated with K-n (e.g., K-1, K-2 ... K-6), where n stands for the number of 5-carbon isoprenoid units in the aliphatic side chain.
- Vitamin K_2 or menaquinone, is mainly the form produced by bacterial fermentation (Figure 4.11b). Like vitamin K_1, vitamin K_2 can be divided into subtypes indicated with MK-n, where n stands for the number of 5-carbon isoprenoid units in the aliphatic side chain (e.g., menaquinone-4 or MK-4, menaquinone-7 or MK-7). MK-4 is synthesized in the liver from ingested menadione or changed to a biologically active menaquinone by intestinal microorganisms (Suttie, 2013).
- Vitamin K_3 or menadione is produced by chemical synthesis (Figure 4.11c). This form, partially water-soluble and highly stable, has no side chain and is the form normally used in compound feeds for animal nutrition.

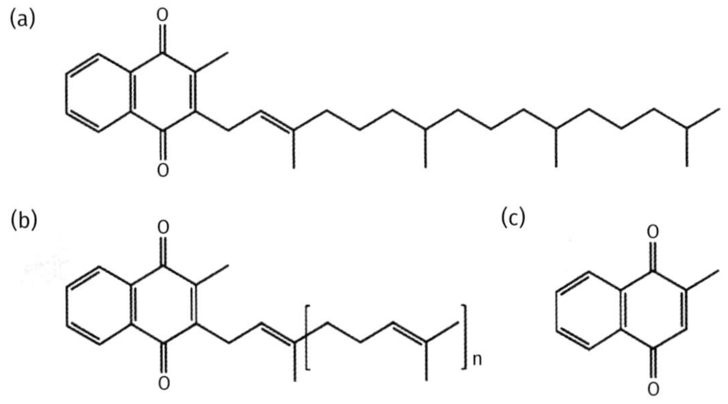

Figure 4.11 Vitamers of vitamin K: (a) vitamin K_1 or phylloquinone; (b) vitamin K_2 or menaquinone; (c) vitamin K_3 or menadione (source: Krossøy et al., 2011)

Vitamin K_1 is a golden yellow, viscous oil. Vitamin K_1 is slowly degraded by atmospheric oxygen but rapidly destroyed by sunlight. The feed industry does not utilize vitamin K_1 due to cost and lack of a stabilized form.

Natural sources

Vitamin K_1 is synthesized in chloroplast of fresh dark-green plants, e.g., alfalfa, and occurs in plant tissues and oils. Light is important for its formation, and parts of plants that do not normally form chlorophyll contain little vitamin K. However, the natural loss of chlorophyll as the leaves yellow does not bring about a corresponding change in vitamin K. Cereals and oil cakes contain only small amounts of vitamin K, which means fish and crustaceans would receive little vitamin K from diets based on grains and oilseed meals.

Vitamin K_2 can be found in all by-product feedstuffs of animal origin, including fish meal and fish liver oils, especially after they have undergone extensive bacterial putrefaction. Vitamin K_2 is also produced by the bacterial flora in birds and mammals. The extent to which vitamin K_2 is synthesized by the gut microflora have not been studied extensively in aquatic species. In other animals, the site of synthesis is the lower gut, an area of poor absorption. Thus, availability to the host is limited unless the animal practices coprophagy, like shrimp for instance, in which case the synthesized vitamin K could be highly available if shrimp microflora was effectively producing it.

Commercial forms

Vitamin K supplementation in animal diets is provided by the synthetic product, namely vitamin K_3, in the form of various bisulfite complexes, which are more stable and potent (Huyghebaert, 1991). The various products used by the feed industry and their respective content in menadione are listed in Table 4.3.

Metabolism
Absorption, conversion to active forms, and transport

Like all fat-soluble vitamins, vitamin K is absorbed in association with dietary fats and requires the presence of bile salts and pancreatic juice for adequate uptake from the alimentary tract. The absorption of vitamin K into the enterocytes depends on its incorporation into mixed micelles, and the optimal formation of these micellar structures requires the presence of both bile and pancreatic juice. Thus, any malfunction of the fat absorption mechanism (e.g., biliary obstruction, malabsorption syndrome) reduces the availability of vitamin K (Ferland, 2006). Unlike phylloquinone and the menaquinones, synthetic menadione salts are relatively water-soluble and absorbed satisfactorily from low-fat diets.

Table 4.3 Menadione salts used for diet supplementation (source: Oviedo-Rondón *et al.*, 2023b)

Vitamin K_3 salt	Menadione (K_3) concentration (%)	Amount of menadione salt to provide 1 g of menadione (K_3) (g)
Menadione sodium bisulfite (MSB)	50	2
Menadione dimethylpyrimidinol bisulfite (MPB)	45.4	2.2
Menadione nicotinamide bisulfite (MNB)	43	2.3
Menadione sodium bisulfite complex (MSBC)	33	3

In aquatic, avian, and terrestrial animals, the vitamers K are absorbed mostly in the proximal intestine. Evidence suggest the absorption of vitamins K_1 and K_2 is negligible and selective in fish (Udagawa, 2000). In poultry and other animals, the measured efficiency of vitamin K absorption ranges from 5 to 80%, depending on the form of the vitamin administered (Combs and McClung, 2022; Oviedo-Rondón *et al.*, 2023a,b). Some reports have indicated that menadione is completely absorbed, probably due to the aqueous solubility of the menadione salts. This information is yet to be demonstrated and described quantitatively in fish and shrimp.

An energy-dependent process absorbs ingested vitamin K_1 from the proximal portion of the small intestine in rodents (Hollander, 1973). In contrast to the active transport of phylloquinone, vitamins K_2 and K_3 are absorbed from the small intestine by a passive noncarrier-mediated process, although specific lipid transporters may be involved according to recent studies on human and laboratory animals (Combs and McClung, 2022; Matsuo *et al.*, 2023). In humans, vitamin K_2 with long side chains can also be absorbed from the ileum while vitamin K_2 with short side chains can be taken up from the distal intestine by a passive process (Mladěnka *et al.*, 2022). Little is known on intestinal sites of absorption of vitamin K_2 in aquatic species.

It has been shown that male animals are more susceptible to dietary vitamin K deprivation than females, apparently due to a stimulation of phylloquinone absorption by estrogens, and the administration of estrogens increases absorption in both male and female animals (Jolly *et al.*, 1977; Suttie, 2013). The difference between male and female in terms of vitamin K absorption is yet to be demonstrated in aquatic species.

In enterocytes, vitamers K are packaged into chylomicrons before entering the portal circulation and reaching the liver. In the liver, vitamers K are bound to lipoproteins to be transported in the plasma. Shearer *et al.* (1970) demonstrated the association of phylloquinone with serum lipoproteins in human, but little is known of the existence of specific carrier proteins in fish and other animals.

Vitamins K_1, K_2, K_3, and their various subtypes are active compounds that can modified into other active forms. The side chain of vitamers K can be modified in the liver and other tissues (e.g., heart, kidney, brain). For instance, the side chains of vitamin K_1 can be removed by dealkylation to produce menadione and then realkylated to form vitamin K_2 with a side chain containing 4 isoprenoids, also referred to as menaquinone-4 (MK-4) (Combs and McClung, 2022). Vitamin K_3 can also be converted into vitamin K_2. Because vitamin K_3 has no side chain, it must be alkylated through an enzymatic process and its main product is MK-4 (reviewed by Krossøy *et al.*, 2011). These active forms act as a cofactor to activate vitamin K-dependent proteins (VKD), which are involved in critical metabolic functions.

Storage and excretion

Although vitamin K is stored mostly in the liver, it remains widely distributed among various tissues (Table 4.4). Phylloquinone is quickly concentrated in the liver, but it does not have a long retention time in this organ (Thierry *et al.*, 1970). The long chains MK-7 to MK-13 are the main forms of menaquinone stored in the liver. Menadione is widely distributed in all tissues.

The inability to rapidly develop a vitamin K deficiency in most species results from the difficulty in preventing the absorption of the vitamin from the diet or intestinal synthesis rather than from significant storage of the vitamin.

Vitamin K_1 is rapidly catabolized in the liver and excreted predominantly in association with the bile in feces, but also in urine (Shearer and Newman, 2014). In human, it has been estimated that ~60–70% of absorbed vitamin K_1 is excreted after each meal (Shearer *et al.*, 1970;

Table 4.4 Average concentrations of vitamin K (ng/g wet tissue) in tissues of different species reared under laboratory conditions or on farms (vitamin K subtypes are in parenthesis)

Species	Tissue	K (ng/g wet tissue)	References
Atlantic salmon (*Salmo salar*)	Eyes	~50 (PK) ~25 (MK-4)	Graff et al. (2010)
	Gills	~100 (PK) ~50 (MK-4)	Graff et al. (2010)
	Heart	~100 (PK) ~70 (MK-4)	Graff et al. (2010)
	Liver	~1,200 (PK) 75–110 (MK-4) 31,000–40,000 (PK) 31,000–40,000 (MK-4)	Graff et al. (2010) Vera et al. (2019)
	Spleen	~50 (PK) ~25 (MK-4)	Graff et al. (2010)
	Digestive tract	~200 (PK) ~40 (MK-4)	Graff et al. (2010)
	Kidney	~300 (PK) ~70 (MK-4)	Graff et al. (2010)
	Gonads	~50 (PK) ~40 (MK-4)	Graff et al. (2010)
	Bone	~50 (PK) ~10 (MK-4)	Graff et al. (2010)
	Fillet	~50 (PK) ~30 (MK-4)	Graff et al. (2010)
	Skin	~100 (PK) ~15 (MK-4)	Graff et al. (2010)
Largemouth bass (*Micropterus salmoides*)	Liver	53–76	Wei et al. (2023)
Ayu (*Plecoglossus altivelis*)	Heart	<5 (PK) ~5 (MK-4) <5 (MK-6)	Udagawa (2000)
	Kidney	<5 (PK) ~8 (MK-4) <5 (MK-6)	Udagawa (2000)
	Liver	~5 (PK) ~10 (MK-4)	Udagawa (2000)
	Gonad	~5 (MK-4)	Udagawa (2000)
Mummichog (*Fundulus heteroclitus*)	Whole-body (larvae)	9–965 (PK) 1–25 (MK-4)	Udagawa (2001)

Note: PK: phylloquinones.

Shearer *et al.*, 1996). The percentage of vitamin K excreted vary depending on vitamers. For instance, liver turnover and excretion appeared higher for vitamin K$_1$ than vitamin K$_2$ in human and rats (Griminger and Donis, 1960; Griminger, 1984; Shearer *et al.*, 1996). Excretion rates of different vitamers K has not been described quantitatively in fish and shrimp.

Biochemical functions

The most well-known biochemical functions of vitamin K in aquaculture relate to blood coagulation and bone health (NRC, 2011; Mai *et al.*, 2022). Vitamin K may also be involved in antioxidative responses, transcription of genes associated with homeostasis of osteoblastic cells, vascular hemostasis, energy metabolism and cellular signaling (Krossøy *et al.*, 2011; Shearer and Newman, 2014; Mai *et al.*, 2022). The main biochemical functions of vitamin K are reviewed in this chapter and supported with practical examples that apply to various species in Chapters 5, 6, and 7.

Vitamin K acts as a cofactor for the enzyme γ-carboxyglutamyl carboxylase (GGCX) that is essential for the conversion of specific glutamate residues into γ-carboxyglutamate (Gla) which, in turn, is needed to produce the so-called VKD also referred to as Gla proteins. It is those Gla proteins that mediate the most well-established metabolic functions of vitamin K. Gla proteins perform multiple metabolic functions implying that vitamin K is pivotal to a broad range of physiological functions. Despite this, the roles of vitamin K – outside the domain of hematology and bone calcification – have not been extensively studied in aquaculture species, likely because clinical signs of deficiencies other than those associated with blood clotting and bone health have not raised serious concerns so far.

Blood clotting

Blood clotting is the first role that was associated with vitamin K. Although blood coagulation time is shorter in fish than mammals, the clotting system between fish, shrimp, birds and mammals is similar and involves 8 VKDs or coagulation factors that are synthesized in the liver – and perhaps other tissues in fish and shrimp – and release into the plasma: prothrombin or factor II, factor VII, factor IX, factor X, protein C, protein S, protein Z, and transthyretin or prealbumin (Tavares-Dias and Oliveira, 2009; Combs and McClung, 2022). The clotting pathway is initiated after factor X has been activated by external or internal damage to the vascular system. This is followed by a cascade of coordinated reactions involving the other VKDs. The clotting pathway ends when factor II converts fibrinogen into its insoluble form, i.e., fibrin, which result in clot formation. Protein S and protein Z serve as regulators of blood clotting or antithrombotic factors. All these VKDs are activated by their association with Ca^{2+} and require phospholipid surfaces that serve as binding sites.

In shrimp, hemolymph clotting stands as the first line of defense of innate immunity (Maningas *et al.*, 2013). Clots can bind to pathogenic bacteria and kill them. The hemolymph clotting system also participates in the production of antimicrobial peptides.

Bone health

Vitamin K prevents bone deformities in fish and enhances calcium retention efficiency in shrimp (Shiau and Liu, 1994a,b; Udagawa, 2001; Roy and Lall, 2007; Lall and Lewis-McCrea, 2007; Richard *et al.*, 2014). Two of the best characterized VKDs associated with bone health are osteocalcin or bone Gla protein and matrix Gla protein, which were initially discovered in bone. These 2 VKDs are also found in fish (Mai *et al.*, 2022). In invertebrates, the presence of osteocalcin is yet to be confirmed, whereas the matrix Gla protein was not detected, which indicated that it is not obligatory for the calcification of exoskeleton (King, 1978).

Osteocalcin is a protein containing 3 Gla residues that give the protein its mineral-binding properties. Osteocalcin accounts for 15–20% of the non-collagen protein in the bone of most vertebrates. In vertebrates, osteocalcin is produced by osteoblasts through synthesis controlled by 1,25-dihydroxy vitamin D. The newly synthesized protein is released into the circulation and can be used to measure bone and other calcified tissues formation. As is true for other non-blood VKD proteins, the physiological role of osteocalcin remains largely unknown.

Nutritional assessment

The typical biomarker of vitamin K status in fish/shrimp is blood/hemolymph clotting time, but other markers have been suggested such as GGCX activity (Krossøy et al., 2010; Mai et al., 2022). Calcium retention efficiency, protein efficiency ratio and VKD protein precursor were suggested as biomarkers of vitamin K status in shrimp (Shiau and Liu, 1994a,b). Vitamin K status may also be assessed by measuring the circulating concentration of each relevant vitamers (e.g., MK-4, MK-7) or by measuring the circulating concentration of uncarboxylated Gla proteins (Oviedo-Rondón et al., 2023a,b).

Direct measurement of K vitamers in tissues is generally accomplished by reversed-phase HPLC or ultra-performance liquid chromatography with fluorescence or mass spectrometric detection. However, many menaquinones are not available as reference compounds. The concentration in circulation reflects recent dietary exposure rather than true status concentrations (Höller et al., 2018). ELISA-based methods for measuring uncarboxylated Gla proteins are currently the most reliable for assessing vitamin K status. These tests can determine tissue-specific proteins such as uncarboxylated osteocalcin (ucOC) for bone.

Deficiency signs

Deficiency signs associated with vitamin K are not always apparent and consistent across studies. When clinical signs of deficiency occur, they appear as impairment of blood coagulation, which leads to hemorrhages, skeletal deformity, and slow growth (Krossøy et al., 2010; NRC, 2011; Mai et al., 2022). Other clinical symptoms include low prothrombin levels, anemia and loss of fin tissue. Vitamin K deficiencies can be induced by insufficient dietary intake, lipid malabsorption and ingestion of vitamin K antagonists like sulphonamides (antibiotics used in aquaculture, e.g., Zhou J. et al., 2022).

Safety

Like deficiency signs, the toxic effects associated with excess vitamin K intake do not occur always and consistently across studies. Dietary vitamin K (menadione sodium bisulfite) at 2,400 mg/kg diet did not affect growth, blood coagulation time and hematocrit concentration of brook trout fingerlings after 20 weeks (Poston, 1971). In contrast, dietary menadione sodium bisulfite at 2,500 mg/kg diet negatively impacted growth and increased the occurrence of vertebral abnormality of mummichog fingerlings after 4 weeks (Udagawa, 2001). These variable responses between studies can be driven by different susceptibility between aquatic species, life stage, vitamin K status, dietary levels of vitamin K tested, synthesis by the microbiota, and differences in the ability of the various vitamin K compounds to evoke a toxic response (Barash, 1978).

There is limited information on the interaction of vitamin K with other fat-soluble vitamins in aquaculture species. Grisdale-Helland et al. (1991) reported negative interactions between vitamin K_1 or menadione sodium bisulfite (equivalent mole basis of 30 mg menadione sodium bisulfite/kg diet) and vitamins A at 50,000 IU/kg diet on growth and survival of salmon fingerlings after 28 weeks. However, differences were marginal between treatments, hematocrit as well as hemoglobin concentration were not affected, and bone integrity was not assessed in this study.

WATER-SOLUBLE VITAMINS

Among the water-soluble vitamins are the B complex vitamins plus choline and vitamin C. Most of them either cannot be synthesized by fish and crustaceans or can but not in sufficient quantities to cover their requirements in different physiological situations. Some can be supplied through the metabolism of the microbiota of the intestinal tract. Still, since absorption is limited to the posterior sections of the digestive tract, this process is not very efficient and depends on subsequent coprophagy. Generally, they are not stored in significant quantities in aquatic species, which prevents toxicity problems but requires regular supplementation in the ration.

The B complex vitamins are essential in the metabolism of carbohydrates, amino acids, lipids, and nucleic acids. Thiamine, riboflavin, niacin, and biotin are necessary for optimal energy metabolism and lipid synthesis. Without adequate dietary pyridoxine, folic acid, and vitamin B_{12}, the synthesis and interconversion of amino acids will be inadequate for the optimum synthesis of proteins (enzymes, tissues, hormones), and growth rates will be reduced. Insufficient dietary folic acid and vitamin B_{12} impairs nucleic acid synthesis and slows down cell division, consequently limiting growth and delaying the cellular immune response to stop pathogens from infecting the animal. Despite these critical functions, B vitamins in fish and shrimp nutrition have not been studied systematically across life stages of a given species reared under variable conditions.

Vitamin B$_1$ (thiamine)
Chemical structure and properties
Thiamine consists of a molecule of pyrimidine and a molecule of thiazole linked by a methylene bridge and contains both nitrogen and sulfur atoms (Figure 4.12a). The metabolically active form is thiamine pyrophosphate (TPP), also referred to as thiamine diphosphate (Figure 4.12b). Benfotiamine (Figure 4.12c) is a promising lipid-soluble vitamin B$_1$ analog in aquaculture

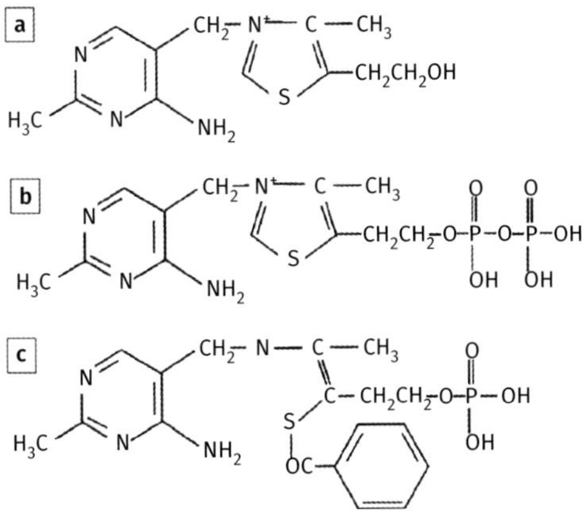

Figure 4.12 Vitamin B$_1$ chemical structures: (a) thiamine; (b) thiamine pyrophosphate (TPP); (c) benfotiamine (source: Beltramo et al., 2008).

because of its better water stability and perhaps higher bioavailability compared to TPP (Beltramo *et al.*, 2008; Xu C. *et al.*, 2018).

Thiamine is isolated in pure form as white, crystalline thiamine hydrochloride. The vitamin has a characteristic sulfurous odor and a slightly bitter taste (Bettendorff, 2013).

Thiamine is mainly found in the form of chloride hydrochloride ($C_{12}H_{17}N_4OSCl.HCl$, molecular mass 337.27 g mol^{-1}) that decomposes at 198°C (Bettendorff, 2013). Thiamine hydrochloride is a hydrophilic molecule: it can form a hydrate even under normal atmospheric conditions by absorbing nearly 1 mole of water and has a solubility of ~1 g/ml water at 25°C. Under ordinary conditions, thiamine hydrochloride is more hygroscopic than mononitrate salt. However, both products should be kept in sealed containers.

Thiamine is sparingly soluble in alcohol and insoluble in fat solvents. It is very sensitive to alkali, in which the thiazole ring opens at room temperature when the pH is above 7. In a dry state, thiamine is stable at 100°C for several hours, but moisture greatly accelerates destruction, and thus it is much less stable to heat in fresh than in dry foods. It is destroyed by ultraviolet light.

Natural sources

Brewer's yeast is the richest known natural source of thiamine (~85 mg/kg). Cereal grains and their by-products, soybean meal, cottonseed meal, and peanut meal, are relatively rich sources of thiamine. Since the vitamin is present primarily in the germ and seed coats, by-products containing the latter are richer than the whole kernel, while highly milled flour is very deficient. In humans, beriberi was prevalent in Orient countries, where polished rice is the dietary staple. Rice may have 5 mg/kg of thiamine, but the content is much lower for polished rice (0.3 mg/kg) and higher for rice bran (23 mg/kg) (Marks, 1975). Wheat germ ranks next to yeast in thiamine concentration. The level of thiamine in grain rises as the protein level rises, depending on species, strain, and use of nitrogenous fertilizers (Bettendorff, 2013).

Nevertheless, the frequent presence of antagonists, such as mycotoxins, and high susceptibility to inactivation by heat must be considered. Reddy and Pushpamma (1986) studied the effects of 1 year of storage and insect infestation on the thiamine content of feeds. Thiamine losses were high in several varieties of sorghum and pigeon pea (40–70%) and lower in rice and chickpea (10–40%). Insect infestation caused further loss. Since thiamine is water-soluble and unstable in heat, large losses may result during certain feed manufacturing processes (McDowell, 2000; Bettendorff, 2013).

Commercial sources

Thiamine sources available for addition to feed are the thiamine chloride hydrochloride (337.28 g/mol; 98%) and thiamine mononitrate (327.36 g/mol; 98%) salts (Figure 4.13). Both are fine granular, white to pale-yellow powders. Because of its lower solubility in

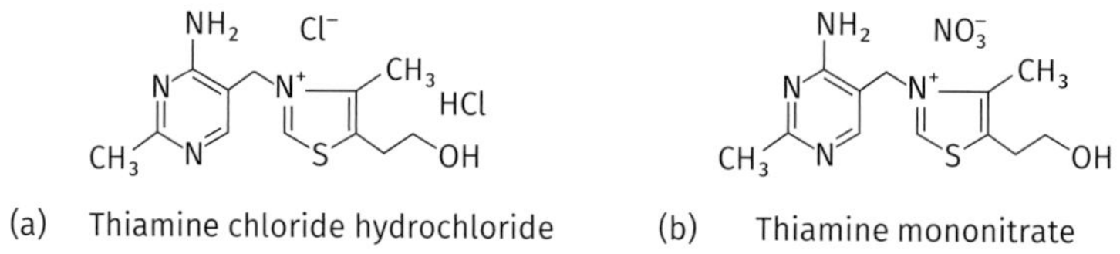

(a) Thiamine chloride hydrochloride   (b) Thiamine mononitrate

Figure 4.13 Vitamin B$_1$ forms used in animal nutrition (source: Oviedo-Rondón *et al.*, 2023a,b)

water, the mononitrate salt has somewhat better stability characteristics in dry products than the hydrochloride (Bettendorff, 2013). Thiamine mononitrate is prepared from thiamine hydrochloride by dissolving the hydrochloride salt in a mildly alkaline solution, followed by precipitation of the nitrate half-salt with a stoichiometric amount of nitric acid.

Metabolism
Absorption and transport
Thiamine appears to be readily digested and released from naturally occurring sources. A precondition for normal thiamine absorption is sufficient stomach hydrochloric acid production. Phosphoric acid esters of thiamine are split (dephosphorylated) in the proximal intestine. The free thiamine formed is soluble in water and easily absorbed. The mechanism of thiamine absorption is not yet fully understood, but active transport and simple diffusion are apparently involved. There is active sodium-dependent transport of thiamine at low concentrations against the electrochemical potential, whereas, at high concentrations, it diffuses passively through the intestinal wall.

Thiamine synthesized by the gut microflora is largely unavailable to animals except by coprophagy. Specific proteins (transporters and carriers) in the cell membrane have binding sites for thiamine, allowing it to be solubilized within the cell membrane. This permits the vitamin to pass through the membrane and ultimately reach the aqueous environment on the other side (Rose, 1990; Bates, 2006). Absorbed thiamine is transported via the portal vein to the liver with the carrier plasma protein.

Thiamine phosphorylation can occur in most tissues, particularly in the liver. Almost 80% of thiamine in animals is phosphorylated in the liver under the action of ATP to form the metabolically active enzyme form TPP, also named diphosphate or cocarboxylase. Of total body thiamine, about 80% is TPP, about 10% is thiamine triphosphate (TTP), and the remainder is thiamine monophosphate (TMP) and free thiamine. These proportions likely apply to aquatic species as well, except for fish eggs where unesterified thiamine (free base) is the main form of vitamin B$_1$ (Brown et al., 1998).

Storage and excretion
Although thiamine is readily absorbed and transported to cells throughout the body, it is one of the most poorly stored vitamins. Most mammals on a thiamine-deficient diet will exhaust their body stores within 1–2 weeks (Ensminger et al., 1983). The thiamine content in individual organs varies considerably, and the vitamin is preferentially retained in organs with high metabolic activity. The stability of thiamine in fish tissues is species and temperature dependent (Brown et al., 1998). In avian and terrestrial species, thiamine is contained in the greatest quantities in major organs such as the liver, heart, brain, and kidneys during deficiencies. Although liver and kidney tissues have the highest thiamine concentrations, approximately 50% of the total thiamine body stores are in muscle tissue (Tanphaichair, 1976). Such information is yet to be confirmed in fish and crustaceans.

Thiamine intakes above current needs are rapidly excreted through the feces. Fecal thiamine may originate from feed, synthesis by microorganisms, or endogenous sources (i.e., via bile or excretion through the mucosa of the large intestine). No data was found on thiamine excretion through urine in aquatic species.

Biochemical functions

Thiamine is one of the critical coenzymes or enzyme cofactors in the metabolism of lysine, branched-chain amino acids, carbohydrates, and lipogenesis (Bettendorff, 2013). The main functions of thiamine are illustrated in Figure 4.14. Primarily, thiamine is important in carbohydrate and energy metabolism. For this reason, thiamine recommendations for aquatic species may increase when the main energy source supplied by the feed is carbohydrates (Morris and Davies, 1995).

The TPP is the active thiamine derivative involved in the tricarboxylic acid cycle (TCA, citric acid, or Krebs cycle). Thiamine is the coenzyme for all enzymatic decarboxylation of α-keto acids. Thus, it functions in the oxidative decarboxylation of pyruvate to acetate, which in turn is combined with coenzyme A (CoA) for entrance into the TCA cycle. Thiamine is essential in two oxidative decarboxylation reactions in the TCA cycle in cell mitochondria and one in the cytoplasm of the cells.

Decarboxylation in the TCA cycle removes carbon dioxide, and the substrate is converted into the compound having the next lower number of carbon atoms:

- pyruvate –> acetyl-CoA + CO_2
- α-ketoglutaric acid –> succinyl-CoA + CO_2.

These reactions are essential for the utilization of carbohydrates to provide energy. Vitamins B_2 (riboflavin), pantothenic acid, and niacin are also involved with thiamine in this biochemical process. Thiamine plays a very important role in glucose metabolism. TPP is a coenzyme in the transketolase reaction that is part of the direct oxidative pathway (pentose phosphate cycle) of glucose metabolism. The pentose phosphate cycle is the only mechanism known for ribose synthesis needed for nucleotide formation. This cycle also reduces nicotinamide adenine dinucleotide phosphate (NADPH), essential for lowering intermediates from carbohydrate metabolism during fatty acid synthesis (Bettendorff, 2013).

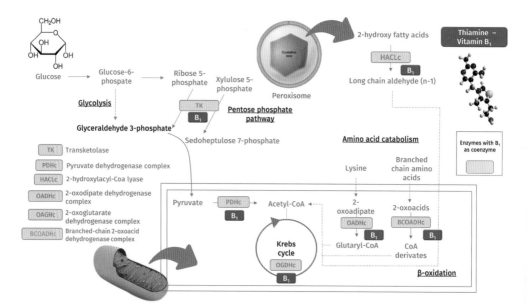

Figure 4.14 Vitamin B_1 roles in metabolism (source: Oviedo-Rondón *et al.*, 2023a,b)

In human, thiamine is, together with vitamin B_6 (pyridoxine) and B_{12} (cobalamin), one of the commonly called "neurotropic" B vitamins, playing special and essential roles both in the central nervous system (CNS) and the peripheral nervous system (PNS) (Muralt, 1962; Cooper et al., 1963; Calderón-Ospina and Nava-Mesa, 2020). These roles have not been studied comprehensively in fish and crustaceans.

Nutritional assessment

Thiamine nutritional status is typically determined by measuring thiamine-dependent erythrocyte transketolase activity (ETKA) in humans and animals (Cowey et al., 1975; Combs and McClung, 2022). However, TPP concentrations in erythrocyte and liver appear to be a more sensitive indicator of thiamine status than transketolase activity of fish (Masumoto et al., 1987). Likewise, TPP concentration in haemolymph was a better indicator of thiamine status than transketolase activity in shrimp (Chen, H.Y., et al., 1994). Muscle thiamine concentration is often measured to assess thiamine status in farmed and wild fish populations (Hemre et al., 2016; Fitzsimons et al., 2021).

Concentrations of total thiamine (phosphorylated + free forms) and free thiamine in eggs can serve to assess nutritional status and predict survival of progeny (Harder et al., 2018; Reed et al., 2023). When measuring thiamine concentrations in tissues to assess thiamine status, one must keep in mind that tissue saturation occurs and that thiamine concentration at saturation varies depending on tissue(s) and species (Mai et al., 2022).

Analytical variability in the determination of thiamine concentration is reported due to standardization and sample stability issues (Sauberlich, 1999; Höller et al., 2018). HPLC-based methods to quantify free thiamine and the phosphorylated form require further improvements.

Deficiency signs

In fish, signs associated with thiamine deficiency often flow from neurological disorders and consist in unusual swimming patterns (i.e., sideways swimming, ataxia followed by passive drifting, lethargy) and hyperirritability (reviewed by NRC, 2011; Hansen et al., 2015; Harder et al., 2018; Mai et al., 2022). Other clinical signs of thiamine deficiency are high mortality at early stages, subcutaneous hemorrhaging, darkening of the skin.

In shrimp, deficiency signs relate to anorexia, growth retardation and low survival (reviewed by NRC, 2011). The effect of thiamine deficiency on nerve cell functions has not been investigated yet in shrimp.

The detailed pathophysiology and biochemistry of thiamine deficiency-induced processes in the brain have been studied in human subjects, animal models, and cultured cells (Gibson and Zhang, 2002; Martin et al., 2003; Ke and Gibson, 2004). Neurodegeneration becomes apparent, initially, as a reversible lesion and later irreversibly brain tissue necrosis. This is explained by the fact that the brain satisfies its energy requirement chiefly by the degradation of glucose and is therefore dependent on biochemical reactions in which thiamine plays a key role.

Thiamine deficiency caused by substances with antithiamine activity, such as thiaminase, have been observed in wild populations. Thiaminase can be present in bacteria, plants, insects, mollusks and fish (Nishimune et al., 2000; Tillitt et al., 2009; Harder et al., 2018; Oonincx and Finke, 2021). It is heat sensitive, but novel ingredients should nevertheless be assessed for their risk to express thiaminase activity if conditions are conducive to this enzyme.

In premixes that include choline and trace minerals, thiamine stability is relatively low, and at ambient temperature, its content may be reduced by up to 50%. This occurs to a greater extent in feed contaminated by mycotoxin-producing fungi such as *Aspergillus* and *Fusarium*, where the B_1 concentration can drop by a factor of up to 10 (Nagaraj *et al.*, 1994). Thiamine content was reduced by 43–50% for 2 cultivars of wheat infested with *Aspergillus flavus* compared to the uncontaminated sound wheat (Kao and Robinson, 1972).

Safety
The effect of excess dosage of thiamine has not been studied comprehensively in fish. However, it appeared that thiamine administered orally or injected in large amounts was not toxic to salmonids. Ketola *et al.* (2008) fed rainbow trout fingerlings with 20,000 mg thiamine hydrochloride/kg diet – which corresponds to levels ≥2,000× above the requirements reported in NRC (2011) and Mai *et al.* (2022), and ≥650× above the recommendations of Liu *et al.* (2022) – for 2 weeks and injected 180 mg thiamine hydrochloride in the yolk sac of Atlantic salmon fry with no negative impacts on survival. On the contrary, survival was superior in groups that received excess thiamine.

Vitamin B$_2$ (riboflavin)
Chemical structure and properties
Riboflavin exists in 3 forms in nature: free dinucleotide riboflavin and the 2 coenzyme derivatives, flavin mononucleotide (FMN) and flavin adenine dinucleotide (FAD) (Pinto and Rivlin, 2013).

Riboflavin is an odorless, bitter, orange-yellow compound that melts at about 280°C. The molecular structure of riboflavin and its metabolically functional forms are shown in Figure 4.15.

Riboflavin is only slightly soluble in water but readily soluble in dilute basic or strongly acidic solutions. It is quite stable to heat, and losses are ≤10% during steam pelleting and extrusion (Mai *et al.*, 2022). Aqueous solutions are unstable to visible and UV light, and instability is increased by heat and alkalinity. Both light and oxygen have been found to induce riboflavin degradation (Becker *et al.*, 2003). When dry, riboflavin is appreciably less affected by light.

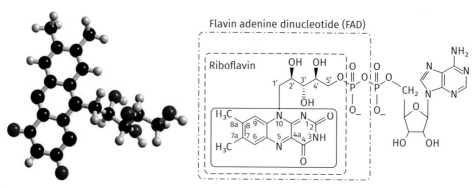

Figure 4.15 Vitamin B$_2$ chemical structure: riboflavin (within the dashed line ---), flavin mononucleotide (within the dot-dash line ---) and flavin adenine dinucleotide (whole structure) (source: Oviedo-Rondón *et al.*, 2023b)

Natural sources

Vitamin B$_2$ was first isolated from egg albumin in 1933 and subsequently detected in milk and liver. Riboflavin can be found in appreciable quantities in microalgae, green plants and by-products of animal origin. Riboflavin is also synthesized by certain yeast, fungi, bacteria, and possibly microalgae (Brown and Farmer, 1994; Averianova *et al.*, 2020). Vitamin B$_2$ synthesis by gut microbiota has not been investigated in aquatic species.

The possibility that dietary riboflavin is more bioavailable from animal products than from plant sources like it is in poultry (Oviedo-Rondón *et al.*, 2023a) – as the flavin complexes in plants are more stable to digestion than in animal sources – has not been demonstrated in fish and shrimp. Salmon feeds with ≤11% of protein-rich marine ingredients and not supplemented with vitamin premix contain ~2.7 mg riboflavin/kg diet (Hemre *et al.*, 2016).

Commercial sources

Riboflavin is commercially available to the feed, food, and pharmaceutical industries as a feed-grade crystalline compound produced mostly by microbial fermentation, although chemical synthesis is also possible (Averianova *et al.*, 2020). The vitamin is formulated in spray-dried powders containing 80% riboflavin. Commercial spray-dried powders show reduced electro-staticity and hygroscopicity for better feed flowability and distribution in feed mash compared to high-potency, USP, or feed-grade crystalline powders (Adams, 1978). Riboflavin 5'-phosphate sodium salt (75–79% riboflavin) is available for applications requiring a water-dispersible source of riboflavin.

Metabolism

Absorption and transport

Like with most B vitamins, what is known about the metabolism of riboflavin comes from studies with mammals and humans. Riboflavin covalently bound to protein is released by non-specific proteolytic digestion. Phosphatases hydrolyze phosphorylated forms (FAD, FMN) of riboflavin in the upper gastrointestinal tract to free the vitamin for absorption. Mucosal cells absorb free riboflavin via an active, sodium-dependent, saturable transport system involving specific riboflavin transporters in all parts of the small intestine (Pinto and Rivlin, 2013; Combs and McClung, 2022). When consuming high doses of riboflavin, absorption can also take place by passive diffusion processes (Combs and McClung, 2022). Absorption is facilitated by bile acids. Riboflavin is then phosphorylated to FMN in mucosal cells by the enzyme flavokinase (Rivlin, 2006). The FMN then enters the portal system, is bound to plasma albumin, transported to the liver, and converted to FAD, the form most present in plasma and tissues.

Transport of flavin by blood plasma involves loose associations with albumin and tight associations with some globulins (McCormick, 1990). A genetically controlled riboflavin-binding protein comparable to that of avian species is present in serum and eggs of fish (Malhotra *et al.*, 1991).

Thyroid hormones, particularly triiodothyronine (T$_3$), regulate the activities of the flavin biosynthetic enzymes, the synthesis of the apoflavoproteins, and the formation of covalently bound flavins (Pinto and Rivlin, 2013). These axes are yet to be described in aquaculture species.

Storage and excretion

Riboflavin is stored predominantly in the form FMN. The main storage sites (in order of impor-tance) are the kidney, spleen, liver, and heart in fish, but hepatic and heart tissues saturation often served to estimate status and requirements (Table 4.5). The concentration of riboflavin

Table 4.5 Examples of average vitamin B$_2$ concentrations (µg/g wet tissue) in tissues of salmonid species reared under laboratory conditions

Species	Tissue	B$_2$ (µg/g wet tissue)	References
Rainbow trout (*Oncorhynchus mykiss*)	Liver	22.2–28.3 4.4–11.8	Woodward (1982) Amezaga and Knox (1990)
	Heart	8.7–12.2 3.4–8.0	Woodward (1982) Amezaga and Knox (1990)
	Spleen	15.3–46.2	Woodward (1982)
	Head kidney	29.1–98.5	Woodward (1982)
	Posterior kidney	47.9–184.2	Woodward (1982)
Coho salmon (*Oncorhynchus kisutch*)	Liver	5.1–14.1	Yu H. R. *et al.* (2022)

in these tissues correlates with dietary intake until the point of saturation is reached. In animals fed riboflavin-deficient diets, hepatic cells have a relatively greater maximal uptake of riboflavin (Rose *et al.*, 1986). Pinto and Rivlin (2013) suggested that physiological mechanisms facilitate the transfer of riboflavin from mature female stores to the eggs. Intakes of riboflavin above current needs are rapidly excreted in the feces. Minor quantities of absorbed riboflavin are excreted in urine and bile salts.

Biochemical functions

Riboflavin is involved in multiple metabolic reactions essential to life and for maintaining health conditions. Riboflavin, in its phosphorylated forms (FMN and FAD) or as a constituent of the flavoproteins, is a coenzyme and cofactor of more than 100 enzymes, namely flavoenzymes, involved in the transfer of electrons in redox reactions, in the metabolism of carbohydrates and amino acids and the synthesis and oxidation of fatty acids and transporting proteins (Pinto and Rivlin, 2013). Flavoproteins assist in the generation of ATP.

Riboflavin is also critical for the metabolism of micronutrients (Combs and McClung, 2022). It is needed for activating pyridoxine (B$_6$) and folate (B$_9$) and enhancing the bioavailability of iron and zinc. In the methionine and homocysteine metabolism, riboflavin plays a role as a coenzyme in the same way that cobalamin, folate, and pyridoxine do (Pinto and Rivlin, 2013). Riboflavin not only is a key link in the utilization of dietary folates and cobalamins but also regulates the methylation of DNA and RNA within the single-carbon metabolism.

Riboflavin helps maintaining healthy conditions by supporting the glutathione redox cycle for the neutralization of ROS, promoting phagocytotic activities of neutrophils and monocytes, participating in erythropoiesis, aiding in the biosynthesis of cholesterol and thyroid hormones (T3 and T4). Riboflavin also protects the cornea, lens and retina from ROS. In addition, FMN plays a structural role in the cornea. Finally, riboflavin is involved in neural development and functions. Figure 4.16 shows the most important functions of riboflavin.

If riboflavin levels are low, the respiration process becomes less efficient, and 10–15% more feed is required to meet energy needs of terrestrial livestock species (Christensen, 1983). It is thus timely and relevant to study the impact of dietary riboflavin levels on feed intake and efficiency in aquaculture to establish recommendations that improve feed utilization and contribute to sustainability.

In poultry, recommendations for dietary riboflavin rise when the quantity of fat or protein in the feed increases (Oviedo-Rondón *et al.*, 2023a). Although there is no apparent reason for

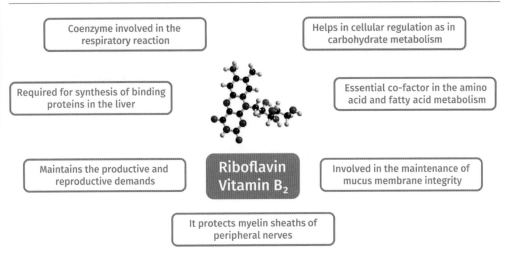

Figure 4.16 Vitamin B$_2$ functions (source: Oviedo-Rondón *et al.*, 2023b)

this not to apply in aquaculture species, limited information exists to recommend dietary riboflavin inclusion as a function of macronutrient levels and, when studies were somewhat designed to address these possible relationships, diets contained high levels of fishmeal (≥47%) and squid meal (≥9%) and were not representative of current plant-based diets (Brønstad *et al.*, 2002).

Nutritional assessment

Although riboflavin concentration can be determined accurately in different tissues, the assessment of the riboflavin nutritional status is not simple, and it seems that the sensitivity to changes in riboflavin intake is quite low (Hemre *et al.*, 2016). The typical functional test used the erythrocyte glutathione reductase activity coefficient (EGRac) assay, which allows for determining the degree of tissue saturation with riboflavin (Sauberlich, 1999). In shrimp, whole-body riboflavin concentration, but not haemolymph glutathione reductase, stood as a sensitive indicator of riboflavin status (Chen and Hwang, 1992). Hepatic D-amino acid oxidase represents another reliable indicator of riboflavin status in fish (Woodward, 1985; Deng and Wilson, 2003). It appears that hepatic and serum malondialdehyde contents along with activities of certain enzymes associated with oxidative stress (e.g., hepatic superoxide dismutase, catalase) can also serve as markers to assess riboflavin status in fish and shrimp (Yu *et al.*, 2022; Sanjeewani and Lee, 2023). Lysozyme activity and intestinal villi length of shrimp were also correlated with dietary levels of riboflavin (Figure 4.17) (Sanjeewani and Lee, 2023).

The riboflavin vitamers – free riboflavin and the coenzymes FMN and FAD – can be analyzed directly in serum (or homogenized erythrocytes) and other tissues using liquid chromatography coupled to tandem mass spectrometry (LC-MS/MS), and data can be compared with EGRac.

Deficiency signs

Several signs associated with riboflavin deficiency are somewhat like those of thiamine deficiency in fish: ataxia lethargy, hyperirritability, hemorrhaging, change of skin color (reviewed by NRC, 2011; Hansen *et al.*, 2015; Mai *et al.*, 2022). Other clinical signs of riboflavin deficiency are anorexia (low feed intake), reduced growth, high mortality, severe fin erosion, dwarfism (increased condition factor) and eye pathologies. However, cataract or ocular opacity is not consistent across studies (Woodward, 1984; Deng and Wilson, 2003).

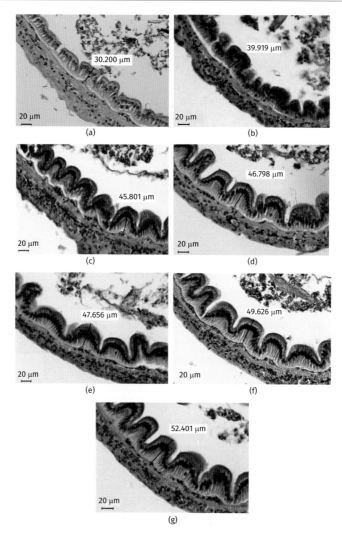

Figure 4.17 Effect of dietary levels of vitamin B$_2$ on intestinal villi length of Pacific white leg shrimp (*Litopenaeus vannamei*): (a) 1.4 mg/kg, (b) 6.4 mg/kg, (c) 17.9 mg/kg, (d) 25.3 mg/kg, (e) 37.1 mg/kg, (f) 46.0 mg/kg, (g) 55.1 mg/kg (source: Sanjeewani and Lee, 2023).

In shrimp, deficiency signs relate to growth retardation, light coloration, irritability, deformities between abdominal segments and short-head dwarfism (Chen and Hwang, 1992; Sanjeewani and Lee, 2023).

Safety

A large body of evidence in the animal nutrition literature has accumulated that supplementation with riboflavin over nutritional requirements has very little toxicity (Pinto and Rivlin, 2013). Most data from rats suggest that dietary levels between 10 and 20 times the requirement (possibly 100 times) can be tolerated safely (NRC, 1987). Dietary supplementation with 600 mg riboflavin/kg diet did not impact growth of rainbow trout fingerlings at different sizes and water temperatures (Hughes, 1984). However, growth rate of coho salmon was affected by dietary supplementation with riboflavin above requirement level (Yu *et al.*, 2022). A similar pattern was observed in shrimp as well (Sanjeewani and Lee, 2023).

Lack of toxicity is probably because the transport system necessary for riboflavin absorption across the gastrointestinal mucosa becomes saturated, limiting riboflavin absorption. Also, the capacity of tissues to store riboflavin and its coenzyme derivatives appears to be limited when excessive amounts are administered. When massive amounts of riboflavin are administered orally, only a small fraction of the dose is absorbed, the remainder being excreted in the feces.

Niacin (vitamin B₃)
Chemical structure and properties
Niacin or vitamin B_3 exists in 2 vitamer forms, namely nicotinic acid (pyridine-3-carboxylic acid) and nicotinamide (or niacinamide; pyridine-3-carboxylic acid amide). These 2 vitamers are required for the biosynthesis of the coenzymes nicotinamide adenine dinucleotide (NAD) and nicotinamide adenine dinucleotide phosphate (NADP) that are involved in the cellular respiration processes (Kirkland, 2013).

The empirical formula is $C_6H_5O_2N$. Nicotinic acid and nicotinamide correspond to 3-pyridine carboxylic acid and its amide (Figure 4.18). Both are white, odorless, crystalline solids soluble in water and alcohol. They are resistant to heat, air, light, and alkali and thus are stable in feeds. Niacin is also stable in the presence of the usual oxidizing agents: however, it will undergo decarboxylation at a high temperature in an alkaline medium.

Natural sources
Niacin is widely distributed in feedstuffs of both plant and animal origin. Good sources are animal and fish by-products, distiller grains and yeast, various distillation, and fermentation solubles, cereals, and certain oilseed meals.

Niacin is stable under normal conditions, but its bioavailability can vary substantially depending on its chemical form and ingredient processing (Wall and Carpenter, 1988). Most of the niacin is bound with a peptide or a carbohydrate in plant ingredients, whereas it is mostly in the free forms NAD and NADP in animal by-products (Combs and McClung, 2022). As a result, niacin bioavailability from plants ranged from 29 to 60% and reached 100% for animal by-products fed to channel catfish (Ng *et al.*, 1998). Extrusion of corn increased niacin bioavailability from 29 to 45% (Ng *et al.*, 1998). Niacin bioavailability is also dependent upon crop maturity at harvest. For instance, corn harvested immaturely ("milk stage") gave niacin bioavailability values from 74 to 88 µg/g in rat growth assays. In contrast, corn harvested at maturity gave assay values of 16–18 µg/g (Carpenter *et al.*, 1988).

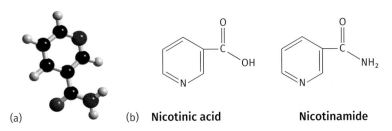

(a)   (b) **Nicotinic acid**   **Nicotinamide**

Figure 4.18 Chemical structure of the 2 forms of niacin, namely (a) nicotinic acid and (b) nicotinamide (source: Oviedo-Rondón *et al.*, 2023b)

Two types of bound niacin were initially described: (1) a peptide with a molecular weight of 12,000–13,000, the so-called niacinogens; and (2) a carbohydrate complex with a molecular weight of approximately 2,370 (Darby *et al.*, 1975). The name niacytin has been used to designate this latter material from wheat bran. Using a microbiological assay, Ghosh *et al.* (1963) reported that 85 to 90% of the total nicotinic acid in cereals is in a bound form. Therefore, in calculating the niacin content of formulated diets, probably all the niacin from cereal grain sources should be ignored or at least given a value no greater than one-third of the total niacin.

Most animals and humans can synthesize niacin from the essential amino acid tryptophan to varying degrees (Combs and McClung, 2022). However, the biosynthesis of niacin from tryptophan appeared limited or absent in fish and shrimp (Shiau and Suen, 1994; NRC, 2011; Xia *et al.*, 2015).

Commercial sources

Nicotinic acid and niacinamide are both available commercially as fine granular-powder formulations containing 99% activity. An additional source of supplemental niacin could be the vitamin K supplement menadione nicotinamide bisulfite (MNB), with a content ≥31% nicotinamide. In chicks and swine, results suggest MNB is fully effective as a source of niacin and vitamin K (Oduho *et al.*, 1993; Marchetti *et al.*, 2000).

Metabolism
Absorption and transport

Like most vitamins, the knowledge on metabolism of niacin relies heavily on studies conducted with human and terrestrial species. Nicotinic acid and nicotinamide are rapidly absorbed from the stomach and the intestine at either physiological or pharmacologic doses (Nabokina *et al.*, 2005; Jacob, 2006). In the gut mucosa nicotinic acid is converted to nicotinamide (Stein *et al.*, 1994). Niacin in foods occurs mostly in its coenzyme forms, i.e., NAD and NADP. Pyrophosphatase activity in the upper small intestine metabolizes NAD and NADP to yield nicotinamide, which is then hydrolyzed to form nicotinamide riboside and eventually free nicotinamide, which seems to be absorbed as such without further hydrolysis in the gastrointestinal tract (Kirkland, 2013).

Intestinal absorption of both nicotinic acid and nicotinamide at low concentrations appears via sodium-dependent high-affinity transporters. At high concentration, both vitamers can be absorbed via passive diffusion. Once absorbed from the lumen into the enterocyte, nicotinamide may be converted via the Preiss–Handler pathway to NAD or released into the portal circulation. Although some nicotinic acid moves into the blood in its native form, the enterocyte's bulk of nicotinic acid is converted to NAD. The intestinal mucosa contains niacin conversion enzymes such as NAD glycohydrolase (Henderson and Gross, 1979). As required, NAD glycohydrolases in the enterocytes release nicotinamide from NAD into the plasma, as the principal circulating form of niacin, for transport to tissues that synthesize NAD as needed.

Blood transport of niacin is associated mainly with red blood cells. Erythrocytes effectively take up nicotinic acid and nicotinamide by facilitating diffusion, converting them to nucleotides to maintain a concentration gradient. However, niacin rapidly leaves the bloodstream and enters the kidney, liver, and peripheral tissues.

Storage and excretion

Although niacin coenzymes are widely distributed in the body, storage is marginal (Table 4.6). The liver or hepatopancreas is the site of the greatest niacin concentration in the body.

Table 4.6 Examples of average vitamin B_3 concentrations (µg/g wet tissue) in tissues of fish and crustaceans species reared under laboratory conditions

Species	Tissue	B_3 (µg/g wet tissue)	References
Atlantic salmon (*Salmo salar*)	Whole-body	24–54	Hemre *et al.* (2016)
Indian catfish (*Heteropneustes fossilis*)	Liver	29–40	Mohamed and Ibrahim (2001)
Pacific whiteleg shrimp (*Litopenaeus vannamei*)	Whole-body	164–219	Hasanthi and Lee (2023)
Oriental river prawn (*Macrobrachium nipponense*)	Hepatopancreas	60–74	Wang J.W. *et al.* (2022)

The tissue content of niacin and its analogs, NAD and NADP, is a variable factor dependent on the diet and several other factors, such as strain, sex, age, and treatment of animals (Hankes, 1984). The liver is a central processing organ for niacin. The liver receives nicotinamide and some nicotinic acid via the portal circulation and nicotinamide released from other extrahepatic tissues. In the liver, nicotinic acid and nicotinamide are metabolized to NAD or to yield compounds for urinary excretion, depending on the niacin status of the organism. Urine is the primary pathway of excretion of absorbed niacin and its metabolites.

Biochemical functions

There are 14 known metabolic reactions in which niacin participates, forming part of the NAD and NADP coenzymes. Both coenzymes stand as the most pivotal electron transport carriers in the Krebs cycle. Therefore, they are essential in the metabolism of carbohydrates, amino acids, and fatty acids and for yielding energy in fish and crustaceans (NRC, 2011; Mai *et al.*, 2022).

Like the riboflavin coenzymes, the NAD and NADP containing enzyme systems play an important role in biological oxidation-reduction, including up to ~500 reactions in the metabolism of carbohydrates, fatty acids, and amino acids, due to their capacity to serve as hydrogen transfer agents (Combs and McClung, 2022). Hydrogen is effectively transferred from the oxidizable substrate to oxygen through a series of graded enzymatic hydrogen transfers. Nicotinamide-containing enzyme systems constitute one such group of hydrogen transfer agents.

Important metabolic reactions catalyzed by NAD and NADP are summarized as follows (McDowell, 2000; Kirkland, 2013):

- carbohydrate metabolism:
  - glycolysis: anaerobic and aerobic oxidation of glucose
  - TCA or Krebs cycle
- lipid metabolism:
  - glycerol synthesis and breakdown
  - fatty acid oxidation and synthesis
  - steroid synthesis
- protein metabolism:
  - degradation and synthesis of amino acids
  - oxidation of carbon chains via the TCA cycle
- rhodopsin synthesis.

Another metabolic function of niacin refers to its role in DNA protection and antioxidative physiological responses. Niacin, riboflavin, and the antioxidant coenzyme Q_{10} are associated with poly (ADP-ribose) synthesized in response to DNA strand breaks and involved in the post-translational modification of nuclear proteins (Kirkland, 2013). The poly ADP-ribosylated proteins function in DNA repair, DNA replication, and cell differentiation (Carson et al., 1987; Premkumar et al., 2008). These functions may be important in tissues with high turnover rates like the skin, intestines, and CNS (Kirkland, 2013; Gasperi et al., 2019). Zhang, J. Z., et al. (1993) suggested that a severe niacin deficiency may increase the susceptibility of DNA to oxidative damage, likely due to the lower availability of NAD. In line with this, dietary supplementation with niacin enhanced antioxidant defense mechanisms and innate immunity of shrimp (Hasanthi and Lee, 2023).

Nutritional assessment

Niacin and its metabolites can be measured in whole blood. However, the main gap is the lack of reliable ranges, even in humans, making the assessment extremely complicated (Höller et al., 2018). Whole-body niacin correlated with dietary level (36–110 mg/kg) in Atlantic salmon, but no tissue saturation was observed in the study by Hemre et al. (2016). Hepatic niacin concentrations leveled off in Indian catfish at 20 mg/kg (Mohamed and Ibrahim, 2001). In shrimp, tissue saturation was observed at the whole-body level when dietary niacin reached 120–160 mg/kg (Hasanthi and Lee, 2023).

Niacin can be determined via gas chromatography (GC), HPLC, or simultaneously via LC-MS/MS, and using microbiological assays (Lawrance, 2015).

Deficiency signs

Certain signs associated with niacin deficiency are comparable to those of thiamine and riboflavin deficiency in fish: abnormal swimming (e.g., ataxia, lethargy) and hemorrhaging [reviewed by NRC (2011), Hansen et al. (2015) and Mai et al. (2022)]. Other clinical signs of niacin

Figure 4.19 Fin erosion caused by vitamin B_3 deficiency in rainbow trout (source: Poston and Wolfe, 1985)

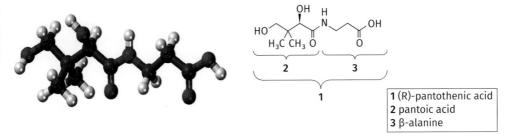

Figure 4.20 Chemical structure of pantothenic acid (source: Oviedo-Rondón *et al.*, 2023b)

deficiency are anorexia (low feed intake), fin erosion, anemia and edema. Severe caudal fin erosion caused by niacin deficiency are shown in Figure 4.19.

In shrimp, slow growth was observed at low dietary niacin concentrations (Xia *et al.*, 2015). No other niacin deficiency signs have been reported in shrimp species so far.

Safety
Research indicates that nicotinic acid and nicotinamide have a high safety margin in fish and crustaceans (NRC, 2011; Mai *et al.*, 2022). Signs associated with excessive niacin intake were reported in brook trout fed 10,000 mg niacin/kg diet and included reduction in body lipid and increase in liver lipid contents (Poston, 1969a).

Pantothenic acid (vitamin B$_5$)
Chemical structure and properties
In popular literature, pantothenic acid is often referred to as vitamin B$_5$, though the origin of this designation is obscure (Rucker and Bauerly, 2013). Pantothenic acid consists of pantoic acid joined to β-alanine by an amide linkage (Figure 4.20). This vitamin is an essential component of the coenzyme A (CoA) and the acyl carrier protein (ACP), which are the 2 metabolic active forms of pantothenic acid.

The free acid of the vitamin is a viscous, pale-yellow oil readily soluble in water and ethyl acetate. It crystallizes as white needles from ethanol and is reasonably stable to light and air. The oil is extremely hygroscopic and easily destroyed by acids, bases, and heat. Maximum heat stability occurs at pH 5.5 to 7.0.

Pantothenic acid is optically active (characteristic of rotating a polarized light). It may be prepared either in the pure dextrorotatory (d) form or the racemic mixture (dl) form. The racemic form has approximately one-half the biological activity of d-calcium pantothenate. The latter represents the commercial form used in fish nutrition. Only the dextrorotatory form, d-pantothenic acid, is effective as a vitamin.

Natural sources
Pantothenic acid, as indicated by its Greek prefix, is widely distributed in feedstuffs of animal and plant origin. Peanut meal, yeast, rice bran, wheat bran, brewer's yeast, fish solubles, fishmeal and rice polishings are good sources of the vitamin for animals. However, corn gluten meal, blood meal and meat and bone meal are low in pantothenic acid (<5 mg/kg).

Pantothenic acid is predominantly bound to CoA and ACP in nature (Combs and McClung, 2022). The d-calcium pantothenate has good stability in feedstuffs during extrusion and storage (Mai *et al.*, 2022). Gadient (1986) considers pantothenic acid slightly sensitive to heat, oxygen, or light but very sensitive to moisture. As a general guideline, pantothenic acid activity in normal pelleted feed over 3 months at room temperature should be 80–100%. Although this vitamin is found in practically all feedstuffs, complementary supplementation is advisable in rations for fish and crustaceans to optimize health and performance.

Commercial sources

Pantothenic acid is available as a commercially synthesized product for feed, known as d- or dl-calcium pantothenate. Because animals can biologically utilize only the d-isomer of pantothenic acid, nutrient requirements for the vitamin are routinely expressed in the d-form. One gram of d-calcium pantothenate is equivalent to 0.92 g of d-pantothenic acid activity. Therefore 1.087 g of d-calcium pantothenate is required to get the activity of 1 g of d-pantothenic acid.

Sometimes a racemic mixture (i.e., equal parts d- and dl-calcium pantothenate) is offered to the feed industry. The racemic mixture of 1 g of the dl-form has 0.46 g of d-pantothenic acid activity. Products sold based on racemic mixture content can be misleading and confusing to a buyer not fully aware of the biological activity supplied by d-calcium pantothenate. To avoid confusion, the label should clearly state the grams of d-calcium pantothenate or its equivalent per unit weight and the grams of d-pantothenic acid. Moreover, because of its hygroscopic and electrostatic properties, the racemic mixture can create handling problems.

Losses of pantothenic acid may occur in premixes that are alkaline (pH>7.0) and acidic (pH<5.0) (Combs and McClung, 2022). Verbeeck (1975) reported calcium pantothenate to be stable in premixes with or without trace minerals, regardless of the mineral form.

Metabolism
Absorption and transport
Pantothenic acid in the bound forms CoA and ACP must be hydrolyzed and dephosphorylated in the intestinal lumen to yield pantetheine (pantothenic acid joined to a cysteamine group) that is absorbed as-is or broken down to release pantothenic acid before its absorption. Pantothenic acid, its salt (d-calcium pantothenate), and the alcohol (pantothenol) are absorbed primarily in the small intestine by a specific transport system that is saturable and sodium ion-dependent (Fenstermacher and Rose, 1986; Miller et al., 2006). Passive diffusion occurs at high concentrations of pantothenic acid in the lumen (Combs and McClung, 2022).

After absorption, pantothenic acid in its free form is transported to various tissues in the plasma. Most cells take it up via another active-transport process. Passive uptake of free pantothenic acid takes place in erythrocytes and brain tissue (Spector, 1986). Pantothenic acid is converted to CoA and other compounds within tissues, apart from erythrocytes (Sauberlich, 1985).

Storage and excretion
Livestock does not appear to have the ability to store appreciable amounts of pantothenic acid: organs such as the liver and kidneys have the highest concentrations in avian and terrestrial species. The liver or hepatopancreas, heart and gills appear to be important tissues for pantothenic acid storage in aquatic species (Table 4.7). The concentration of pantothenic acid can be high and variable in erythrocytes (Masumoto et al., 1994).

The excretion of pantothenic acid has not been studied in aquaculture species. Urinary excretion is the major route of body loss of absorbed pantothenic acid in avian and terrestrial livestock species, and excretion is prompt when the vitamin is consumed in excess.

Table 4.7 Examples of average vitamin B$_5$ and coenzyme A (CoA) concentrations (µg/g wet tissue) in tissues of fish and shrimp species reared under laboratory conditions

Species	Tissue	B$_5$ (µg/g)	CoA (µg/g wet tissue)	References
Rainbow trout (*Oncorhynchus mykiss*)	Liver	21–23*	–	Masumoto *et al.* (1994)
	Heart	16–22*	–	Masumoto *et al.* (1994)
	Gills	9–19*	–	Masumoto *et al.* (1994)
	Erythrocyte	219–5,480**	–	Masumoto *et al.* (1994)
Atlantic salmon (*Salmo salar*)	Gills	2–6	–	Hemre *et al.* (2016)
Grouper (*Epinephelus malabaricus*)	Liver	23–48	9–23	Lin *et al.* (2012)
Blunt snout bream (*Megalobrama amblycephala*)	Liver	18–38	10–17	Qian *et al.* (2015)
Grass shrimp (*Penaeus monodon*)	Hepatopancreas	4–13	107–184	Shiau and Hsu (1999)

Notes: * Total pantothenic acid (free + bound); ** only free pantothenic acid was measured in erythrocytes.

Most pantothenic acid is excreted as a free vitamin. Pantothenic acid excretion increased with increased crude protein intake.

Biochemical functions

Pantothenic acid is involved in several metabolic pathways important in intermediary metabolism in all tissues. Its main functions have been relatively well defined in mammals and human and include (Rucker and Bauerly, 2013):

- utilization of nutrients
- synthesis of fatty acids, phospholipids, cholesterol, and steroid hormones
- synthesis of neurotransmitters, steroid hormones, porphyrins, and hemoglobin
- participation in the citric acid cycle or Krebs cycle, as a constituent of acetyl-coenzyme A (CoA) and other enzymes and coenzymes
- energy-yielding oxidation of fatty acids, glucose, and amino acids
- involved in the production of antibodies, the adrenal glands' activity, and the acetylation of choline for nerve impulse transmission
- relationship with vitamin B$_{12}$; if the latter is deficient, it accentuates the lack of pantothenic acid
- interactions with folic acid, biotin, and copper.

CoA's most important function is acting as a carrier mechanism for carboxylic acids (Miller *et al.*, 2006; Rucker and Bauerly, 2013). When bound to CoA, such acids have a high potential for transfer to other groups, and such carboxylic acids are normally referred to as "active." The most important of these reactions is the combination of CoA with acetate to form "active acetate" with a high-energy bond that renders acetate capable of further chemical interactions. Combining CoA with two-carbon fragments from fatty acids, glucose, and certain amino acids to form acetyl-CoA is essential in their complete metabolism because the coenzyme enables these fragments to enter the TCA cycle. For example, acetyl-CoA is utilized directly by combining with oxaloacetic acid to form citric acid, entering the TCA cycle. Coenzyme A, along with ACP, functions as a carrier of acyl groups in enzymatic reactions involved in (1) synthesizing fatty acids, phospholipids, cholesterol, and other sterols, (2) oxidizing fatty acids, pyruvate, and αα-ketoglutarate, and (3) allowing biological acetylations. Decarboxylation of αα-ketoglutaric acid in the TCA cycle yields succinic acid, which is then converted to the "active" form by linkage with CoA. Active succinate and glycine are involved in the first step of heme biosynthesis.

In the form of acetyl-CoA, acetic acid can also combine with choline to form acetylcholine, a chemical transmitter at the nerve synapse, and can be used to detoxify various drugs such as sulfonamides, which are used in animal production including aquaculture. Pantothenic acid also participates in adaptive immunity by stimulating the synthesis of antibodies, increasing animal resistance to pathogens. The positive effect of pantothenic acid on immunity and oxidative stress resistance has been shown in fish, but the modes of action are yet to be described (Wen *et al.*, 2010; Qian *et al.*, 2015).

Nutritional assessment

Common biomarkers to assess pantothenic acid status in fish are concentrations of free B_5 in gills, heart, erythrocytes and, to a lesser extent, liver in fish (Masumoto *et al.*, 1994). Pantothenic acid and CoA concentrations in liver and hepatopancreas have been used also as biomarkers in blunt snout bream and shrimp (Shiau and Hsu, 1999; Qian *et al.*, 2015). Masumoto *et al.* (1999) reported that concentrations of free branched-chain amino acids (BCAA, i.e., Ile, Leu, Val) were 2 times higher in plasma of rainbow trout fed pantothenic acid-deficient diets. Therefore, plasma free BCAA could serve an indicator of pantothenic acid status in fish. Lysozyme activity in the gill tissue also served as indicator to assess status and estimate requirement for pantothenic acid in grass carp (*Ctenopharyngodon idella*) (Li, L., *et al.*, 2015a,b).

Early assays to measure concentrations of pantothenic acid used B_5-dependent microorganisms such as *Lactobacillus plantarum* for quantification (Sauberlich, 1999). This assay is prone to various interferences, and therefore more specific radioimmunoassay (RIA) or ELISA tests have been developed (Sauberlich, 1999) and used to measure it in blood plasma to determine body status (Höller *et al.*, 2018).

Deficiency signs

The primary lesions of pantothenic acid deficiency appear to involve the gills where hyperplasia and fusion of epithelial cells of lamellae occur in fish [reviewed by NRC (2011), Hansen *et al.* (2015) and Mai *et al.* (2022)]. Severe caudal fin erosion caused by niacin deficiency are shown in Figure 4.21.

Certain signs associated with pantothenic acid deficiency are comparable to those of thiamine, riboflavin, and niacin deficiency in fish: abnormal swimming (e.g., lethargy) and hemorrhaging. Other clinical signs of pantothenic acid deficiency are anorexia, skin lesion, anemia and hemorrhages followed by high mortality. In shrimp, Shiau and Hsu (1999) observed an increase in irritability, light coloration and thin and soft shells when fed pantothenic acid-deficient diets.

In poultry, pantothenic acid deficiency severely depletes hatchability (Oviedo-Rondón *et al.*, 2023a). The effect of dietary intake of pantothenic acid and its status in mature fish and crustaceans may also affect embryo development and survival, but this possibility has not been studied in aquaculture species yet.

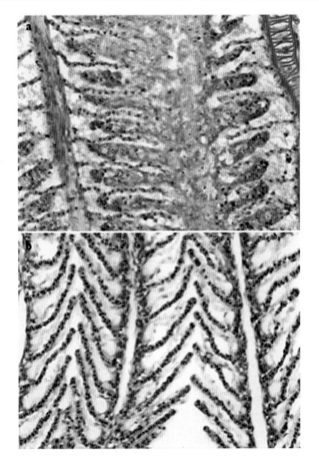

Figure 4.21 Gill disorder caused by vitamin B$_5$ deficiency in hybrid striped bass. Upper photo: swelling and fusion of lamellae; Lower photo: normal gill lamellae (source: Raggi *et al.*, 2016)

The most common antagonist of pantothenic acid is Ω-methyl-pantothenic acid which has been used to produce a vitamin deficiency in humans (Hodges *et al.*, 1958). Other antivitamins include pantoyltaurine, phenylpantothenate hydroxocobalamin (c-lactam), an analog of vitamin B$_{12}$, and antimetabolites of the vitamin containing alkyl or aryl ureido and carbamate components in the amide part of the molecule (Fox, 1991; Brass, 1993; Rucker and Bauerly, 2013). The negative impact of these antagonists can be reversed by higher dietary inclusion of pantothenic acid.

Safety

Pantothenic acid is generally regarded as nontoxic. Excesses are mostly excreted in the urine or feces. Pantothenic acid at 75 mg/kg depressed growth and innate immunity while increasing malondialdehyde and activities of markers associated with inflammation in gills of grass carp (*Ctenopharyngodon idella*) (Li L. *et al.*, 2015a,b). The dietary levels of pantothenic acid considered excess have not been defined in various aquaculture species and at different life stages. Calcium pantothenate, sodium pantothenate, and panthenol are not mutagenic in bacterial tests.

Vitamin B$_6$ (pyridoxine)
Chemical structure and properties

The term vitamin B$_6$ refers to a group of 3 pyridine derivatives named according to the functional group in the position 4 (Dakshinamurti and Dakshinamurti, 2013):

- pyridoxol or pyridoxine (PN), the alcohol form
- pyridoxal (PL), the aldehyde form
- pyridoxamine (PM), the amine form (Figure 4.22a).

Pyridoxine is the predominant plant form, whereas pyridoxal and pyridoxamine are vitamin forms generally found in animal products. Three additional vitamin B$_6$ forms are the phosphorylated forms pyridoxine 5'-phosphate (PNP), pyridoxal-5'-phosphate or codecarboxylase (PLP) and pyridoxamine 5'-phosphate (PMP). The natural, free forms of the vitamers could be converted to the key coenzymatic form, PLP, by the action of 2 enzymes, a kinase, and an oxidase.

Various forms of vitamin B$_6$ found in animal tissues are interconvertible, with vitamin B$_6$ metabolically active mainly as PLP and to a lesser degree as PMP (Figure 4.22c). Vitamin B$_6$ is stable in response to heat, acid, and alkali exposure; however, light is highly destructive, especially in neutral or alkaline media. The free base and the commonly available hydrochloride salt are soluble in water and alcohol.

Natural sources

Vitamin B$_6$ is widely distributed in feedstuffs. Most vitamin B$_6$ in animal products is pyridoxal and pyridoxamine phosphates, whereas, in plants and seeds, the usual form is pyridoxine (McDowell, 2000). The vitamin B$_6$ present in cereal grains is concentrated mainly in bran, with the rest containing only small amounts.

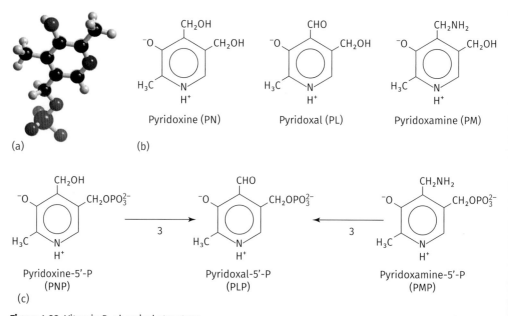

Figure 4.22 Vitamin B$_6$ chemical structure

Most of the ingredients are good sources of this vitamin, but its bioavailability varies between 70 and 80% in human and 40 and 65% in poultry (McDowell and Ward, 2008; Combs and McClung, 2022). The levels of vitamin B_6 contained in feedstuffs are also affected by processing, refining, and storage with losses as high as 70% being reported (Shideler, 1983) and with a commonly accepted range of 0–40% (Birdsall, 1975). Pyridoxine is far more stable than either pyridoxal or pyridoxamine. Therefore, the processing losses of vitamin B_6 with plant-derived foods (which mostly contain pyridoxine) are low compared to animal products (which mostly contain pyridoxal and pyridoxamine) that are losing large quantities (Lewis et al., 2015; Yang, P., et al., 2020). Losses of vitamin B_6 from natural sources increase in neutral and alkaline conditions particularly when exposed to light (Combs and McClung, 2022; Mai et al., 2022).

Pyridoxine-5′-β-D-glucoside (PNG), a conjugated form of vitamin B_6, is abundant in various plant-derived foods (McCormick, 2006). This form of B_6 may account for up to 50% of the total vitamin B_6 content of oilseeds, such as soybeans and sunflower seeds.

There are several vitamin B_6 antagonists, which either compete for reactive apoenzyme sites or react with PLP to form inactive compounds. The presence of a vitamin B_6 antagonist in linseed meal is of particular interest to animal nutritionists. This substance was identified in 1967 as linatine, l-((N-γ-L-glutamyl) amino)-D-proline and was found to have antibiotic properties (Parsons and Klostermann, 1967). Pesticides (e.g., carbaryl, propoxur, or thiram) can be antagonistic to vitamin B_6. Feeding a diet enriched with vitamin B_6 prevented disturbances in the active transport of methionine in rats intoxicated with pesticides (Witkowska et al., 1992; Dakshinamurti and Dakshinamurti, 2013). Similarly, pyridoxine reduced stress physiological responses of fish exposed to the pesticide endosulfan and thermal stress (Akhtar, et al., 2010; Kumar et al., 2016).

Commercial sources

Commercially, vitamin B_6 is available as fine crystalline powder of pyridoxine hydrochloride 99% and dilutions. Supplemental vitamin B_6 is reported to have higher bioavailability and stability than the naturally occurring vitamin.

Metabolism

Absorption and transport

Digestion of vitamin B_6 would first involve splitting the vitamin, as it is bound to the protein portion of foods. Vitamin B_6 compounds are absorbed from the diet in dephosphorylated forms. Vitamin B_6 is absorbed mainly by passive diffusion. Using a rodent model, Sakurai et al. (1992) reported that a physiological dose of pyridoxamine was rapidly transformed to pyridoxal in the intestinal tissues and then released in the form of pyridoxal into the portal blood.

After absorption, vitamin B_6 compounds quickly appear in the liver, where they are mostly phosphorylated into PLP, considered the most active vitamin form in metabolism. Both niacin (as NADP-dependent enzyme) and riboflavin (as the flavoprotein pyridoxamine phosphate oxidase) are important for the conversion of vitamin B_6 forms and phosphorylation reactions (Kodentsova et al., 1993).

Under normal conditions, most of the vitamin B_6 in the blood is present as PLP that is linked to proteins, largely albumin in the plasma and hemoglobin in the red blood cells (McCormick, 2006). Although other tissues also contribute to vitamin B_6 metabolism, the liver is responsible for forming PLP found in plasma. Like PLP, pyridoxal is associated primarily with plasma albumin and red blood cell hemoglobin (Mehansho and Henderson, 1980). Pyridoxal phosphate accounts for 60% of plasma vitamin B_6. Researchers do not agree on whether pyridoxal or PLP is the transport form of B_6.

Storage and excretion

Vitamin B_6 is widely distributed in various tissues, mainly as PLP or pyridoxamine phosphate. Only small quantities of vitamin B_6 are stored in the body (Table 4.8).

In mammals and humans, pyridoxic acid is the major excretory metabolite of the vitamin, eliminated via urine. Also, small quantities of pyridoxol, pyridoxal, pyridoxamine, and phosphorylated derivatives are excreted in urine (Henderson, 1984). Excess vitamin B_6 is excreted via urine (Mai *et al.*, 2022).

Biochemical functions

Vitamin B_6, primarily as PLP and, to a lesser extent, as PMP, plays an essential role in the amino acid, carbohydrate, and fatty acid metabolism, and the energy-producing citric acid cycle, with over 60 enzymes known to depend on vitamin B_6 coenzymes.

Pyridoxal phosphate functions in practically all reactions involved in amino acid metabolism, including transamination, decarboxylation, deamination, and the cleavage or synthesis of amino acids. The largest group of vitamin B_6-dependent enzymes are transaminases. Aminotransferase is involved in the interconversion of a pair of amino acids into their corresponding keto acids, e.g., amino groups transferred from aspartate to α-ketoglutarate forming oxaloacetate and glutamate. Non-oxidative decarboxylation involves PLP as a coenzyme, e.g., converting amino acids into biogenic amines, such as histamine, serotonin, GABA, and taurine.

The recommended levels of vitamin B_6 supply increase with protein content or with amino acid imbalance in poultry since more enzymes are needed to metabolize the excess amino acids, which depend on vitamin B_6 (Daghir and Shah, 1973). To our knowledge, this phenomenon has been studied in fish only by Hardy *et al.* (1979) and is yet to describe in crustaceans.

Vitamin B_6 participates in several functions that include (Marks, 1975; McCormick, 2006; Dakshinamurti and Dakshinamurti, 2013):

- deaminases: for serine, threonine, and cystathionine
- glycogen phosphorylase catalyzes glycogen breakdown to glucose-1-phosphate (pyridoxal phosphate does not appear to be a coenzyme for the enzyme but rather affects the enzyme conformation)
- synthesis of epinephrine and norepinephrine from either phenylalanine or tyrosine – both norepinephrine and epinephrine are involved in carbohydrate metabolism and other body reactions

Table 4.8 Examples of average total vitamin B_6 concentrations (µg/g wet tissue) in fish and crustacean species reared under laboratory conditions

Species	Tissue	B_6 (µg/g wet tissue)	References
Atlantic salmon (*Salmo salar*)	Whole-body	0.3–1.2	Albrektsen *et al.* (1993)
	Muscle	0.2–1.8 1.0–3.6 1.75–6.15 ~6–7 4.9–7.8	Albrektsen *et al.* (1993) Albrektsen *et al.* (1994) Hemre *et al.* (2016) Espe *et al.* (2020) Hamre *et al.* (2020)
	Liver	3.4–5.5 ~5–7	Albrektsen *et al.* (1993) Espe *et al.* (2020)
Major carp (*Catla catla*)	Liver	1.72–4.17	Khan and Khan (2021)
Prawn (*Penaeus japonicus*)	Whole-body	0.08–0.40	Deshimaru and Kuroki (1979)

- transmethylation involving methionine
- incorporation of iron in hemoglobin synthesis
- formation of antibodies – B$_6$ deficiency inhibits the synthesis of globulins that carry antibodies
- innate immunity and inflammation – dietary level of vitamin B$_6$ was linked to and increase phagocytic activity of leukocytes, enhanced lysozyme activity and protection against inflammation (Akhtar *et al.*, 2010; Morris *et al.*, 2010; Feng *et al.*, 2010).

Nutritional assessment

The 6 interconvertible forms of vitamin B$_6$ can be measured in plasma to assess nutritional status. Nonetheless, PLP is the vitamer normally used, and its measurement in erythrocytes seems to provide a more appropriate evaluation (Sauberlich, 1999). Considering the complex metabolic pathways in which vitamin B$_6$ is involved, assessing the ratios between metabolites, or the possibility of quantifying numerous amino acids and metabolites related to PLP-dependent pathways, may provide a better insight into nutritional status (Höller *et al.*, 2018). Alanine and aspartate amino transferase have been used as markers of pyridoxin status in fish and shrimp (reviewed by NRC, 2011). Li E. *et al.* (2010) used aminotransferase activities to estimate requirement and assess status of vitamin B$_6$ in shrimp. The fact that enzyme activities typically level off beyond requirements makes it difficult to assess status with certainty when normal ranges are unknown or highly variable.

Deficiency signs

Several signs associated with pyridoxine deficiency are associated with neurological disorders and some other clinical signs are comparable to those of thiamine, riboflavin, niacin and pantothenic acid deficiency in fish: abnormal swimming (e.g., erratic swimming, convulsion, hyperirritability), anorexia, poor growth [reviewed by NRC (2011), Hansen *et al.* (2015) and Mai *et al.* (2022)]. In shrimp, signs associated with pyridoxine deficiencies reported in the literature related to poor growth and survival, but not to nervous disorders (Deshimaru and Kuroki, 1979; Boonyaratpalin, 1998; Shiau and Wu, 2003).

Safety

Insufficient data is available to support estimates of the maximum dietary tolerable levels of vitamin B$_6$ for aquaculture species. It is suggested, primarily from dog and rat data, that nutritional levels at least 50 times the dietary requirements are safe for most species (NRC, 1987).

Signs of toxicity, which occur most obviously in the PNS, include changes in movement and peripheral sensation, ataxia, muscle weakness, and incoordination at levels approaching 1,000 times the requirement in rats, poultry, and human (Krinke and Fitzgerald, 1988; Leeson and Summers, 2008; Dakshinamurti and Dakshinamurti, 2013).

Biotin (vitamin B$_7$)
Chemical structure and properties

The chemical structure of biotin includes a sulfur atom in its ring (like thiamine) and a transverse bond across the ring (Figure 4.23). Biotin is a bicyclic compound, a monocarboxylic acid with sulfur as a thioether linkage. One of the rings contains a ureido group (-N-CO-N-), and the other is a tetrahydrothiophene ring. The tetrahydrothiophene ring has a valeric acid side chain (short-chain and saturated fatty acid). With its rather unique structure, biotin contains 3 asymmetric carbon atoms; therefore, 8 different isomers are possible. Of these isomers, only d-biotin has vitamin activity (Mock, 2013).

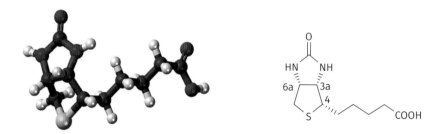

Figure 4.23 D-biotin chemical structure (source: Oviedo-Rondón *et al.*, 2023b)

Biotin crystallizes from water as long, white needles. Its melting point is 232–233°C. Free biotin is soluble in dilute alkaline solutions and hot water and practically insoluble in fats and organic solvents. Biotin is quite stable under ordinary conditions. It is destroyed by nitric acid, other strong acids, strong bases, and formaldehyde and is inactivated by oxidative rancidity reactions (Scott *et al.*, 1982). It is gradually destroyed by UV radiation.

Structurally related biotin analogs can vary from no activity to partial replacement of biotin activity to antibiotin action. Mild oxidation converts biotin to sulfoxide, and strong oxidation converts it to sulfone. Strong agents result in sulfur replacement by oxygen, resulting in oxybiotin and desthiobiotin. Oxybiotin has some biotin activity (Halver, 2002).

Natural sources

Biotin is present in common fish and shrimp feedstuffs: however, corn, wheat, other cereals, meat, and fish are relatively poor sources of biotin. Of all the vitamins present in feed ingredients of plant origin, biotin is the one that presents the most variable content, being affected by numerous environmental factors. For example, 59 samples of corn analyzed for biotin varied between 56 and 115 µg/kg, and 62 samples of meat meal ranged from 17 to 323 µg/kg (Frigg, 1987). Compared to cereal grains, oilseed meals are better sources of total biotin. Soybean meal, for instance, contains a mean biotin content of 270 µg/kg with a range of 200–387 µg/kg (Frigg and Volker, 1994). Milling wheat or corn reduced biotin concentrations (Bonjour, 1991).

Biotin is present in feedstuffs and yeast in both bound and free forms: therefore, it is important to know the form of biotin, i.e., bound or unbound, and its overall content in the feed. Much of the bound biotin to protein or lysine is unavailable to animal species as covalent bonds hinder its digestion and availability for the animal.

Other possible factors influencing biotin availability (and requirement) are dietary lipids (Walton *et al.*, 1984). In poultry, fiber interfered with biotin intestinal absorption (Misir and Blair, 1984; Oloyo, 1991), but this negative interaction has not been reported in fish and shrimp so far.

Biotin synthesis by the gut microbiota is suspected, but more research is warranted to demonstrate its importance and define the factors affecting its production.

Biotin is unstable in oxidizing conditions and, therefore, is destroyed by heat, especially under conditions that support simultaneous lipid peroxidation, by solvent extraction and improper storage conditions. Steam pelleting does not affect the stability of biotin much (0–10% degradation), but extrusion does to a greater extent (10–30% degradation) (Mai *et al.*, 2022).

Commercial sources

Biotin is commercially available as a 100% crystalline product or as various dilutions, pre-mixed, and low potency spray-dried preparations. The d-form of biotin is the biologically active form. It is the form that occurs in nature and is also the commercially available form. A 2% spray-dried biotin product is also commercially available in feed or drinking water.

Metabolism
Absorption and transport

In most species that have been investigated, physiological concentrations of biotin are absorbed in the small intestine by a sodium-dependent active-transport process, which is inhibited by dethiobiotin and biocytin (Said and Derweesh, 1991). Biotinidase, present in pancreatic juice and intestinal mucosa, catalyzes the hydrolysis of biocytin (a bound form of biotin) to biotin and free lysine during the luminal phase of proteolysis. (Said *et al.*, 1988; Said, 2011).

Biotin is absorbed intact in the first third to half of the small intestine (Bonjour, 1991). Biotin exits the enterocyte across the basolateral membrane. This transport is also carrier mediated. However, this carrier is Na$^+$ independent, electrogenic, and cannot accumulate biotin against a concentration gradient (Said, 2011). Biotin transport is regulated by multiple factors, including biotin nutritional status, enterocyte maturity, anatomic location, and ontogeny (Said, 2011). Biotin transport is more active in the villus cells than in the crypt cells.

Biotin appears to circulate in the bloodstream free and bound to a serum glycoprotein, which also has biotinidase activity and catalyzes biocytin's hydrolysis. In humans, 81% of biotin in plasma was free and the remainder bound (Mock and Malik, 1992; Mock, 2013). In the plasma of chickens, 2 biotin-binding proteins have been detected, which appear to be functionally different. Information on biotin transport, tissue deposition, and storage in animals and humans is very limited. Mock (1990) reported that biotin is transported as a free water-soluble component of plasma, is taken up by cells via active transport, and is attached to its apoenzymes. Said *et al.* (1992) reported that biotin is transported via a specialized, carrier-mediated transport system into the human liver. This system is Na$^+$ gradient-dependent and transports biotin via an electroneutral process (Mock, 2013). Not much is known about the transport of biotin in fish.

Storage and excretion

Biotin is stored in all tissues. Intracellular distribution of biotin corresponds to known locations of biotin-dependent carboxylase enzymes, especially the mitochondria (Mock, 2013). Examples of biotin concentrations in fish and shrimp tissues are given at Table 4.9.

In terrestrial animals, investigations of biotin metabolism are difficult to interpret, as biotin-producing microorganisms are present in the intestinal tract distal to the cecum. The amount of biotin excreted in urine and feces often exceeds the total dietary intake, whereas urinary biotin excretion is usually less than intake. Efficient conservation of biotin and the recycling of biocytin released from the catabolism of biotin-containing enzymes may be as important as the intestinal bacterial synthesis of the vitamin in meeting biotin requirements (Bender, 1992). 14C-labeled biotin showed the major portion of intraperitoneally injected radioactivity to be excreted in the urine and none in the feces or as expired carbon dioxide (Lee *et al.*, 1973). Biliary excretion of biotin and metabolites in rats and pigs is negligible (Zempleni *et al.*, 1997).

Table 4.9 Examples of average total vitamin B$_7$ concentrations (µg/g wet tissue) in fish and crustacean species reared under laboratory conditions

Species	Tissue	B$_7$ (µg/g wet tissue)	References
Rainbow trout (*Oncorhynchus mykiss*)	Liver	0.03–0.68	Woodward and Frigg (1989)
Atlantic salmon (*Salmo salar*)	Whole-body	75–81	Hemre *et al.* (2016)
	Liver	6–12	Mai *et al.* (2022)
Hybrid tilapia (*Oreochromis niloticus X O. aureus*)	Whole-body	0.03–0.69	Shiau and Chin (1999)
Spotted snakehead (*Channa punctatus*)	Liver	2.1–7.2	Zehra and Khan (2019)
Grass shrimp (*Penaeus monodon*)	Hepatopancreas	0.08–1.10	Shiau and Chin (1998)

Biochemical functions

Biotin is a coenzyme essential for gluconeogenesis, lipogenesis, and the elongation of essential fatty acids. It converts carbohydrates to protein and vice versa and transforms protein and carbohydrate into fat. Biotin is important for the normal functioning of the reproductive and nervous systems and the thyroid and adrenal glands.

Biotin is known also as a regulator of blood glucose levels, but this role in fish and shrimp remains unclear, especially under stressful conditions. As a component of 5 carboxylating enzymes, it can transport carboxyl units and fix carbon dioxide (bicarbonate) in tissue and compounds in the Krebs cycle (Camporeale and Zempleni, 2006; Mock, 2013; Mai *et al.*, 2022).

The 5 biotin-dependent carboxylases are:

- propionyl-CoA carboxylase (PCC)
- methylcrotonyl-CoA carboxylase (MCC)
- pyruvate carboxylase (PC)
- acetyl-CoA carboxylase 1 (ACC1)
- acetyl-CoA carboxylase 2 (ACC2).

All except ACC2 are mitochondrial enzymes. In carbohydrate metabolism, biotin functions in both carbon dioxide fixation and decarboxylation, with the energy-producing citric acid cycle dependent upon the presence of this vitamin. The hydrolysis of ATP drives the reaction to ADP and inorganic phosphate. Specific biotin-dependent reactions in carbohydrate metabolism are:

- carboxylation of pyruvic acid to oxaloacetic acid
- conversion of malic acid to pyruvic acid
- interconversion of succinic acid and propionic acid
- conversion of oxalosuccinic acid to α-ketoglutaric acid.

In protein metabolism, biotin enzymes are important in protein synthesis, amino acid deamination, purine synthesis, and nucleic acid metabolism. Biotin is required for transcarboxylation in the degradation of various amino acids.

The vitamin deficiency in mammals hinders the normal conversion of the deaminated chain of leucine to acetyl-CoA. Depleting hepatic biotin reduces the hepatic activity of methylcrotonyl-CoA carboxylase, which is needed for leucine degradation (Mock and Mock, 1992; Mock, 2013). Likewise, the ability to synthesize citrulline from ornithine is reduced in liver homogenates from biotin-deficient rats. The urea cycle enzyme ornithine transcarbamylase was significantly lower in the livers of biotin-deficient rats (Maeda et al., 1996).

Acetyl-coenzyme A (CoA)-carboxylase catalyzes the addition of carbon dioxide to acetyl-CoA to form malonyl CoA, the first reaction in the synthesis of fatty acids. Biotin is required for normal long-chain unsaturated fatty acid synthesis and is important for essential fatty acid metabolism. Deficiency in rats and chicks inhibited arachidonic acid (20:4) synthesis from linoleic acid (18:2) while increasing linolenic acid (18:3) and its metabolite (22:6) (Watkins and Kratzer, 1987).

Evidence has emerged that biotin plays unique roles in cell signaling, epigenetic control of gene expression, and chromatin structure (Rodríguez-Meléndez and Zempleni, 2003). Manthey et al. (2002) reported that biotin affects the expression of biotin transporters, biotinylations of carboxylases, and metabolism of interleukin-2 in Jurkat cells. These roles have not been studied extensively in aquaculture species.

Finally, Zhao S. et al. (2012) reported that biotin had a positive impact on digestive enzyme activities, gut microbiota and gut morphology of Jian carp, which may translate into enhanced nutrient digestibility.

Nutritional assessment

Carboxylase assays that measure the saturation of biotin-dependent enzymes (e.g., liver PC) with biotin stand as the most reliable and accurate approach to assess nutritional status in human and animals, including fish and possibly crustaceans (Woodward and Frigg, 1989; Shiau and Chin, 1998; Halver, 2002; Combs and McClung, 2022; Mai et al., 2022). Biotin and its metabolites can be analyzed in plasma using different methods, including microbiological, GC, avidin binding, colorimetric, polarographic, and isotope dilution assays (Sauberlich, 1999).

Deficiency signs

Again, several signs of biotin deficiency are comparable to those of thiamine, riboflavin, niacin and pantothenic acid deficiency in fish: abnormal swimming (e.g., convulsion), anorexia, poor growth and increased mortality (reviewed by NRC, 2011; Hansen et al., 2015; Mai et al., 2022). Other clinical signs of biotin deficiency relate to histopathological damages in the gills, liver, intestine, kidney and erythrocytes, as well as skin depigmentation and darkening. In shrimp, signs associated with biotin deficiencies reported in the literature were reduced feed intake, poor growth and survival (e.g., Shiau and Chin, 1998; Reddy et al., 1999).

Safety

Insufficient data is available to support estimates of the maximum dietary tolerable levels of biotin for aquaculture species. Biotin toxicity is unlikely to occur in aquatic species (Yossa et al., 2015). It is suggested, primarily from poultry data, that nutritional levels 4 to 10 times above the dietary requirements are safe (NRC, 1987).

Folic acid (Vitamin B₉)
Chemical structure and properties

The terms folate, folacin, and folic acid can be used interchangeably and refer to many compounds that possess folic acid's biological activity. Folacin is the generic descriptor for the original vitamin folic acid and related compounds that qualitatively show folic acid activity.

Folic acid is structurally one of the most complex vitamins after vitamin B_{12}. More than 100 forms have been identified in animals (Combs and McClung, 2022). The pure substance is designated pterylomonoglutamic acid. The basic folate molecule is 5,6,7,8-tetrahydropteroyl-glutamate, also referred to as tetrahydrofolate (THF) monoglutamate, which consists of a 2-amino-4-hydroxy-pteridine (pterin) moiety linked via a methylene group at the C-6 position to a p-aminobenzoyl-glutamic acid. Its chemical structure contains 3 distinct parts: glutamic acid, a para-aminobenzoic acid (PABA) residue, and a pteridine nucleus (Figure 4.24).

Folic acid is a yellowish-orange crystalline powder that is tasteless, odorless, and insoluble in alcohol, ether, and other organic solvents. It is slightly soluble in hot water in the acid form but quite soluble in the salt form. It is fairly stable to air and heat in neutral and alkaline solution but unstable in acid solution. It is readily degraded by light and UV radiation, and heating can considerably reduce folic acid activity, particularly under oxidative conditions (Gregory, 1989).

Natural sources

Oilseed meals, animal by-products and yeasts are sources of folic acid normally considered in animal nutrition (NRC, 2011; Combs and McClung, 2022). Only limited amounts of free folic acid occur in natural products. Folic acid is found in natural sources predominantly in its reduced form as polyglutamyl derivatives of tetrahydrofolic acid (FH_4 or THF). Evidence suggests folate is also synthesized endogenously by the gut microbiota (Duncan et al., 1993).

Much of the naturally occurring folic acid in feedstuffs is conjugated with varying numbers of different glutamic acid molecules, reducing its absorption efficiency. The bioavailability of monoglutamate folic acid is substantially greater than polyglutamyl forms (Clifford et al., 1990; Gregory et al., 1991). However, the bioavailability of folates in feedstuffs is difficult to assess and variable because of various nutritional factors and their interactions (Combs and McClung, 2022). As an example of bioavailability variability, Babu and Srikantia (1976) reported that the availability of folic acid in 7 separate food items ranged from 37 to 72% in the monoglutamate form, whiles others have concluded that folic acid bioavailability in various foods generally exceeded 70% (Clifford et al., 1990).

A considerable loss of folic acid (35–50%) occurs during feed manufacturing and storage (Mai et al., 2022). Folic acid is sensitive to light and heating, particularly in acid solutions. Under aerobic conditions, the destruction of most folic acid forms is significant with heating.

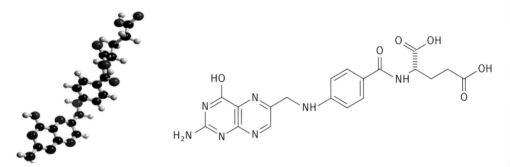

Figure 4.24 Folic acid chemical structure (source: Oviedo-Rondón et al., 2023b)

Commercial sources

Spray-dried folic acid and dilutions of crystalline folic acid are the most widely used product forms in animal feeds. Several lines of evidence indicate higher bioavailability of added folic acid than naturally occurring folates in many foods, which have approximately 50% lower availability (Gregory, 2001). Synthesized folic acid is the monoglutamate form.

Although folacin is only sparingly soluble in water, sodium salt is quite soluble and is used in injections and feed supplements (McGinnis, 1986).

Metabolism
Absorption and transport

Polyglutamate forms of folic acid are digested via hydrolysis to mono- or di-glutamate before transport across the intestinal mucosa. The enzyme responsible for the hydrolysis of polyglutamates is a carboxy peptidase known as folate conjugase (Baugh and Krumdieck, 1971). Most likely, several conjugase enzymes are responsible for the hydrolysis of the long-chain folate polyglutamates to the monoglutamate, which then enter the mucosal cell (Rosenberg and Neumann, 1974). Kesavan and Noronha (1983) suggested from rat results that luminal conjugase is a secretion of pancreatic origin and that the hydrolysis of polyglutamate forms of folic acid occurs in the lumen rather than at the mucosal surface or within the mucosal cell.

Monoglutamate is absorbed predominantly in the small intestine, mostly by an active process involving sodium. It is estimated that 20–30% of monoglutamates are absorbed through passive diffusion into the enterocyte and is occurring especially at high folate intakes (Combs and McClung, 2022).

After hydrolysis and absorption from the intestine, dietary folates are transported in erythrocytes as monoglutamate derivatives, predominantly as 5-methyltetrahydrofolate (5-MTHF), also referred to as 5-methyl-FH$_4$.

The monoglutamate derivatives then enter cells by specific transport systems involving folate receptors and transporters. The polyglutamates, the major folic acid form in cells, are built up by an enzyme, namely folate polyglutamate synthetase, in a stepwise manner. Polyglutamates keep folic acid within the cells since only the monoglutamate forms are transported across membranes, and only monoglutamates are found in plasma and urine (Wagner, 2001).

Storage and excretion

Folic acid is widely distributed in tissues, largely in the conjugated polyglutamate forms of folic acid, generally containing 3 to 7 glutamyl residues linked by peptide bonds. The natural coenzymes are abundant in every tissue examined (Wagner *et al.*, 1984). Specific folate-binding proteins (FBPs) that bind folic acid mono- and polyglutamate exist in many tissues and body fluids, including liver, kidney and small intestinal brush border membranes (Combs and McClung, 2022). These FBPs have also been suggested to play a role in folic acid transport analogous to the intrinsic factor in the absorption of vitamin B$_{12}$. Examples of folic acid concentrations in fish and shrimp tissues are given in Table 4.10.

What is known about folic acid excretion comes to a great extent from the human and mammal literature (Combs and McClung, 2022). Folic acid is predominantly excreted via feces and bile. Fecal folic acid concentrations are high, often higher than intake, meaning synthesis of this vitamin by the gut microbiota takes place. Bile contains high levels of folic acid due to enterohepatic circulation, with most biliary folic acid reabsorbed in the intestine (Bailey *et al.*, 2013). Urinary excretion of folic acid represents a small fraction of total excretion.

Table 4.10 Examples of average total vitamin B_9 concentrations (µg/g wet tissue) in fish and crustacean species reared under laboratory conditions

Species	Tissue	B_9 (µg/g wet tissue)	References
Atlantic salmon (*Salmo salar*)	Whole-body	0.2–0.3	Taylor *et al.* (2019)[1]
	Liver	9.6–13.5 7.4–8.5 ~10–15 ~5–8	Hemre *et al.* (2016) Taylor *et al.* (2019)[1] Espe *et al.* (2020) Skjærven *et al.* (2020)
	Muscle	~0.1–0.2 ~0.05–0.06	Espe *et al.* (2020) Skjærven *et al.* (2020)
	Eggs	~0.4–0.8	Skjærven *et al.* (2020)
Zebrafish[2] (*Danio rerio*)	Liver	~4.4–19.9	Kao *et al.* (2014)
	Brain	~0.4–0.8	Kao *et al.* (2014)
Grass shrimp (*Penaeus monodon*)	Hepatopancreas	1.2–1.5	Shiau and Huang (2001)

Notes: [1]Diploid; [2]non-transgenic.

Biochemical functions

The principal functions of folic acid are related to:

- the synthesis of protein and purines, and pyrimidines, which make up the nucleic acids needed for cell division (Bailey *et al.*, 2013)
- the interconversions of various amino acids
- the maturation process of red corpuscles and the functioning of the immune system.

This means that there are multiple coenzyme forms in transferring one-carbon activity (Bailey *et al.*, 2013). In forms 5, 6, 7, 8-tetrahydrofolic acid (THF), folic acid is indispensable in transferring single-carbon (C_1) units in various reactions, a role analogous to that of pantothenic acid in the transfer of two-carbon units (Bailey and Gregory, 2006; Bailey *et al.*, 2013). The one-carbon teams can be formyl, methylene, or methyl groups. Some biosynthetic relationships of one-carbon units are shown in Figure 4.25.

The major *in vivo* pathway providing methyl groups involves the transfer of a one-carbon unit from serine to tetrahydrofolate to form 5,10-methylenetetrahydrofolate, which is subsequently reduced to 5-methyltetrahydrofolate. Methyl tetrahydrofolate then supplies methyl groups to remethylate homocysteine in the activated methyl cycle, providing methionine for synthesizing the important methyl donor agent S-adenosylmethionine (Krumdieck, 1990; Jacob *et al.*, 1994; Bailey *et al.*, 2013).

The important physiological function of THF consists of binding the C_1 units to the vitamin molecule, thus transforming them to "active formic acid" or "active formaldehyde." These are interconvertible by reduction or oxidation and transferable to appropriate acceptors. Folic acid polyglutamates work at least as well as or better than the corresponding monoglutamate forms in every enzyme system examined (Wagner, 2001). It is now accepted that the pteroylpolyglutamates are the acceptors and donors of one-carbon units in amino acid and nucleotide metabolism, while the monoglutamate is merely a transport form.

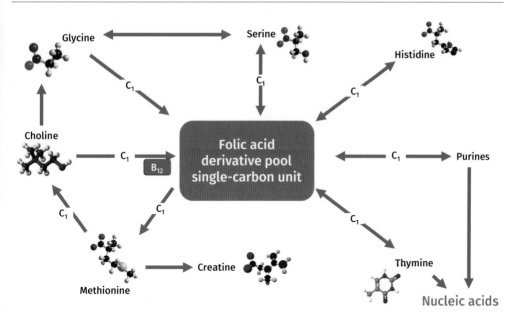

Figure 4.25 Folic acid metabolism requires single-carbon units (Source: Oviedo-Rondón *et al.*, 2023b)

Specific reactions involving single-carbon transfer by folic acid compounds are:

- purine and pyrimidine synthesis
- interconversion of serine and glycine
- glycine–carbon as a source of C_1 units for many syntheses
- histidine degradation
- synthesis of methyl groups for such compounds as methionine, choline, and thymine (a pyrimidine base).

As folacin is involved in the interconversion of serine and glycine, in the metabolism of histidine, and in the addition of methyl groups to compounds such as methionine, choline, and thiamine, inadequate levels of other methyl group donors – such as vitamin B_{12}, serine, methionine, betaine, and choline – increase folic acid requirements (Bailey *et al.*, 2013). This interaction with other nutrients is especially important for methionine, which tends to be the first limiting amino acid in protein-rich ingredients from soy. Logically, high protein levels in the diet raise the dietary recommendations for folate.

Folic acid is also essentially involved in all the reactions of labile methyl groups. The metabolism of labile methyl groups plays an important role in methionine biosynthesis from homocysteine and choline from ethanolamine. Folic acid has a sparing effect on the requirements of choline.

Folic acid is also needed to maintain the immune system. The blastogenic response of T-lymphocytes to certain mitogens is decreased in folic acid-deficient humans and animals, and the thymus is preferentially altered (Dhur *et al.*, 1991). More specific information on the role of folic acid with regards to immunity is presented in Chapters 5, 6, and 7.

Nutritional assessment

Serum/plasma folate is a classical indicator of current folate status and is used as a first-line clinical indicator of folate deficiency (Sauberlich, 1999). Folate liver content stands as another reliable indicator in fish (Mai *et al.*, 2022). Relative weight of liver or hepatopancreas (hepatosomatic index) has also been suggested as a marker of folate status (Shiau and Huang, 2001; NRC, 2011; Mai *et al.*, 2022).

Folates can be measured using HPLC-MS/MS, electrochemical or fluorescence-based techniques, radio- and immuno-based assays, and the traditional *Lactobacillus casei* growth assay. This microbiological approach measures all biologically active folate species, including di- and tri-glutamates of the species, but cannot differentiate between the species (Höller *et al.*, 2018).

As folate and cobalamin (B_{12}) jointly participate in one-carbon metabolism and thus have close biological links, both are usually measured concurrently since the deficiency will interact with the blood status markers of the other (Höller *et al.*, 2018).

Deficiency signs

With a folic acid deficiency, there is a reduction in the biosynthesis of nucleic acids essential for cell formation and function. Hence, vitamin deficiency leads to impaired cell division and alterations in protein synthesis. These effects are most noticeable in rapidly growing tissues such as red blood cells, leukocytes, intestinal mucosa and embryos. In the absence of adequate nucleoproteins, normal maturation of primordial red blood cells does not occur, and hematopoiesis is inhibited at the megaloblast stage. As a result of this megaloblastic arrest for normal red blood cell maturation, the first sign of folic acid deficiency is represented by characteristic macrocytic anemia.

Several signs of folic acid deficiency are comparable to those of thiamine, riboflavin, niacin, pantothenic acid and biotin deficiency in fish: abnormal swimming (e.g. lethargy), anorexia and poor growth (reviewed by NRC, 2011; Hansen *et al.*, 2015; Mai *et al.*, 2022). Other clinical signs of folic acid deficiency relate to anemia and red blood cell anomalies, infarction of the spleen, pale gills, dark skin coloration and decrease resistance to disease. In shrimp, signs associated with folic acid deficiencies were reduced feed intake, poor growth and survival (Reddy *et al.*, 1999; Shiau and Huang, 2001).

The effects of folic acid deficiency on humoral immunity have been more thoroughly investigated in animals than humans. The antibody responses to several antigens have been shown to decrease. As *de novo* synthesis of methyl groups requires the participation of folic acid coenzymes, the effect of folic acid deficiency on pancreatic exocrine function was examined in rats (Balaghi and Wagner, 1992; Balaghi *et al.*, 1993). Pancreatic secretion was significantly reduced in the deficient group compared with the pair-fed control groups after 5 weeks. The negative effect of low dietary folate levels on synthesis of methyl groups would be interesting to investigate in fish and crustaceans.

Safety

Folic acid is generally considered a nontoxic vitamin (NRC, 1987; Mai *et al.*, 2022). EFSA FEEDAP Panel (2012) concluded that there is no need to prescribe a maximum level of folic acid in animal feeds and that folic acid does not pose a risk to the environment. Toxic levels of folic acid for fish and crustaceans are unknown. In poultry, renal hypertrophy has been observed when birds were fed with up to 5,000 times the normal intake of folic acid (Leeson and Summers, 2008).

Vitamin B$_{12}$ (cobalamin)
Chemical structure and properties
Nutritionists now consider vitamin B$_{12}$ to be the generic name for a group of compounds with vitamin B$_{12}$ activity in which the cobalt atom is in the center of the corrin nucleus (cobalt-containing corrinoids). Cobalamin has a molecular weight of 1,355 and is the most complex structure and heaviest compound of all vitamins (Figure 4.26). The empirical formula of vitamin B$_{12}$ is C$_{63}$H$_{88}$O$_{14}$N$_{14}$PCo.

Vitamin B$_{12}$ is a dark-red crystalline hygroscopic substance, freely soluble in water and alcohol but insoluble in acetone, chloroform, and ether. Oxidizing and reducing agents and exposure to sunlight tend to destroy its activity. Vitamin B$_{12}$ is considered relatively stable and unaffected by temperature during processing of animal feeds (Riaz et al., 2009), although up to 60% losses has been reported after 3 months of storage at room temperature in aquaculture feeds (Mai et al., 2022). In food processing, vitamin B$_{12}$ was affected by die temperature during extrusion of fortified rice (Bajaj and Singhal, 2019). Therefore, the stability of vitamin B$_{12}$ is likely affected by several factors other than just temperature (e.g., moisture, pH, storage duration) and further studies are required to understand and optimize its stability in aquaculture nutrition.

Adenosylcobalamin and methylcobalamin are naturally occurring forms of vitamin B$_{12}$ in feedstuffs and animal tissues. Cobalamin is not a naturally occurring form of the vitamin but is the most widely used form of cobalamin in clinical practice because of its relative availability and stability (Green and Miller, 2013).

Natural sources
Feedstuffs of animal origin – fishmeal, poultry by-product meal, blood meal, etc. – are reasonably good sources of vitamin B$_{12}$. The kidney and liver are excellent sources. Ruminant tissues are richer in vitamin B$_{12}$ than those of monogastric species. Vitamin B$_{12}$ presence in the tissues of animals follows from ingestion of vitamin B$_{12}$ in animal feeds or from intestinal synthesis.

Plant products are practically devoid of B$_{12}$. The vitamin B$_{12}$ reported in higher plants in small amounts may result from synthesis by soil microorganisms, excretion of the vitamin onto

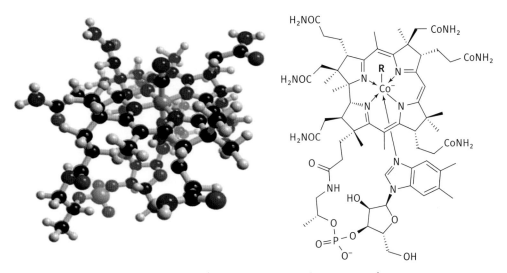

Figure 4.26 Vitamin B$_{12}$ chemical structure (source: Oviedo-Rondón et al., 2023b)

the soil, with subsequent absorption by the plant. The root nodules of certain legumes contain small quantities of vitamin B_{12}. Certain seaweed species (algae) have been reported to contain appreciable amounts of vitamin B_{12} (up to 1 µg/g of solids). Dagnelie *et al.* (1991) reported that vitamin B_{12} from algae is largely unavailable.

Microorganisms synthesize vitamin B_{12} in the intestinal tract of rainbow trout (Sugita *et al.*, 1991). However, cobalamin produced in the distal intestine is likely not bioavailable because the receptors necessary for absorbing the vitamin are typically found in the proximal intestine, upstream of the site of corrinoid production (Seetharam and Alpers, 1982).

Commercial sources

Commercial sources of vitamin B_{12} are produced from fermentation products, and it is available as cobalamin, the most stable form of this vitamin. Little is known about the bioavailability of orally ingested B_{12} in feeds.

Metabolism

Absorption and transport

Absorption of vitamin B_{12} is predominantly via active transport. The absorptive site for vitamin B_{12} is the proximal intestine. Vitamin B_{12} is also secreted into this section of the small intestine and then reabsorbed further down the intestinal tract. Passing vitamin B_{12} through the intestinal wall is a complex procedure that has not been studied extensively in aquatic species (Mai *et al.*, 2022). In most species, the absorption of vitamin B_{12} involves certain carrier compounds able to bind the vitamin molecule (McDowell, 2000) and requires the following conditions (Green and Miller, 2013):

- production of the intrinsic factor (lacking in fish, Hansen *et al.* (2013)) for absorption of vitamin B_{12} through the small intestine
- functional pancreas (trypsin secretion) required for release of bound vitamin B_{12} before combining the vitamin with the intrinsic factor
- functional small intestine with receptor and absorption sites.

The absorption of vitamin B_{12} appears to be slow: peak blood concentrations of the vitamin are not achieved for some 6 to 8 hours after an oral dose.

Gastric juice defects are responsible for most cases of food-vitamin B_{12} malabsorption in monogastric animals (Carmel, 1994). Factors that diminish vitamin B_{12} absorption include protein, iron, and vitamin B_6 deficiencies, and dietary tannic acid (Hoffmann-La Roche, 1984).

There are structural differences in vitamin B_{12} intrinsic factors among species. Intrinsic factors have been demonstrated in humans, monkeys, pigs, rats, cows, ferrets, rabbits, hamsters, foxes, lions, tigers, and leopards. They have not yet been detected in guinea pig, horses, sheep, chickens, and fish. Similarly, there are species differences in vitamin B_{12} transport proteins (Polak *et al.*, 1979; Green and Miller, 2013). Vitamin B_{12} is bound to transcobalamin (TC) and haptocorrin (HC) for transport in the blood, with about 20% being attached to TC and the rest to HC. The TC-bound cobalamin is the form most actively transported into tissues.

Expression of genes encoding for proteins involved in vitamin B_{12} absorption and transport, but not the vitamin B_{12} binding intrinsic factor, have been observed in zebrafish (Hansen *et al.*, 2013).

Storage and excretion

It is estimated that ~60% of the whole-body vitamin B_{12} is stored in the liver (Combs and McClung, 2022). Other storage sites include the muscle, kidney, heart, spleen, and brain (Green and Miller, 2013). Even though vitamin B_{12} is water-soluble, Kominato (1971) reported a tissue half-life of 32 days in rats, indicating considerable tissue storage. Levels of vitamin B_{12} in tissues of salmon and zebrafish are reported in Table 4.11. Vitamin B_{12} is excreted via the bile and urine.

Biochemical functions

Although the most important tasks of vitamin B_{12} concern the metabolism of nucleic acids and proteins, it also functions in the metabolism of fats and carbohydrates. Overall, protein synthesis is impaired in vitamin B_{12}-deficient animals (Friesecke, 1980). Moreover, the promotion of red blood cell synthesis and the maintenance of nervous system integrity are functions attributed to vitamin B_{12} (McDowell, 2000). Vitamin B_{12} is metabolically related to other essential nutrients, such as choline, methionine, and folic acid. (Savage and Lindenbaum, 1995; Stabler, 2006).

A summary of vitamin B_{12} functions includes (Figure 4.27):

- purine and pyrimidine synthesis
- transfer of methyl groups (1C metabolism)
- formation of proteins from amino acids
- carbohydrate and fat metabolism
- vitamin B_{12} is an important cofactor in the following functions:
  - the maintenance of normal DNA synthesis: failure of this metabolic pathway can lead to megaloblastic anemia
  - the regeneration of methionine for the dual purposes of maintaining protein synthesis and methylation capacity
  - the avoidance of homocysteine accumulation, an amino acid metabolite implicated in vascular damage, thrombosis, and several associated degenerative diseases, including coronary artery disease, stroke, and osteoporosis (Green and Miller, 2013).

Gluconeogenesis and hemopoiesis are critically affected by cobalt deficiency, and carbohydrate, lipid, and nucleic acid metabolism are all dependent on adequate B_{12} and folic acid

Table 4.11 Examples of average total vitamin B_{12} concentrations (µg/g wet tissue) in fish and crustacean species reared under laboratory conditions

Species	Tissue	B_{12} (µg/g wet tissue)	References
Atlantic salmon (*Salmo salar*)	Muscle	0.02–0.05 ~0.05–0.12 ~0.01 0.07	Hemre *et al.* (2016) Espe *et al.* (2020) Skjærven *et al.* (2020) Mai *et al.* (2022)
	Liver	0.40–0.48 ~0.50–0.75 ~0.4 0.6	Hemre *et al.* (2016) Espe *et al.* (2020) Skjærven *et al.* (2020) Mai *et al.* (2022)
	Eggs	~0.05–010	Skjærven *et al.* (2020)
Zebrafish (*Danio rerio*)	Whole-body	~31–73	Hansen *et al.* (2013)

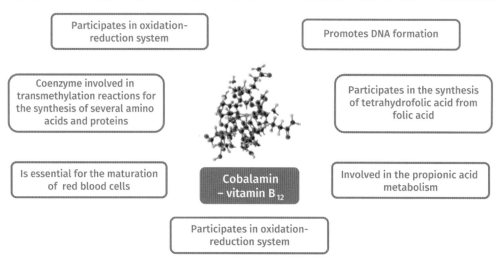

Figure 4.27 Vitamin B₁₂ functions (source: Oviedo-Rondón *et al.*, 2023b)

metabolism. Vitamin B₁₂ is involved in the metabolism of fatty acids, the synthesis of proteins, and reactions involving the transfer of methyl and hydrogenated/hydrogen groups (Green and Miller, 2013).

Vitamin B₁₂ is a cofactor of 2 important metabolic reactions in cells, one involving mitochondrial adenosylcobalamin and the other cytoplasmic, mostly related to methylcobalamine. In the mitochondrial reaction, B₁₂ in the form of 5′-deoxyadenosylcobalamin is required for the enzyme methylmalonyl-CoA mutase, a vitamin B₁₂-requiring enzyme (5′-deoxyadenosyl-cobalamin) that catalyzes the conversion of methylmalonyl-CoA to succinyl-CoA (Green and Miller, 2013). This is an intermediate step in transforming propionate to succinate during the oxidation of odd-chain fatty acids and the catabolism of ketogenic amino acids. In animal metabolism, propionate of dietary or metabolic origin is converted into succinate, entering the tricarboxylic acid (Krebs) cycle.

In the cytoplasmic reaction, B₁₂ in the form of methylcobalamin is required in the folate-dependent methylation of the sulfur amino acid homocysteine to form methionine catalyzed by the enzyme methionine synthase. Apart from being necessary for adequate protein synthesis, methionine is also a key precursor for the maintenance of methylation capacity through the synthesis of the universal methyl donor S-adenosylmethionine. Additionally, the methionine synthase reaction is finally necessary for normal DNA synthesis.

The methyl group transferred to homocysteine during methionine synthesis is donated by the folate derivative methyl tetrahydrofolate (methyl-THF), forming THF. THF is later transformed to 5,20-methylenetetrahydrofolate (methylene-THF) by a one-carbon transfer during serine conversion to glycine. Methylene-THF can be reduced again to form methyl-THF. Still, it also serves as the critical one-carbon source for the *de novo* synthesis of thymidylate from deoxyuridylate, required for DNA replication.

Finally, an additional function of vitamin B₁₂ relates to immune function. In mice, vitamin B₁₂ deficiency affected immunoglobulin production and cytokine levels (Funada *et al.*, 2001). The effect of vitamin B₁₂ on immunity has not been demonstrated yet in fish and shrimp.

Nutritional assessment

TC-bound cobalamin, the form most actively transported into tissues, can be measured and is considered a relevant marker of vitamin B_{12} status in animal species. In aquatic animals, tissue concentration has been used to assess vitamin B_{12} status. Plasma level stands as a potential non-lethal marker for assessing vitamin B_{12} status (Mai et al., 2022). A newer method, not yet tested in fish or shrimp, estimates holotranscobalamin (holoTC) as a fraction of vitamin B_{12} carried by TC in serum and, therefore, available for tissue uptake (Höller et al., 2018).

Historically, vitamin B_{12} was measured using microbiological assays such as the *Lactobacillus delbrueckii* method, which was later adapted for high-throughput use (Sauberlich, 1999). Measuring the total vitamin B_{12} concentration in serum is the first-line clinical test for determining vitamin B_{12} deficiency. The current assays are mostly based on the competitive binding of the serum vitamin to intrinsic factor, followed by radiometric or fluorescence-based detection. (Höller et al., 2018).

Deficiency signs

Several signs of vitamin B_{12} deficiency are comparable to those of other B vitamins: anorexia and poor growth [reviewed by NRC (2011), Hansen et al. (2015) and Mai et al. (2022)]. Clinical signs of vitamin B_{12} deficiency can be confounded with those of folic acid and relate to anemia and fragmented erythrocytes in fish and shrimp.

When B_{12} is deficient, 1C metabolism can be compromised after tissue stores have been exhausted. Synthesis and transformation of methyl-THF can be impaired through interdiction of the methionine synthase reaction and impact negatively DNA replication (Green and Miller, 2013).

Deficiency of vitamin B_{12} will induce folic acid deficiency by blocking the utilization of folic acid derivatives: folic acid remains trapped as methylfolate and thus becomes metabolically useless. This explains why the hematologic damage of vitamin B_{12} deficiency is indistinguishable from that of folacin deficiency, resulting in an inadequate quantity of methylene-THF to participate adequately in DNA synthesis. Vitamin B_{12} deficiency can be generated with the addition of high dietary levels of propionic acid.

Safety

Adding vitamin B_{12} to feeds in amounts far above requirement or absorbability appears to be without hazard. Dietary levels of at least several hundred times the requirement are considered safe for most species (NRC, 1987). EFSA FEEDAP Panel (2023) concluded that vitamin B_{12} is considered safe in animal feeds and that there is no concern for end consumer safety. Toxic levels of vitamin B_{12}. for fish and crustaceans are unknown. In avian species, signs of toxicity were observed, but were likely confounded with toxic effects of fermentation residues, inadvertently included with B_{12} during manufacture (Leeson and Summers, 2008).

Vitamin C
Chemical structure and properties

Vitamin C, also called ascorbic acid, has 4 stereoisomers: L-ascorbic acid, D-ascorbic acid, L-isoascorbic acid and D-isoascorbic acid or erythorbic acid. The stereoisomer L-ascorbic acid is widely found in nature, whereas D-ascorbic acid, L-isoascorbic acid and erythorbic acid are synthetic products. Only the stereoisomer L-ascorbic acid is biologically active. Bioactive vitamin C includes 2 forms (Figure 4.28):

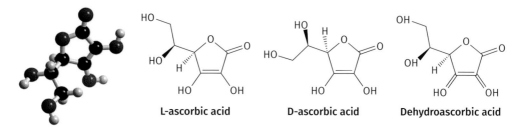

Figure 4.28 Vitamin C chemical structure (source: Oviedo-Rondón *et al.*, 2023b)

- L-ascorbic acid: reduced form
- L-dehydroascorbic acid or dehydroascorbic acid: oxidized form.

In nature, the reduced form of ascorbic acid may reversibly oxidize to the dehydroxidized form, i.e., dehydroascorbic acid (Johnston *et al.*, 2013), and dehydroascorbic acid is irreversibly oxidized to the inactive diketogulonic acid. The latter can be further oxidized to oxalic acid and L-threonic acid. Since this change takes place readily, vitamin C is susceptible to destruction through oxidation, accelerated by heat and light. Reversible oxidation-reduction of ascorbic acid with dehydroascorbic acid is vitamin C's most important chemical property and the basis for its known physiological activities and stabilities (Moser and Bendich, 1991).

Vitamin C is a white to yellow-tinged crystalline powder. It crystallizes, out of the water, like square or oblong crystals. It is slightly soluble in acetone. A 0.5% solution of ascorbic acid in water is strongly acid with a pH of 3. The vitamin is more stable in acid than in an alkaline medium.

Vitamin C is the least stable and, therefore, most easily destroyed of all the vitamins. Several chemical substances, such as air pollutants, industrial toxins, heavy metals, and some pharmacologically active compounds, are antagonistic to vitamin C and can lead to increased vitamin requirements (Johnston *et al.*, 2013).

Natural sources
The main sources of vitamin C are fruits and green plants, but some foods of animal origin contain more than traces of the vitamin. Vitamin C occurs in significant quantities in animal organs, such as the liver and kidney, but in only small amounts in meat. However, vitamin C is very low in cereals and oilseeds used to make feed ingredients in aquaculture nutrition. Post-harvest storage values vary with time, temperature, damage, and enzyme content (Zee *et al.*, 1991; Johnston *et al.*, 2013).

Commercial sources
Ascorbic acid is commercially available as:

- 100% crystalline L-ascorbic acid
- Sodium ascorbate
- 97.5% L-ascorbic acid – ethyl cellulose-coated (EC)
- 35% phosphorylated Na/Ca salt of L-ascorbic acid (Ca or Na ascorbyl-2-phosphate; $C_6H_9O_9P$ molecular mass 256.11 g/mol)
- 50% phosphorylated Na salt of L-ascorbic acid (Ca or Na ascorbyl-2-phosphate; $C_6H_6O_9Na_3 \cdot H_2O$ molecular mass 358.08 g/mol).

When providing supplemental ascorbic acid in heat-treated feeds, it is strongly advisable to use a stabilized form like EC-coated or phosphorylated forms to minimize losses of activity. Among the phosphorylated Na/Ca salt of L-ascorbic acids, the monophosphate form contains more ascorbic acid on a molecular weight basis and is widely used in aquaculture feeds (NRC, 2011).

Metabolism
Absorption and transport
Vitamin C is absorbed like carbohydrates (monosaccharides) in the proximal intestine. Intestinal absorption in vitamin C-dependent animals (e.g., primates and most species of fish) requires sodium-dependent vitamin C transporters (SVCT1 and 2) and glucose transporters (GLUTs) (Johnston, 2006; Johnston *et al.*, 2013; Mai *et al.*, 2022).

Absorbed vitamin C circulates in the blood predominantly in the form of L-ascorbic acid. A small fraction (10–20%) of L-ascorbic acid is oxidized and transformed into dehydroascorbic acid. Circulating vitamin C readily equilibrates with the body pool of the vitamin.

Uptake of vitamin C by different tissues takes place like in the small intestine, i.e. via SVCTs for L-ascorbic acid and GLUTs for dehydroascorbic acid. Once in the cells, ascorbic acid is first converted to dehydroascorbic acid by several enzymes or non-enzymatic processes and can then be reduced back to ascorbic acid in cells (Johnston *et al.*, 2013).

The bioavailability of dietary vitamin C depends on its chemical form, species, nutritional status as well as feed processing and storage conditions (NRC, 2011; Mai *et al.*, 2022).

Storage and excretion
Ascorbic acid is widely distributed throughout the tissues in fish and crustaceans (Table 4.12). In experimental animals and aquatic species, the highest concentrations of vitamin C are found in adrenal tissues (present in the head kidney), and high levels are also found in the liver. Vitamin C also tends to localize around healing wounds. Tissue levels are decreased by virtually all forms of stress, which also create a demand for higher dietary intake to optimize animal health.

Ascorbic acid is excreted mainly in the urine, with small amounts in feces. Urinary excretion of vitamin C depends on body stores, intake, and renal function.

Biochemical functions
Ascorbic acid is involved in fundamental biological and metabolic processes, and its function is related to its reversible oxidation and reduction characteristics. Thus, its action is important in:

- calcification processes
- immune response
- adaptation to stress
- maintenance of electrolytic balance.

Vitamin C's biochemical and physiological functions have been copiously reviewed over the last decade (Johnston *et al.*, 2013; Dawood and Koshio, 2018; Mai *et al.*, 2022). Vitamin C is well known for its antioxidant properties. Other important roles of vitamin C in salmonids include reproduction, conversion of vitamin D_3 to its active form, biosynthesis of collagen, iron absorption, glucocorticoid synthesis, wound healing, immune defense, and stress/disease resistance

Table 4.12 Examples of average total vitamin C concentrations (μg/g wet tissue) in fish and crustacean species reared under laboratory conditions

Species	Tissue	C (μg/g wet tissue)	References
Rainbow trout (*Oncorhynchus mykiss*)	Blood	34.4–51.0 27.8–51.7	Halver (2002) Verlhac *et al.* (1998)
	Head kidney	125–247 148.1–208.2	Halver (2002) Verlhac *et al.* (1998)
	Liver	95.2–149.0	Verlhac *et al.* (1998)
Coho salmon (*Oncorhynchus kisutch*)	Liver	27.1–83.4	Xu *et al.* (2022a)
Atlantic salmon (*Salmo salar*)	Blood	22.3–37.8	Halver (2002)
	Head kidney	89–321	Halver (2002)
Tilapia (*Oreochromis niloticus*)	Liver	33.6–104.7	Wu F. *et al.* (2015)
	Muscle	14.7–31.9	Wu F. *et al.* (2015)
	Blood	3.5–4.5	Ibrahim *et al.* (2020)
Mrigal carp (*Cirrhinus mrigala*)	Liver	12.8–64.9	Zehra and Khan (2012)
Channel catfish (*Ictalurus punctatus*)	Liver	32.6–77.3	Lovell and Lim (1978)
	Head kidney	175.4–295.2	Lovell and Lim (1978)
Grouper (*Epinephelus malabaricus*)	Liver	21.4–80.2	Lin and Shiau (2004)
Pufferfish (*Takifugu obscurus*)	Liver	20.6–103.4	Cheng C. H. *et al.* (2017)
	Muscle	6.3–32.9	Cheng C. H. *et al.* (2017)
Whiteleg shrimp (*Litopenaeus vannamei*)	Whole-body	2.5–21.5	He H. and Lawrence (1993a)
Kuruma prawn (*Penaeus japonicus*)	Muscle	46.2–67.3	Alava (1993)

(Sandnes *et al.*, 1984; Eskelinen, 1989; Navarre and Halver, 1989; Waagbø *et al.*, 1992; Wahli *et al.*, 2003; NRC, 2011; Mai *et al.*, 2022). The main biochemical functions of vitamin C are summarized in Figure 4.29.

Antioxidant and immune role (stimulation of phagocytic activity)

One of the most interesting properties of vitamin C is its ability to act as a reducing agent or electron donor. It reacts rapidly with free radicals and works synergistically with vitamin E, hence accounting for the observed sparing effect on this vitamin (Jacob, 1995). In the process of sparing fatty acid oxidation, tocopherol is oxidized to the tocopheryl free radical. Ascorbic acid can donate an electron to the tocopheryl free radical, regenerating the reduced antioxidant form of tocopherol.

Ascorbic acid is reported to have a stimulating effect on the phagocytic activity of leukocytes, the function of the reticuloendothelial system, and the formation of antibodies. Ascorbic acid levels are very high in phagocytic cells, with these cells using free radicals and other highly reactive oxygen-containing molecules to help kill pathogens that invade the body.

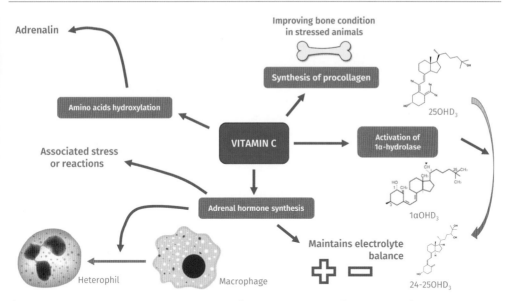

Figure 4.29 Some vitamin C roles on metabolism (source: Oviedo-Rondón *et al.*, 2023b)

Ascorbic acid minimizes the oxidative stress associated with activated phagocytic leukocytes' respiratory burst, thereby controlling the inflammation and tissue damage associated with immune responses (Chien *et al.*, 2004).

Conversion of vitamin D₃ to its active form

Vitamin C, because of its relationship to hydroxylation enzymes, has a direct effect on C-1 hydroxylation of 25OHD₃ to the active form 1,25(OH)₂D₃ (Suter, 1990; Cantatore *et al.*, 1991). Vitamin C has been shown to influence the developmental process in the growth plate for bone growth (Farqhuarson *et al.*, 1998).

Biosynthesis of collagen

The beneficial effects of ascorbic acid in the collagen biosynthesis are extensively documented and represent the most clearly established role for vitamin C. Collagens are the tough, fibrous, intercellular materials (proteins) that are the principal components of skin and connective tissue, cartilage, as well as the organic matrix of bones and teeth and the ground substance between cells.

Beneficial effects result from ascorbic acid synthesizing "repair" collagen and wound healing. Failure of wounds to heal and gum and bone changes resulting from vitamin C undernutrition are direct consequences of reducing insoluble collagen fibers. Ascorbic acid is a cofactor in extracellular matrix metabolism because it affects collagen, laminin, various cell-surface integrins, and elastin. Vitamin C is a cofactor for enzymes key to the post-translational modification of matrix proteins (Johnston *et al.*, 2013).

In the case of vitamin C deficiency, the impairment of collagen synthesis appears to be due to lowered ability to hydroxylate lysine and proline. In addition to the relationship of ascorbic acid to hydroxylase enzymes, Franceschi (1992) suggests that vitamin C is required for the differentiation of connective tissue such as muscle, cartilage, and bone derived from mesenchyme (embryonic cells capable of developing into connective tissue). It is proposed that the collagen matrix produced by ascorbic acid-treated cells provides a permissive environment for tissue-specific gene expression. A common finding is that vitamin C can alter the expression of multiple genes as cells progress through specific differentiation programs (Ikeda *et al.*, 1997).

Absorption of minerals (iron)
Ascorbic acid has a role in metal ion metabolism due to its reducing and chelating properties. This results in enhanced absorption of minerals from the diet and their mobilization and distribution throughout the body. Ascorbic acid reduces the ferric iron at the acid pH in the stomach and forms complexes with iron ions that stay in solution at alkaline conditions in the proximal intestine.

Control of glucocorticoid synthesis
Vitamin C controls the synthesis of glucocorticoids and norepinephrine in the adrenal gland of the anterior kidney. The protective effects of vitamin C (also vitamin E) on health may partially result from reducing glucocorticoid circulating levels (Nockels, 1990). During stress conditions (e.g., heat stress), glucocorticoids, which suppress the immune response, are elevated. Vitamin C reduces adrenal glucocorticoid synthesis, helping to maintain immunocompetence.

Other functions
Vitamin C is involved in carnitine synthesis. Carnitine is synthesized from lysine and methionine and is dependent on 2 hydroxylases containing ferrous iron and L-ascorbic acid. Vitamin C deficiency can reduce the formation of carnitine, resulting in the disruption of fatty acid β-oxidation and accumulation of triglycerides in tissues.

Vitamin C also participates in many hormone activation processes. Hormones like melanotropins, calcitonin, growth hormone-releasing factors, corticotrophin and thyrotropin, vasopressin, oxytocin, cholecystokinin, and gastrin undergo amidations where ascorbic acid serves as a reductant to maintain copper in a reduced state at the active site of the enzyme (Johnston *et al.*, 2013).

Finally, ascorbic acid may play a role in broodstock nutrition by enhancing egg and fry survival (Eskelinen, 1989; Dawood and Koshio, 2018).

Nutritional assessment
Several biological compartments such as body tissues (e.g. liver, vertebral collagen), whole blood, erythrocytes, leucocytes, and plasma or serum can be used to assess vitamin C status (NRC, 2011; Mai *et al.*, 2022). However, serum or plasma concentration of ascorbate is the most reliable marker. Analysis of ascorbic acid in biological samples is complicated by the high susceptibility of this compound to oxidation which requires the use, for example, of EDTA (Höller *et al.*, 2018). Several approaches have been developed to measure vitamin C in biological materials: HPLC provides an efficient means to quantify vitamin C with good selectivity and sensitivity (Höller *et al.*, 2018).

Deficiency signs
Bone deformities (e.g., scoliosis, lordosis, jaw; Figure 4.30), eye pathology (e.g., cataract, exophthalmia), abnormal skin pigmentation (decolouration or darkening), hemorrhaging (internal, skin and fins), anorexia, lethargy, slow growth, poor disease resistance and high mortality are typical signs of vitamin C deficiency in aquatic species [reviewed by NRC (2011), Dawood and Koshio (2018) and Mai *et al.* (2022)]. High levels of triglycerides, and sometimes cholesterol, in plasma are other signs of vitamin C deficiency (John *et al.*, 1979; Deng *et al.*, 2019).

Figure 4.30 Atlantic salmon with jaw deformity (source: courtesy Hardy, R. W.)

Safety
In general, high intakes of vitamin C are considered low toxicity. No signs of toxicity were observed in tilapia fed 1,100 mg vitamin C/kg diet for almost 6 months (Falcon *et al.*, 2007). In European sea bass at larval stage, dietary vitamin C at 400 mg/kg was associated with bone deformities and fin abnormalities (Darias *et al.*, 2011).

Choline
Chemical structure and properties
Choline (2-hydroxy-*N,N,N*-trimethylethanaminium) (Figure 4.31) is a water-soluble nutrient essential to many animals, including fish and shrimp (de Lima *et al.*, 2018; Shiau and Cho, 2002; Simon, 1999; Won *et al.*, 2019; Zeisel, 1981). Choline is a strong base, hygroscopic, and soluble in ethanol but insoluble in organic solvents (Combs and McClung, 2017). Choline contains 3 methyl groups which allows it to serve as a methyl donor and participate in the one-carbon cycle.

Natural sources
Foods and feedstuff rich in choline (as total choline; wet weight basis) include raw egg yolk (6.8 g/kg), solvent-extracted canola meal (6.7 g/kg), poultry-feather meal (6.0 g/kg), fish meals

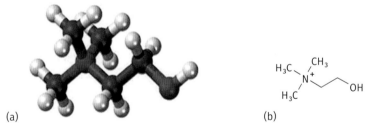

(a)   (b)

Figure 4.31 Choline chemical structure (source: Oviedo-Rondón *et al.*, 2023b)

(2.9-5.2 g/kg), beef by-products (3.3 g/kg), sunflower meals (3.6-3.7 g/kg), soy products (raw, meal, protein isolate/concentrate, flour; 0.6-2.8 g/kg), and raw chicken liver (1.9 g/kg) (NRC, 2011; USDA, 2008). Most raw foods (fresh or frozen) including, fishes, legumes, vegetables, and fruits used to feed cultured fish contain less than 0.5 g choline/kg (Ahmmed *et al.*, 2020; USDA, 2008). In comparison, common protein sources used in formulated aquafeeds obtained from marine or terrestrial origins generally contain at least 1.0 g choline/kg, although ingredients such as casein, corn gluten meal, and kelp meal have much lower choline levels (NRC, 2011). Krill meal and krill oil are particularly rich in choline-containing phospholipids (i.e., phosphatidylcholine and lysophosphatidylcholine) (Ahmmed *et al.*, 2020; Burri and Johnsen, 2015; Valverde *et al.*, 2012; Winther *et al.*, 2011). Based on the molecular weight of free choline (104.17 g/mol) in these phospholipids, krill meal/oil can provide >6.0 g choline/kg. Because the choline content in oils is rarely reported in the literature, oils that do not contain much choline-containing phospholipids such as fish oil, rapeseed oil, and soybean oil presumably have low levels of choline (Tocher *et al.*, 2008).

Commercial sources

Commercially, choline can be mixed with vitamin premixes or separately as lecithin, phosphatidylcholine, or choline salts (e.g., chloride, bitartrate, dihydrogen citrate) (NRC, 2011). The term "lecithin" commonly refers to phosphatidylcholine, however, feed-grade lecithin may contain other phospholipid classes because they are usually extracted and purified from egg yolk, rapeseed, sunflower, and soybean (EFSA FEEDAP Panel, 2016). Furthermore, phosphatidylcholine only contains approximately 13-14% of choline by weight depending on the extraction source. For this reason, chemically synthesized choline salts are richer sources of choline than naturally derived sources. For example, choline chloride (CC) is a widely used supplement that contains 74.6% choline by weight; accordingly, feed-grade CC supplements (50-75% CC) can supply between 373-560 g choline/kg. Supplemental choline can be in both solid or liquid form and its purity varies by the manufacturing method and inclusion levels of other nutrients, hence the molecular weight of choline in these supplements should be taken into consideration when formulating aquafeed to ensure that dietary choline is sufficient to meet the animal's requirement. The liquid form of CC (usually 75% CC) is corrosive and requires special storage and handling equipment. It is not suitable for inclusion in concentrated vitamin premixes but is most economical to add directly to concentrate feed mixtures. Analytical grade CC crystals (>95%) are hygroscopic and should be stored appropriately to minimize additional weight from absorbing moisture.

Metabolism
Absorption, transport

Choline can be acquired from the diet in the form of free choline and other choline-containing molecules that can then be hydrolyzed by pancreatic and intestinal enzymes to form free choline. The water-soluble compounds such as free choline, phosphocholine, and glycerophosphocholine can be absorbed by a saturable carrier system or passive diffusion before entering the portal vein and stored in the liver (Corbin and Zeisel, 2012). Intact choline-containing compounds are mostly absorbed as lysophosphatidylcholine (lysoPC). The lysoPC are re-esterified with free choline to form choline-containing lipids, incorporated into lipoproteins, and then circulated in the lymphatic system where choline is transported to other tissues.

Storage, and excretion
Choline is primarily stored in the liver of fish and hepatopancreas in shrimp. Depletion of choline from the body pool involves metabolizing choline to betaine and trimethylamine and excreting via urine and feces.

Biochemical functions and synthesis

Choline is important for its role in lipid metabolism, mainly attributed to being the head group of phospholipids such as phosphatidylcholine and sphingomyelin. Several comprehensive reviews on the roles of choline-containing phospholipids in teleost fish already exist (Tocher, 1995; 2003; Tocher and Sargent, 1990; Tocher *et al.*, 2008). Although not well studied in fish and shrimp, the phospholipids in fish and mammalian cells share the same structure i.e., hydrophilic phosphate-head group (substitute for choline, ethanolamine, serine etc.) attached to the glycerol backbone at *sn*-3 and the hydrophobic fatty acids at the *sn*-1 and *sn*-2 (Ridgway, 2016; Tocher *et al.*, 2008). Thus, phospholipids in fish and shrimp should also share the same amphipathic properties and functions as they do in mammals. The amphipathic property allows the formation of plasma and organelle membranes and lipoproteins to transport hydrophobic lipids in aqueous environments such as plasma and lymph (Tocher *et al.*, 2008; Vance, 2015).

The biochemistry and biosynthesis of choline and its metabolites have been extensively researched and reviewed for terrestrial animals for several decades (Figure 4.32) (Al-Humadi *et al.*, 2012; Fagone and Jackowski, 2013; Zeisel, 1981; Zeisel and Blusztajn, 1994). At present, there are limited studies that have directly investigated the metabolic pathways of choline in fish and shrimp. Moreover, the role of choline in fish and shrimp has not received the same attention as it has in terrestrial animals. Nonetheless, there is presently little evidence to suggest that choline metabolism in fish and shrimp is different to terrestrial animals (Simon, 1999; Zwingelstein *et al.*, 1998).

Choline metabolism primarily occurs in the liver, and it can be obtained from the diet or biosynthesized *de novo* (Caudill, 2010). Based on the mammalian model, choline is synthesized *via* 2 major metabolic pathways: 1) phosphatidylethanolamine-*N*-methyltransferase (PEMT), and 2) cytidine diphosphocholine (CDP-choline). In the PEMT pathway (Figure 4.32), homocysteine can receive a methyl group from either betaine or 5-methyltetrahydrofolate *via* the folate cycle. To form methionine, homocysteine is catalyzed by vitamin B_{12}–dependent methionine synthase. The enzyme methionine adenosyltransferase activates methionine to *S*-adenosylmethionine (SAM), a primary methyl donor required for more than 60 methylation enzymes (Caudill, 2010). In the following reactions phosphatidylethanolamine (PE) accepts a methyl group from SAM, catalyzed by PEMT; this reaction occurs 3 times before PE forms PC which can be cleaved to form a new choline molecule. The methylation process makes PEMT one of the biggest methyl group consumers of SAM (Fagone and Jackowski, 2013). In addition, the biosynthesis of PC through PE methylation has also been demonstrated in mussel (*Mytilus galloprovincialis*) (Athamena *et al.*, 2011a), rainbow trout, eel (*Anguilla Anguilla*), green crab (*Carcinus maenas*), and Chinese crab (*Eriocheir sinensis*) (Athamena *et al.*, 2011b; Zwingelstein *et al.*, 1998). As described above, the flux of SAM is influenced by folate, vitamin B_{12} and methionine, thus a disturbance in these nutrients can subsequently affect choline metabolism and *vice versa*. In the CDP-choline pathway, choline-containing molecules such as phosphocholine and PC can be cleaved to release free choline (Figure 4.32). Other choline-containing molecules including sphingomyelin, acetyl choline (Ach), glycerophosphocholine, lysophosphatidylcholine, and Platelet-activating factor (PAF) can also release free choline.

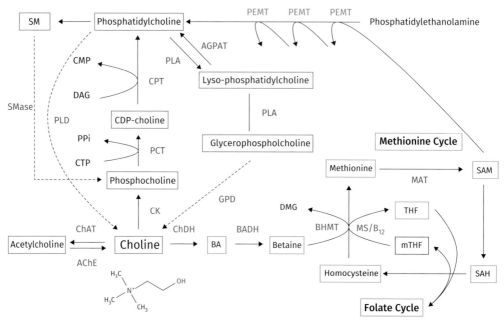

Figure 4.32 Major steps of choline biosynthesis in mammals (source: adapted from Liu A. *et al.* 2020)

Notes: AChE, acetylcholinesterase; ChAT, choline acetyltransferase; ChDH, choline dehydrogenase; BA, betaine aldehyde; BADH, betaine aldehyde dehydrogenase; BHMT, betaine homocysteine methyltransferase; DMG, dimethylglycine; MAT, methionine adenosyltransferase; SAM, *S*-adenosylmethionine; SAH, *S*-adenosylhomocysteine; mTHF, methyl tetrahydrofolate; MS, B$_{12}$, cobalamine (vitamin B$_{12}$); THF, tetrahydrofolate; PEMT, phosphatidylethanolamine-*N*-methyltransferase; CK, choline kinase; CTP, cytidine triphosphate; PCT, phosphocholine cytidylyltransferase; PPi, inorganic pyrophosphate; CDP-choline, cytidine diphosphocholine; DAG, diacylglycerol; CPT, choline phosphotransferase; CMP, cytidine monophosphate; PLD, phospholipase D; AGPAT, acylglycerophosphate acyltransferase; PLA, phospholipase A; GPD, glycerophosphodiesterase; SM, sphingomyelin; SMase; sphingomyelinase.

Supplementing aquafeed with lecithin (predominantly PC) or PC have been reported to improve lipid digestibility in juvenile Atlantic salmon (*Salmo salar*) (Hung *et al.*, 1997), red drum (*Sciaenops ocellatus*) (Craig and Gatlin, 1997), arctic char (*Salvelinu alpinuis* L.) (Olsen *et al.*, 1999), and common carp (*Cyprinus carpio*) (Geurden *et al.*, 2008). Gilthead seabream larvae were also able to assimilate, and transport digested lipid better when fed a PC-supplemented diet, which would likely enhance energy availability and growth performance (Hadas *et al.*, 2003). However, a diet supplemented with CC did not induce the same effect on lipid digestibility in juvenile Atlantic salmon; it presumably took priority in methyl groups donation after absorption (Hung *et al.*, 1997). Wu P. *et al.* (2011) later demonstrated that dietary choline could promote the digestion and absorptive capability of juvenile Jian carp (*Cyprinus carpio* var. Jian) by elevating intestinal trypsin, chymotrypsin, and amylase activities.

In mammals and teleost fish, the liver is one of the main sites for lipid synthesis, catabolism, transportation, and storage (Tocher, 2003). Thus, the liver is often used as an indicator of animal health. In the mammalian model, the result of excess lipid deposition in the liver is often called 'fatty liver disease' (hepatic steatosis), a common liver disease that can cause DNA damage which can eventually progress to liver fibrosis and cirrhosis (Asaoka *et al.*, 2013; Corbin and Zeisel, 2012; da Costa *et al.*, 2006; Rucker *et al.*, 2008). Since choline-containing phospholipids are important components of lipoproteins that are responsible for mobilizing lipids to organs in need of metabolic use or storage (Tocher, 2003), hepatic steatosis induced from choline deficiency is often associated with reduced very-low-density-lipoprotein (VLDL) production in the liver (Li and Vance, 2008; Luo *et al.*, 2016; Sherriff *et al.*, 2016). The liver in some farmed fish can show signs of fatty liver disease such as a swollen pale liver or increased vacuolation in the liver, which is similar to that exhibited by mammals with fatty liver disease (Benedito-Palos *et al.*, 2008; Fernandes Jr. *et al.*, 2016; Shimada *et al.*, 2014; Tao *et al.*, 2018; Wang, X., *et al.*, 2015). However, reports on the effect of choline on liver lipid metabolism in farmed fish appeared to be species dependent.

PAF is another choline-containing phospholipid that is rarely mentioned in fish nutrition but presumably exhibits a similar function as in mammals (Tocher *et al.*, 2008). PAF is a potent pro-inflammatory phospholipid that mediates several immunoregulatory functions, including platelet and neutrophil aggregation, inflammation, and organ injury (Karidis *et al.*, 2006; Stafforini *et al.*, 2003). PAF is synthesized by various inflammatory cells, including platelets, macrophages, neutrophils, monocytes, and eosinophils (Montrucchio *et al.*, 2000). The synthesis of PAF by gill, kidney, liver, and spleen has been demonstrated in rainbow trout (*Oncorhynchus mykiss*) (Turner and Lumb, 1989).

As a methyl donor, choline, and its subsequent methyl donors such as betaine, methionine, and *S*-adenosylmethionine (SAM) are involved in critical cellular processes such as histone, DNA, protein, and phospholipid methylation, gene expression mediation, and the metabolism of hormones, neurotransmitters, lipids, and amino acids to name a few (Mentch and Locasale, 2015; Neidhart, 2016). Luo *et al.* (2016), for the first time, demonstrated that dietary choline strongly influences some mRNA expression of genes that are involved in lipid metabolism in the muscle and liver of yellow catfish (*Pelteobagrus fulvidraco*). In juvenile Jian carp, dietary choline has been demonstrated to regulate gene expressions (i.e., target of rapamycin and eIF4E-binding protein 2) in tissues that are important in protein synthesis and degradation (Wu, P., *et al.*, 2011).

Nutritional assessment and deficiency signs

Growth, feed efficiency, and hepatic lipid content are common indicators used to assess choline status in young fish and shrimp. Choline-deficient fish and shrimp typically have poorer growth, poorer feed efficiency, higher mortality, and changing hepatic lipid content (species dependent). Pathological changes induced by choline deficiency are not widely reported. However, haemorrhagic organs (e.g., kidneys, intestine) were reported in rainbow trout (McLaren *et al.*, 1947) and Channel catfish (Dupree, 1966). White sturgeon exhibited extensive liver fat and fatty cysts, thinning intestinal wall muscle, sloughing of mucosal epithelium and lamina propria, focal degeneration of the exocrine pancreas, and inflammatory cell infiltration in the liver (Hung, 1989). Japanese yellowtail showed dark skin coloration when fed a choline-deficient diet (Hosokawa *et al.*, 2001).

Safety

Supplementing choline at twice the requirement of fish or shrimp does not appear to be toxic. However, choline toxicity (i.e., poor survival) was reported in hybrid striped bass (*Monrone saxatillis* × *M. chrysops*) fed more than 8 g choline bitartrate/kg diet, but not when provided as CC (Griffin *et al.*, 1994). Thus, it was likely the source of choline that resulted in higher mortality (~23%) in hybrid striped bass rather than choline level (Griffin *et al.*, 1994).

Chapter 5

Optimum vitamin nutrition for salmonids

INTRODUCTION

Atlantic salmon (*Salmo salar* L.) aquaculture production is >2.7 million tonnes annually with an estimated value of US$15 billion in 2020 (FAO, 2022). In 2022, the top 4 salmon producing countries were Norway with 1,365,400 gutted weight tonnes (GWT), Chile (678,000 GWT), United Kingdom (148,500 GWT) and Canada (136,000 GWT) (Mowi, 2023).

Despite the critical functions of vitamins, the number of studies on salmonids remains low and calls for greater attention. For instance, among the 15 vitamins reported in the aquaculture literature (Mai *et al.*, 2022; Lall and Dumas, 2022; NRC, 2011), vitamin B_9 (folate) has received the most attention when it comes to estimating Atlantic salmon vitamin requirements (Sandnes, 1994; Waagbø *et al.*, 2001; Hemre *et al.*, 2016). Requirements or recommendations for 5 vitamins (i.e., B_1, B_3, B_5, inositol, vitamin K) were based on a single study (B_1, B_3 and B_5: Hemre *et al.*, 2016; Waagbø *et al.*, 1998; vitamin K: Poston, 1976a). No requirement study was available for vitamins A and D in Atlantic salmon, although vitamin A requirement has been estimated for coho salmon (*Oncorhynchus kisutch*) recently and maximum authorized level exist for vitamin D in fish feeds (EFSA FEEDAP Panel, 2017; Xu, C.M., *et al.*, 2022b). Moreover, future vitamin nutrition papers should provide clear information on the measures taken to control factors affecting vitamin stability, such as the form(s) and formulation of vitamins, vitamin composition of raw materials used in feed formulas, feed processing, storage conditions and duration. Furthermore, experiments at farms are needed to complement academic trials conducted under controlled and optimal laboratory conditions. Thus, previously established recommendations for dietary vitamins require revision as they can be biased and suboptimal for practical applications with modern formulations and production practices. Finally, raw materials are an important driver in sustainable production and increasing public awareness of food production further drives the need to identify and update the vitamin inclusion levels in new generation diets that containing novel ingredients (Albrektsen *et al.*, 2022; Koehn *et al.*, 2022).

Therefore, this chapter (1) highlights new findings, published since NRC (2011), on vitamin nutrition for salmonid aquaculture focused on deficiency signs and benefits for health and welfare, (2) proposes optimal levels of vitamins based on updated literature and industry practice, and (3) recommends future research directions with consideration for vitamin functionalities.

LIPID-SOLUBLE VITAMINS: PHYSIOLOGICAL ROLES, DEFICIENCY SIGNS AND PROPOSED SUPPLEMENTATION LEVELS

Vitamin A

Among the vitamins studied in fish nutrition, vitamin A deserves greater attention especially since salmonid muscle is a good source of vitamin A in human nutrition (Yerlikaya *et al.*, 2022). There are limited studies that quantified the dietary vitamin A requirement of farmed salmonid species despite the critical roles played by this vitamin in fish, such as vision process, cell differentiation, embryonic and larval development, reproduction, and antioxidation (Poston *et al.*, 1977; Furuita *et al.*, 2003; Hemre *et al.*, 2004; Alsop *et al.*, 2008; Toyama *et al.*, 2008; Guimarães *et al.*, 2014; Morshedian *et al.*, 2017; Jiang, W. D., *et al.*, 2019; Xu, C.M., *et al.*, 2022b). The functions and requirements of vitamin A in fish have been reviewed recently by Hernandez and Hardy (2020) and Mai *et al.* (2022). More recently, the vitamin A requirement of post-smolt coho salmon was estimated at 6,422 IU/kg diet (Xu, C.M., *et al.*, 2022b). However, caution must be used since the number of fish per treatment replicate was low (10 salmon/tank), the duration of the study was short (10 weeks), and FCR high (>2.0).

Vitamin A requirement likely varies with fish size. For example, Kitamura *et al.* (1967a) reported that 0.2 g rainbow trout (*Oncorhynchus mykiss*) fed a casein basal diet devoid of supplemental vitamin A exhibited slow growth, high mortality, anemia, eye, and fin hemorrhage, and bent gill opercula. However, 5.9 g rainbow trout fed a casein-based diet without vitamin A did not show vitamin A-deficient signs as described in the previous studies (Poston *et al.*, 1977). Based on weight gain, the dietary vitamin A requirement of rainbow trout fry was more than 2,500 IU (Kitamura *et al.*, 1967a).

While the analysis of variance (ANOVA) is a suitable statistical approach for some requirements and toxicity studies with a dose-response design (Shearer, 2000; Thorarensen *et al.*, 2015), the recommended dietary vitamin A requirement of rainbow trout was derived from input levels of 2,500 and 50,000 IU/kg diet, which might not be a reliable and practical estimate for diet formulation (Kitamura *et al.*, 1967a). Hernandez and Hardy (2020) appeared to have recalculated the dietary requirement (2,500–3,500 IU/kg diet) of rainbow trout fry (Kitamura *et al.*, 1967a). However, the authors did not provide information on the statistical approach, making it difficult to verify the values.

Because vitamin A can be stored (mainly in the liver of immature fish as retinyl palmitate), excess intake of vitamin A can be toxic; the condition is sometimes referred to as "hypervitaminosis A." Nevertheless, farmed salmonids do not appear to risk experiencing hypervitaminosis A under current commercial farming practices (Ørnsrud *et al.*, 2013). For example, maximum tolerable levels varied between 609,000 and 904,000 IU/kg diet for post-smolt Atlantic salmon and rainbow trout fry, respectively (Hilton, 1983; Ørnsrud *et al.*, 2013). These levels are more than 75× current recommendations (Liu, A., *et al.*, 2022). The underlying mechanism of vitamin A toxicity requires elucidation as some abnormalities might be antagonistic effects or secondary infections (Hilton, 1983; Ørnsrud *et al.*, 2004b; Ørnsrud *et al.*, 2013). Ørnsrud *et al.* (2009) reported that injecting Atlantic salmon (35 g) with vitamin A (all-*trans*-retinoic acid) reduced plasma calcitriol levels which led to increased intestinal calcium absorption and predictable changes in mRNA gene expressions involved in bone formation (e.g., matrix Gla protein or mgp, collagen type I alpha 2 chain or col1a2, and alkaline phosphatase or alp). The authors suggested that an increased level of vitamin A might disrupt the vitamin D and calcium homeostasis, thereby affecting bone metabolism.

Detoxification is a newly discovered function of vitamin A in salmonids (Table 5.2). Aryl hydrocarbon receptor (AhR)-binding toxicants (e.g., PAHs) are present in higher concentrations in plant meal and oil compared with traditional marine aquafeed ingredients. With the increasing use of plant ingredients in aquafeed, there have been studies on the interaction between vitamin A and these toxicants (Alsop *et al.*, 2007; Arukwe and Nordbø, 2008; Berntssen *et al.*, 2010). A study on post-smolt Atlantic salmon reported several dietary vitamin A and PAHs interactively induced changes in metabolites and gene expressions related to detoxification, energy, and lipid metabolism (Berntssen *et al.*, 2016). Post-smolt (195 g) salmon fed plant-based diets (2,309 IU vitamin A/kg diet) with added PAHs (100 mg/kg benzo[a]pyrene and 10 mg/kg phenanthrene) showed reduced growth and depleted hepatic vitamin A store. However, supplementing retinyl acetate at 8,721 IU/kg diet partially alleviated the harmful effects (Berntssen *et al.*, 2016).

Furthermore, information on the effect of vitamin A on broodstock and the grow-out stage of salmonids had been limited until the recent decade. Fontagné-Dicharry *et al.* (2010) suggested that broodstock rainbow trout fed practical diets (fishmeal-based and plant-based) should contain vitamin A of 58,600–193,800 IU/kg diet to sustain broodstock fecundity and swim-up fry growth, particularly in a plant-based diet. However, the measured vitamin A levels in the 2 sets of practical diets differed, and the authors did not design the experiments to allow direct statistical comparison. For these reasons, the recommended vitamin A inclusion level in broodstock rainbow trout feed should be interpreted with care.

Considering that more plant and other novel ingredients will be used in the coming years, follow-on research might be necessary to determine whether nutrients (e.g., vitamin D), contaminants (e.g., PAHs) in novel aquafeeds, and abiotic factors can increase vitamin A expenditure. Bayir *et al.* (2022) showed that the expression of most RBP genes was downregulated when canola oil was substituted for fish oil in Atlantic salmon feeds not supplemented with vitamin A. These results suggest that uptake, transport, and metabolism of vitamin A are downregulated, and supplementation above reported requirement values may be necessary. **Therefore, existing literature and current industrial practices advocate moving from the NRC recommendation (2,500 IU/kg diet) to values ranging from 4,000 to 8,000 IU/kg diet (Table 5.1) not only to avoid deficiencies but to optimize performance and metabolic functions (Table 5.2).**

Vitamin D

Vitamin D – traditionally provided by marine ingredients – is among the vitamins most impacted by the changes in feed formulation over the last 3 decades (Table 5.3) as the direct result of fish oil replacement with lipids of oilseed origin (Sissener *et al.*, 2013; Ytrestøyl *et al.*, 2015; Rektsen *et al.*, 2022). It is also worth mentioning that concentrations of vitamin D_3 vary considerably in marine products, from 0.01 to 0.18 mg/kg fishmeal and 0.03 to 3.90 mg/kg fish oil (EFSA FEEDAP Panel, 2017).

A robust experimental design is needed to accurately identify salmonids' dietary vitamin D requirement and provide a more detailed description of its role in calcium homeostasis. The current dietary vitamin D requirement of rainbow trout (≈5.0–6.0 g initial body weight) presented in NRC (2011) was based on a study conducted by Barnett *et al.* (1982b). However, the authors stated that the requirement is more than 1,600 IU/kg diet, possibly ≥2,400 IU/kg diet based on growth response and feed efficiency. It would be prudent to ensure that other salmonid species are not vitamin D deficient to facilitate osmotic regulation, particularly during smoltification (Graff *et al.*, 2004; Lock *et al.*, 2007).

Vitamin D-deficient rainbow trout can synthesize vitamin D in the skin after solar light (290–1,200 nm) and blue light exposure (380–480 nm), presumably from 7-dehydrocholesterol like most terrestrial animals (Pierens and Fraser, 2015). However, based on early studies, salmonids remain dependent on the external source because no evidence supports their ability to synthesize enough vitamin D to meet their requirements (Barnett *et al.*, 1979; Barnett *et al.*, 1982a).

It has become critical to determine the optimal dietary levels of vitamin D because of its possible role in multiple metabolic functions such as antibacterial and antiviral immune responses (Table 5.2; Núñez-Acuña *et al.*, 2018; Soto-Dávila *et al.*, 2020), prevention of skeletal muscle degeneration (George *et al.*, 1981; Barnett *et al.*, 1982a), fatty acid oxidation (Peng, X., *et al.*, 2017), and secretion of insulin and prolactin (Seale *et al.*, 2006; Hideya *et al.*, 2008).

Young salmonids are relatively tolerant to high levels of dietary vitamin D (1,000,000–3,750,000 IU/kg diet; Poston, 1969b; Hilton and Ferguson, 1982; Graff *et al.*, 2002b) (see Lock *et al.* (2010) for a summary of early studies). Thus, it is unlikely that current commercial vitamin D levels in feeds will cause hypervitaminosis D in farmed salmonids. In agreement, the EFSA Panel on Additives and Products or Substances Used in Animal Feed (FEEDAP) concluded that increasing the maximum authorized vitamin D content in fish feeds from 0.075 to 1.50 mg/kg diet (3,000–60,000 IU/kg diet) would not compromise feed safety (EFSA FEEDAP Panel, 2017). Pierens and Fraser (2015) demonstrated that much of the radioactivity was secreted in bile when rainbow trout were injected or fed radioactively labeled vitamin D. Their results indicated that the liver of rainbow trout can efficiently detoxify excess vitamin D by esterifying with fatty acids.

We recommend using liver lipid, plasma triiodothyronine, clinical signs of muscle tetany, and weight gain to assess the vitamin D requirement based on rainbow trout studies (Leatherland *et al.*, 1980; Barnett *et al.*, 1982a,b). Liver content of metabolites, 25-hydroxyvitamin D_3 and 1,25-dihydroxyvitamin D_3 (1,25-D_3) stood as good indicators of vitamin D status in carp and olive flounder (Takeuchi, 2001). The reliability of these metabolites as indicators of vitamin D status requires validation in salmonids. Recently, Rider *et al.* (2023) reported that dietary calcifediol (25-hydroxyvitamin D_3 or 25-OH-D_3) supplementation (687–6,854 µg/kg diet) was safe and improved growth and feed efficiency of rainbow trout. In the study by Prabhu *et al.* (2019), whole-body vitamin D content correlated weakly with dietary vitamin D levels. Nevertheless, the authors recommended that 2,400–3,600 IU vitamin D/kg diet was required for post-smolt Atlantic salmon. In freshwater, the dietary vitamin D requirement of parr could not be identified due to the absence of a relationship between whole-body and dietary vitamin D levels.

Vitamin D levels have been declining in salmon feeds and, as a result, in salmon fillet as well over the last 2 decades as the inclusion of marine ingredients decreased in feed formulas (Sissener *et al.*, 2013). **Based on existing literature and current industrial practices, we suggest moving from the NRC recommendation (1,600–2,400 IU/kg diet) to values ranging from 2,500 to 8,000 IU/kg diet (Table 5.1) not only to avoid deficiencies but to optimize performance and metabolic functions.**

Vitamin E

Vitamin E (α-tocopherol) is a powerful biological antioxidant and plays a critical role as a potentiator of innate and adaptive immunity (Table 5.2) (Blazer and Wolke, 1984; Moccia *et al.*, 1984; Taksdal *et al.*, 1995; Bell *et al.*, 2000; Lygren *et al.*, 2001; Canyurt and Akhan, 2008; El-Sayed and Izquierdo, 2022).

Because vitamin E cannot be synthesized endogenously in aquatic animals (Hamre, 2011; NRC, 2011; Mai *et al.*, 2022), a lack of dietary vitamin E in salmonids can lead to slower growth, anemia, lower disease resistance, and histopathological changes (Hamre, 2011; NRC, 2011; Mai *et al.*, 2022). The dietary vitamin E requirement (all-*rac*-α-tocopherol acetate) suggested by NRC (2011) was 50–60 mg/kg diet for salmonids (Cowey *et al.*, 1983; Hamre and Lie, 1995; Woodall *et al.*, 1964). Hamre and Lie (1995) also suggested that the optimal dietary vitamin E (all-*rac*-α-tocopherol acetate) level could be as high as 120 mg/kg diet for Atlantic salmon. A recent study reviewed recommended vitamin E/kg diet for parr (18.3 g) and post-smolt (228.0 g) Atlantic salmon (Hamre *et al.*, 2016) and suggested the inclusion of 150 mg/kg diet. Hamre *et al.* (2016) adopted a dose-response design by adding graded levels of premix to a plant-based practical diet. Because the premix contains multiple nutrients such as vitamin E, vitamin C, and Se, the results would be the fish's response (e.g., body α-tocopherol, muscle glutathione, and GSH peroxidase 1) to a suite of changing nutrients, rather than vitamin E per se. There is no indication that using plant ingredients in salmonids feed would increase the dietary vitamin E requirement (Hamre *et al.*, 2016).

Dietary vitamin E supplementation above requirements (50–60 mg/kg diet as reported by NRC (2011) and Mai *et al.* (2022)) improves resistance against oxidative stress. Increasing dietary vitamin E above 78 mg/kg diet not only increased the growth of Caspian trout, but also induced immune responses (e.g., lysozyme activity and immunoglobin M), and antioxidant responses, such as superoxide dismutase and glutathione peroxidase (Saheli *et al.*, 2021). As a result, the dietary vitamin E (all-*rac*-α-tocopherol acetate) requirements of Caspian trout (*Salmo caspius*; 9.7 g initial body weight) were estimated at 78.7–82.2 mg/kg diet (Saheli *et al.*, 2021). Yin *et al.* (2022) observed that oxidative stress depleted vitamin E reserves in salmon tissues and reported that Atlantic salmon fed a diet with 210 mg/kg diet had better protection against oxidation than fish fed with 60 mg/kg diet. To ensure that fish growth or health are not compromised, it would be beneficial to identify the optimal dietary vitamin E inclusion level using practical formulation under stressful and suboptimal farming conditions. (Gabaudan and Hardy, 2000; Tavčar-Kalcher and Vengušt, 2007).

Based on these recent studies, it is suggested to increase dietary levels of vitamin E above the NRC recommendation (50–60 mg/kg diet) to values ranging between 200 and 600 mg/kg diet to avoid deficiencies, improve performance, immune status, and decrease nutrient loss (Table 5.1 and 5.2).

INTERACTION WITH OTHER NUTRIENTS WITH SPECIAL ATTENTION TO IMMUNITY

The relationship between vitamin E and selenium was first demonstrated in Atlantic salmon; both nutrients were required to prevent muscular dystrophy (Poston *et al.*, 1976). After that, vitamin E and its interaction with other micronutrients (e.g., vitamin C, folic acid) had received further attention in fish nutrition, and in particular for salmonids over the last decade (e.g., Mirvaghefi *et al.*, 2015; Naderi *et al.*, 2019; Hamre *et al.*, 2022).

Research on optimizing nutrition to strengthen immunity and prevent diseases, particularly under suboptimal or stressful farming conditions, is now more critical than ever. Vitamin E, selenium, and vitamin C are crucial components of an interacting network of antioxidants, strengthening certain immune responses and disease resistance capacity. A study conducted with healthy fish under optimal rearing conditions reported that lymphocyte counts of rainbow trout (2.1 g) increased significantly when fed a fishmeal-based diet supplemented with vitamin E and C at 200 or 1,000 mg/kg diet (Rahimi *et al.*, 2015). It would be relevant to reassess these outcomes under suboptimal conditions.

Interestingly, Atlantic salmon challenged against the SRS pathogen, *Piscirickettsia salmonis*, showed a 65% reduction in cumulative mortality when fed a fishmeal-based diet containing arachidonic acid (ARA) and vitamin E at 280 and 730 mg/kg diet, respectively (Dantagnan *et al.*, 2017). This could be related to the fact that lymphocyte counts significantly increase in rainbow trout fingerlings fed a diet supplemented with 1,000 ppm of vitamin E (internal DSM research, not published data). Dantagnan *et al.* (2017) found no interactive effect on production performance and immune responses such as respiratory burst and phagocytic activity apart from lysozyme activity. Furthermore, Harsij *et al.* (2020) demonstrated when rainbow trout (11.9 g) was challenged by a sub-lethal dose of ammonia (0.024 mg/L), a diet supplemented with a combination of vitamin E (60 mg/kg diet), vitamin C (200 mg/kg diet), and selenium nanoparticles (0.2 mg/kg diet) could enhance the antioxidant capacity and immune responses as well as production performance.

A group of researchers conducted an experiment to evaluate the effects of dietary vitamin E, selenium nanoparticles, and their combination on various production parameters and health indicators under high stocking density (80 kg/m) of rainbow trout (42.6 g) (Naderi *et al.*, 2017a, b, c; Naderi *et al.*, 2019). An interpretation of all results found that supplementing 500 mg/kg diet vitamin E and 1 mg/kg diet of selenium nanoparticle enhanced growth and maintained the immune status close to fish reared at lower stocking density (20 kg/m) (Naderi *et al.*, 2017a; Naderi *et al.*, 2019). The series of publications offered practical information relevant to feed formulations, but perhaps some of the most interesting findings were the effects of vitamin E and selenium on liver proteomics. Naderi *et al.* (2017b) shed light on the role of vitamin E and selenium in changing the protein gene expressions and potentially facilitating protein metabolism in salmonids.

In addition, lipid oxidation in fish fillets can be decreased by using high doses of dietary vitamin E to increase shelf life and, ultimately, contribute to the reduction of food waste (Hamre *et al.*, 1998; Zhang, L., *et al.*, 2007; Faizan *et al.*, 2013; Menoyo *et al.*, 2014). Such an approach illustrates that the concept of optimal vitamin nutrition can further support the sustainable growth of the aquaculture industry by improving flesh quality and nutritional functionality and concomitantly helping to reduce food loss and waste.

Vitamin K

Vitamin K is considered essential to salmonids but has received limited attention in the past 50 years. Estimates of dietary requirements of salmonids remain vague, with inconsistent responses to low dietary vitamin K (Kitamura *et al.*, 1967b; Salte and Norberg, 1991; Krossøy *et al.*, 2009). NRC (2011) currently recommends less than 10 mg/kg diet of vitamin K for Atlantic salmon. Recommendations for vitamin K inclusion levels in the fish nutrition literature ranged from 0.1 to 10 mg/kg diet (Poston, 1976a; Woodward, 1994; Kaushik *et al.*, 1998; Halver, 2002; Krossøy *et al.*, 2009).

In addition to the study variations such as life stages, experimental designs, vitamin K form (e.g., phylloquinone (lipid-soluble) and menadione salts (water-soluble)), the instability of vitamin K during feed production and storage might have contributed to the requirement variations (Krossøy *et al.*, 2011; NRC, 2011; Mai *et al.*, 2022). Krossøy *et al.* (2011) gave a comprehensive review of the role of vitamin K and dietary requirements in fish. Considering that no further research on salmonids has been conducted in the last decade, we echo the statements of Krossøy *et al.* (2011) and call for more research on salmonids focusing on defining the precise dietary requirement and the bioavailability of different vitamin K forms. Like what Sivagurunathan *et al.* (2023) have done with a warmwater species, salmonid nutrition research should incorporate the interaction of vitamin K with other micronutrients (e.g., vitamin A and D) and culturing stress factors (e.g., disease and temperature) when feeding practical diets containing novel ingredients.

Therefore, existing literature and current industrial practices suggest that moving from the NRC recommendation (below 10 mg/kg diet) to safer values ranging from minimum 8 up to

12 mg/kg diet (Table 5.1) not only to avoid deficiencies but to improve performance, immune status and decrease feed loses (Table 5.2).

Astaxanthin

Astaxanthin is currently referred to as a "vitamin-like" micronutrient or molecule in the literature, owing to the requirement for astaxanthin in first feeding rainbow trout for growth and survival in first-feeding rainbow trout (Ahmadi *et al.*, 2006) and Atlantic salmon (Christiansen *et al.*, 1994; NRC, 2011; Mai *et al.*, 2022). Salmonids do not synthesize astaxanthin *de novo* and therefore require supplementation in the diet. In addition to being an essential pigmentation source in farmed salmonids for the industry, astaxanthin possesses strong antioxidant activity and is often referred to as "super vitamin E" (Lim *et al.*, 2018; Mai *et al.*, 2022). Astaxanthin plays a particularly important role in early life and is required for optimal growth and survival up to first-feeding stage. For these reasons, astaxanthin is listed as a lipid-soluble vitamin.

Clinical assessments of astaxanthin 'deficiency' remain undefined either because this pigment is not essential or its roles require further studies, especially in salmonids at growing stage. Instead, dietary inclusion levels (12.5–200 mg/kg diet) were recommended to enhance pigmentation and meet functional purposes such as antioxidant activity, stress tolerance, immune responses, and reproductive performance (Christiansen *et al.*, 1994; Christiansen and Torrissen, 1996; Bazyar Lakeh *et al.*, 2010; Noori and Razi, 2018: Mai *et al.*, 2022).

Astaxanthin is a lipid-soluble carotenoid well known as the principal carotenoid pigment in wild and farmed salmonids' flesh (Storebakken and No, 1992). In wild rainbow trout, astaxanthin represents 38–87% of total carotenoids (Schiedt *et al.*, 1986) and stands as the main carotenoid in farmed Atlantic salmon (Ytrestøyl *et al.*, 2004). Carotenoids are deposited in the free non-esterified form in the muscle of salmonids (Storebakken and No, 1992).

Recently, Elbahnaswy and Elshopakey (2023) and Lim *et al.* (2018) provided comprehensive reviews on applications of astaxanthin in aquafeed, beyond its importance as a pigment. Astaxanthin can fulfill different metabolic functions depending on life stage. Substantial studies point to the benefits of astaxanthin in improving growth, fillet quality and reproduction performances, stress tolerance, immune responses, and their related gene expressions in aquaculture species (NRC, 2011; Mai *et al.*, 2022). For instance, astaxanthin at 90 mg/kg diet fed to rainbow trout (initial body weight 28 g/fish) for 56 days significantly increased the expression of genes associated with antioxidative, antiinflammatory and nonspecific immunity responses (Zhao, W., *et al.*, 2023). Similar results were obtained with larger rainbow trout (initial body weight 200–300 g/fish) fed either 100 mg astaxanthin/kg diet (Zhao, W., *et al.*, 2022) or 47 mg astaxanthin/kg diet (Kheirabadi *et al.*, 2022).

Astaxanthin digestibility and retention efficiency depend on its source, chemical form, and raw materials in the formulation. The non-esterified form (free) of astaxanthin is absorbed in the digestive tract of fish more efficiently than the esterified forms (NRC, 2011). The non-esterified form represents 100% of astaxanthin found in nature-identical products formulated for use in aquaculture feeds and red yeast (*Phaffia rhodozyma*) (EFSA FEEDAP Panel, 2005). Astaxanthin from crustaceans (e.g., krill) and microalgae (e.g., *Haematococcus pluvialis*) is predominantly in the mono- and di-ester forms (EFSA FEEDAP Panel, 2005). Astaxanthin concentrations are relatively low in yeast (≈4,000 ppm in *Phaffia* sp.), bacteria (~22,000 ppm in *Paracoccus* sp.), microalgae (≈55,000 ppm in *Haematococcus pluvialis*), krill oil (200–500 ppm), and krill meal (≈120 ppm) compared to synthetic formulations where it reaches 100,000 ppm or 10% (Yamaguchi *et al.*, 1986; European Commission, 2002; EFSA FEEDAP Panel, 2007; EFSA NDA Panel, 2014). The low concentration of astaxanthin from microorganisms can cause issues in nutrient-dense diets like those formulated for salmon, where the room available for "nutrient-sparse" ingredients is narrow. Bioavailability of astaxanthin from microorganisms may also be compromised by cell wall constituents. Dietary components other than pigments can also contribute to astaxanthin retention and fillet color. For instance, astaxanthin digestibility increases with lipid digestibility and dietary cholesterol level (Chimsung *et al.*, 2014; Courtot *et al.*, 2022). Dietary levels of EPA and DHA correlated positively with fillet pigmentation/redness in 2 long-term studies with Atlantic salmon fed 40–60 mg astaxanthin/kg diet (Hatlen *et al.*, 2022; Ytrestøyl *et al.*, 2023). Bioactive compounds from marine microalgae may enhance astaxanthin deposition in fillet of salmon (Sørensen *et al.*, 2023). Nutritional factors responsible for enhancing astaxanthin digestion, absorption and deposition need to be described in more detail under commercial formulations used today.

The role of astaxanthin on disease and stress resistance is worth exploring further, particularly in fish at early stages and in broodstock. Amar *et al.* (2012) examined the resistance of rainbow trout fry (0.6 g) fed various carotenoids against infectious hematopoietic necrosis virus (IHNV), a disease that is detrimental to salmonid farms by causing up to 100% mortality depending on multiple factors (Winton, 1991; Lapatra, 1998; Mochizuki *et al.*, 2009). The study of Amar *et al.*, 2012 showed promising results in reducing mortality (22%) when fish fed astaxanthin (100 mg/kg diet) were challenged by a low dose of IHNV (Figure 5.1).

The effects of water temperature, stress, and disease on astaxanthin metabolism and retention deserve further study. Anecdotal evidence suggests that handling of salmon for lice and gill treatments could impair flesh pigmentation. The mechanism behind this possibility is not known, but it could result from impaired absorption or enhanced *in vivo* oxidation.

Finally, new findings on the requirements and functions of astaxanthin in salmonids may call for relaxing the upper limits on dietary inclusion of astaxanthin prescribed in current regulations, e.g., 80 ppm in Canada and 100 ppm in Europe (e.g., European Commission, 2020; Canada Minister of Justice, 2021).

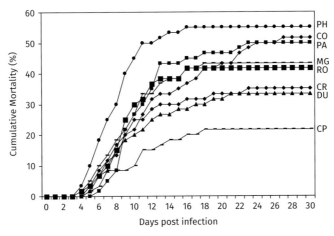

Figure 5.1 Effect of dietary astaxanthin sources at 0 (-) and 500 (+) mg/kg diet on cumulative mortality of juvenile rainbow trout after exposure to infectious hematopoietic necrosis virus (source: adapted from Amar *et al.*, 2012)

Notes: CO: Control, unsupplemented; CP: Carophyll® Pink (astaxanthin); CR: Carophyll® Red (canthaxanthin); DU: *Dunaliella salina* (β-carotene); MG: Marigold, *Tagetes erecta* (lutein/zeaxanthin); PA: Paprika, *Capsicum annuum* (capsanthin); PH: *Phaffia rhodozyma*; RO: Rovimix® β-carotene.

WATER-SOLUBLE VITAMINS: PHYSIOLOGICAL ROLES, DEFICIENCY SIGNS AND PROPOSED SUPPLEMENTATION LEVELS

Vitamin B₁ (Thiamine)

Well-defined dietary thiamine requirement is lacking in salmonids and calls for more research to fill this knowledge gap and identify how thiamine functions in salmonids when fed different novel ingredients. The coenzyme form of thiamine, TPP, is a key player in carbohydrate metabolism (enzymatic decarboxylation reactions); hence it has a profound effect on functions that heavily rely on using glucose as energy (e.g., neural functions) (Mai *et al.*, 2022). However, thiamine metabolism and its physiological functions are not well described in fish and warrant more research in this area to optimize commercial feed formulations.

In salmonids, dietary thiamine deficiency can lead to anorexia, changes in skin coloration, mortality, reproduction disorders, and abnormal behavior (e.g., hyperirritability and ataxia) likely linked to neurological disorders (McLaren *et al.*, 1947; Halver, 1957; Kitamura *et al.*, 1967b; Masumoto *et al.*, 1987; Harder *et al.*, 2018) (Table 5.2). While thiamine deficiency symptoms have been known in some farmed and wild salmonid species (NRC, 2011; Harder *et al.*, 2018), there is no dietary thiamine requirement quantified for Atlantic salmon and the requirements of brook trout, brown trout, and rainbow trout can vary from 1 to 15 mg kg⁻¹ diet (Phillips *et al.*, 1946; Morito *et al.*, 1986; Halver, 2002). One study on lake trout (*Salvelinus namaycush*) alevins reported that supplementing thiamine alone at 20 mg/kg diet or combined with magnesium (700 mg/ kg diet) to a casein-based diet significantly improved weight gain and survival (Lee B.J. *et al.*, 2012). The study indicated that early mortality syndrome is undoubtedly associated with thiamine deficiency in fish embryos, and the interdependent role of thiamine and magnesium in broodstocks should be investigated.

Furthermore, given the role of thiamine in carbohydrate metabolism, few studies have suggested that thiamine requirement might increase when using more ingredients containing high starch content, such as plant ingredients (e.g., Morris and Davies, 1995; Hansen *et al.*, 2015). However, using muscle thiamine content as an indicator, Hemre *et al.* (2016) suggested that dietary thiamine was sufficient when feeding parr and post-smolt Atlantic salmon a plant-based diet containing 6.2 mg thiamine/kg diet.

In conclusion, existing literature and current industrial practices suggest moving from the NRC recommendation (1 to 10 mg/kg diet) to values ranging from a minimum 15 up to 30 mg/kg diet to avoid deficiencies and improve production performance (Table 5.1 and 5.2).

Riboflavin, niacin, pantothenic acid, and inositol

The essentiality and requirements of riboflavin (vitamin B₂), niacin (nicotinic acid or vitamin B₃), pantothenic acid (vitamin B₅), and inositol (myo-inositol is the active form) have been reported in some salmonid species. However, these vitamins are understudied in farmed salmonids, with almost no reports since the 1980s. Consequently, we referred to Hansen *et al.* (2015), Shirmohammad *et al.* (2016), Cui *et al.* (2022), Vera *et al.* (2020), and Mai *et al.* (2022) for a summary of the biochemistry, deficiency symptoms, and dietary requirements of these 4 B complex vitamins in salmonids. Waagbø *et al.* (1998) concluded that an unsupplemented fish-meal-based diet containing ca. 300 mg inositol/kg diet was sufficient for Atlantic salmon fry, given that no improved growth and pathological signs of deficiency were observed. In comparison, McLaren *et al.* (1947) suggested that rainbow trout required 250–500 mg inositol/kg diet to avoid deficiency symptoms and increase weight gain.

The authors' reviews indicated that previous requirements might not be reliable and warrant revision. Recently, the recommended dietary riboflavin, niacin, and pantothenic acid supplementation in a plant-based diet for parr and post-smolt Atlantic salmon were 10–12, 65, and 22 mg/kg diet, respectively (Hemre *et al.*, 2016). These values should be interpreted with caution because the experiment adopted a design with graded levels of a unique premix, which could have been a confounder. A similar approach was used by Vera *et al.* (2020). Based on the limited information in the fish literature, it is recommended to investigate further the metabolic functions and modes of action of the 4 B complex vitamins in salmonids.

Vitamin B$_6$ (pyridoxine)

Despite the critical role of vitamin B$_6$ and recent research in salmonid nutrition, its functions and interactions with other nutrients are understudied. In animal tissues, vitamin B$_6$ (pyridoxine) functions as coenzymes (i.e., pyridoxal phosphate (PLP) and pyridoxamine phosphate (PMP)) that participate in the protein, lipid, and carbohydrate metabolism (Combs and McClung, 2017; Mai *et al.*, 2022). Although no study investigated the ability of salmonids to synthesize vitamin B$_6$ *de novo*, the essentiality and deficiency symptoms of vitamin B$_6$ have been demonstrated in rainbow trout (McLaren *et al.*, 1947; Kitamura *et al.*, 1967b; Smith C.E. *et al.*, 1974), Chinook salmon (Halver, 1957), and coho salmon (*O. kisutch*) (Coates and Halver, 1958). In addition to low feed intake and high mortality, vitamin B$_6$ deficiency often led to behavioral changes (e.g., abnormal swimming, hyperirritability, and ataxia) presumably due to the disruption of neurotransmitter synthesis (i.e., 5-hydroxytryptamine and serotonin) that is dependent on PLP. Other reported deficiency signs in salmonids were organ damages such as kidney and intestinal lesions.

The current dietary vitamin B$_6$ requirement for salmonids suggested by NRC (2011) and Halver (2002) were 4–7 mg/kg diet and 3–6 mg/kg diet, respectively. Lall and Weerakoon (1990) estimated the dietary vitamin B$_6$ requirement of Atlantic salmon at 5 mg/kg diet. Dietary B$_6$ requirement was estimated at 2–8 mg/kg diet for the Atlantic salmon fry fed a fishmeal-based diet (Albrektsen *et al.*, 1993) and 10–16 mg/kg diet for parr and post-smolt Atlantic salmon fed a plant-based diet (Hemre *et al.*, 2016).

Beyond the growth and health of salmonids, an important aspect of farmed fish is its flesh quality for human consumption. Owing to the role that vitamin B$_6$ can play in lipid metabolism, Maranesi *et al.* (2005) found that supplementing 50 mg/kg diet to a fishmeal-based diet significantly increased the percentage of docosahexaenoic (DHA) in rainbow trout (110 g) fillet. In contrast, a study on rainbow trout (30 g) fed plant-based diets (e.g., defatted soybean meal, canola oil, and linseed oil) containing up to 19 mg/kg diet reported no significant changes in the fillet fatty acid profile (Senadheera *et al.*, 2012). It was likely that the vitamin B$_6$ in the experimental diets was not enough to trigger a response because the study by Maranesi *et al.* (2005) did not detect significant changes in fillet DHA up to 25 mg/kg diet supplementation of vitamin B$_6$. Nevertheless, Senadheera *et al.* (2012) demonstrated that vitamin B$_6$ could stimulate the activity of the fatty acid elongase and Δ-6 and Δ-5 desaturase enzymes. Considering that vitamin B$_6$ is a key player in protein and amino acid metabolism, perhaps the role of vitamin B$_6$ in deamination and transamination has priority over fatty acids metabolism when dietary vitamin B$_6$ is not in excess. Accordingly, it might be necessary to fortify commercial diets with vitamin B$_6$ beyond the NRC (2011) and Halver (2002) recommendations.

Taken together, the discussed data and current industrial practices suggest moving from the NRC recommendation (3–6 mg/kg diet) to values ranging from 20–35 mg/kg diet (Table 5.2).

However, more research is required to elucidate the metabolic functions of vitamin B$_6$ in salmonids and how the vitamin interacts with nutrients including those that participate in the one-carbon metabolism (e.g., choline, methionine, folate, and B$_{12}$). It would be relevant to the salmonid industry to investigate if higher dietary B$_6$ supplementation improves fillet fatty acids profile when feeding diets containing novel ingredients. In addition, there is a need to determine if nutrients involved in the one-carbon metabolism can spare a portion of vitamin B_6 and vice versa. Accordingly, the dietary vitamin B_6 inclusion level in salmonid feeds should be revised, especially regarding finishing feed containing novel ingredients.

Biotin (vitamin B_7)

Biotin is an essential nutrient for fish that serves as a coenzyme for carboxylases critical to the glucose, amino acid, and fatty acid metabolism in all animals (Combs and McClung, 2017; Mai et al., 2022). A comprehensive review of biotin physiology and nutrition in fish was provided by Yossa et al. (2015). Despite this, our understanding of biotin nutrition remains limited in fish, including salmonids, due to a lack of studies.

Early studies showed that common symptoms of dietary biotin deficiency were histopathological changes and lesions in organs such as gills, liver, and kidney. These symptoms were present in lake trout (Poston and Page, 1982), rainbow trout (Woodward and Frigg, 1989), and Atlantic salmon (Maeland et al., 1998). The NRC (2011) recommended dietary biotin requirement is 0.15 mg/kg diet for rainbow trout and 1.0 mg/kg diet for Oncorhynchus spp. but recommendations are absent for Atlantic salmon. According to the literature, the biotin requirement ranged from 0.05 to 1.0 mg/kg diet for the salmonids (Poston, 1976b; McLaren et al., 1947; Castledine et al., 1978; Poston and Page, 1982; Woodward and Frigg, 1989; Maeland et al., 1998). The low dietary requirement supposedly resulted in the limited progress in biotin research in fish nutrition. Recently, using muscle biotin content and biotin-dependent genes as markers, the background biotin level (0.3 mg/kg diet) appeared to be sufficient in a plant-based diet for Atlantic salmon (Hemre et al., 2016). This estimate was in line with the NRC (2011) recommendation.

While biotin deficiency symptoms have been induced experimentally in major salmonid species, biotin deficiency is unlikely to occur under commercial conditions because biotin is present in many feed ingredients, unless fed with diets containing biotin antagonists (e.g., avidin) or if the fish is starved for several weeks (Hansen et al., 2015; Yossa et al., 2015). Biotin is also produced by intestinal bacteria in some animals but has yet to be demonstrated in salmonids (Sugita et al., 1992; Yossa et al., 2011; Combs and McClung, 2017).

Future studies could examine (1) if biotin plays any role in preventing tissue lesions such as the gills during infestation by parasites (e.g., sea lice), and (2) the possible contribution of biotin from intestinal bacteria. Echoing Yossa et al. (2015), there is much to explore in biotin nutrition in fish and its novel roles in reproduction, disease resistance, and fatty acid metabolism beyond the dietary requirements. The dietary biotin inclusion level may need to be revised depending on outcomes from such studies. **In conclusion, existing literature and current industrial practices suggest moving from the NRC recommendation (0.15 to 1.0 mg/kg diet) to values ranging from a minimum 1.0 up to 1.2 mg/kg diet to improve production performance (Table 5.2).**

Folate (vitamin B_9)

Research beyond identifying folate deficiency and dietary requirements in salmonids remains limited. Folate (folacin or vitamin B_9) functions as a coenzyme (tetrahydrofolate; THF) that catalyzes several essential steps of the one-carbon metabolism that are fundamental to the methylation of DNA, amino acids, and lipids in animals (Naderi and House, 2018; Mai et al., 2022). Folate is a collective term for compounds exhibiting the biological activity of folic acid. Folic acid is commonly supplemented in aquafeed due to higher stability compared with the natural form. The essentiality of folate has been demonstrated in the salmonids (Kitamura *et al.*, 1967b; Smith, 1968; Smith and Halver, 1969) and the deficiency symptoms were summarized by NRC (2011) and Mai *et al.* (2022).

The NRC (2011) recommended dietary folate requirement ranges from 1.0–2.0 mg/kg diet for *Oncorhynchus* spp. (Table 5.1). This requirement was recently augmented to 3.3 mg/kg diet for Atlantic salmon (Hemre *et al.*, 2016) fed with a plant-ingredient based diet. While production performance was not affected, supplementation of 6 mg folic acid/kg diet to a commercial diet was reported to improve some immune responses (e.g., lysozyme activity and IgM production) (Soheil *et al.*, 2013). Although more precise estimates and robust experimental designs are required, the latter studies indicate that the folate inclusion level in modern salmonid feeds should be revised. Considering that folate is actively involved in the one-carbon metabolic cycle, future studies should (1) investigate the fundamental functions of folate in salmonids, (2) describe how dietary folate increases transmethylation and availability of other essential nutrients that participate in the one-carbon metabolism such as methionine and choline in fish, and (3) explore its possible role in epigenetics to improve the efficiency of nutrient utilization and immunity (Clare *et al.*, 2019). Like biotin, there are indications that some freshwater fish species could obtain folate from intestinal bacteria (Duncan *et al.*, 1993), thus, future studies can assess the contribution of folate from intestinal bacteria in salmonids.

The existing literature and current industrial practices suggest that salmonid feed should contain 3–10 mg folate/kg diet to avoid deficiency symptoms, improve feed efficiency, and improve immune responses (Table 5.1 and 5.2).

Vitamin B$_{12}$ (cobalamin)

The dietary level of vitamin B$_{12}$ have been impacted greatly by the change in feed formulas over time, which resulted in variable contents of this vitamin in salmon fillets (see Figure 4.6; Rektsen *et al.*, 2022). Together with folate, vitamin B$_6$, methionine, and choline, vitamin B$_{12}$ coenzymes (methylcobalamin and adenosylcobalamin) are important participants in the one-carbon metabolism in animals, including fish (Hansen *et al.*, 2015; Espe *et al.*, 2020; Mai *et al.*, 2022). Vitamin B$_{12}$ is required for red blood cell and myelin synthesis, and several critical roles have yet to be demonstrated in fish (Smith A.D. *et al.*, 2018; Calderón-Ospina and Nava-Mesa, 2020; Mai *et al.*, 2022). Known vitamin B$_{12}$ deficiency symptoms in salmonids are reduced feed intake and megaloblastic anemia (Halver, 1957; Kitamura *et al.*, 1967b). A summary of vitamin B$_{12}$ nutrition in fish is provided by Hansen *et al.*, (2015).

While vitamin B$_{12}$ has essential metabolic roles in salmonids, the dietary requirement is low. The NRC (2011) recommended a dietary vitamin B$_{12}$ requirement of 0.02 mg/kg diet for *Oncorhynchus* spp. but did not establish requirement values for Atlantic salmon and rainbow trout. A dietary level of vitamin B$_{12}$ in the premix at ≥0.17 mg/kg diet was recommended in a plant-based diet for Atlantic salmon while no adverse effects were detected between 0.07 and 0.72 mg/kg diet (Hemre *et al.*, 2016; Vera *et al.*, 2020). Given that the synthesis of vitamin B$_{12}$ from intestinal bacteria has been demonstrated in rainbow trout (Sugita *et al.*, 1991), the vitamin B$_{12}$ requirement is perhaps partially supplied by intestinal bacteria; hence, the low dietary requirement. Furthermore, there is evidence of dietary cobalt replacing the dietary need for vitamin B$_{12}$ in fish (Hansen *et al.*, 2015). On this basis, additional vitamin B$_{12}$ supplementation might increase if dietary cobalt is insufficient. Accordingly, a vitamin B$_{12}$ content of 0.17–0.25 mg/kg diet would be adequate for securing good growth and health in salmonid farms. However, future research should investigate the interaction between cobalt and vitamin B$_{12}$ as well as the intestinal bacterial production of vitamin B$_{12}$ in farmed salmonids at different life stages. **Taken together, the discussed data and current industrial practices suggest moving from the NRC recommendation (0.02 mg/kg diet) to values ranging from 0.17–0.25 mg/kg diet**

(Table 5.2). More studies are also needed to understand the vitamin B_{12} interdependency with other one-carbon metabolism nutrients (e.g. methionine, folate).

Choline

Choline is sometimes described as a 'vitamin-like' nutrient in the literature when its essentiality is not demonstrated for the fish species studied. Although the essentiality and requirement of choline have yet to be determined for all farmed salmonid species, a clear dietary need for choline has been demonstrated for juvenile rainbow trout (McLaren et al., 1947; Poston, 1991), chinook salmon (O. tshawytscha) (Halver, 1957), and lake trout (Salvelinus namaycush) (Ketola, 1976). For example, previous studies reported that chinook salmon and rainbow trout fed a choline-deficient diet exhibited hemorrhagic kidney and intestine (McLaren et al., 1947; Halver, 1957), anemia, projected eyes, and extended abdomens (Kitamura et al., 1967b). However, apart from poorer body weight (BW) gain and/or feed conversion ratio, other studies did not mention any gross pathology changes in rainbow trout (1.4 and 3.2 g) (Rumsey, 1991) and lake trout (5 g) (Ketola, 1976) fed apparent choline-deficient diets.

Using weight gain and hepatic lipid content as the primary response criteria, the studies mentioned above estimated the dietary choline requirements of salmonids (0.1–12.0 g) ranged from approximately 50–2,850 mg/kg diet. NRC (2011) recommended the dietary choline inclusion level of 800 mg/kg diet for rainbow trout and other Oncorhynchus spp. but did not establish an inclusion level for Atlantic salmon.

Until recent years there has been limited research that examines the blood chemistry, histopathology changes, gene expression, and other analyses that could have provided a deeper understanding of the effects of choline on the health of farmed salmonids. A lack of research in these areas and over-reliance on information from closely related species have become a missed opportunity to avoid health-related challenges in farmed fish.

With a significant increase of plant ingredients in salmonid feeds over the last 2 decades, there have been recurring reports of intestinal morphological changes and 'floating feces' in farmed Atlantic salmon since the early 2000s (e.g., Penn, 2011). The reports indicated that farmed Atlantic salmon developed intestinal steatosis causing "lipid malabsorption syndrome" (LMS). Common characteristics of LMS in animals include the pale and foamy appearance of the pyloric caeca and pyloric intestine due to excessive lipid accumulation in the enterocyte, most likely associated with impaired ability to transport lipids (Hansen et al., 2020a). Because choline is one of the key players in lipid transport and metabolism, a group of researchers has conducted a series of studies to quantify the dietary choline requirement and to better understand the role of choline in preventing LMS in Atlantic salmon fed plant-based diets (Hansen et al., 2020a,b; Krogdahl et al., 2020).

Hansen et al. (2020a) demonstrated that choline supplementation to a plant-based diet containing 944 mg choline/kg diet (measured total choline of 4,250 mg/kg diet) could prevent signs of LMS in post-smolt salmon (362 g) (Figure 5.2).

Using the degree of whiteness and vacuolated enterocyte as primary responses, a dietary choline inclusion level of 3,350 mg/kg diet (upper 95% confidence interval) was recommended for post-smolt Atlantic salmon (456 g) to avoid LMS development (Hansen et al., 2020b). Similar recommendations of ≥2,600 (Kaur et al., 2023) and 3,500 mg choline/kg diet (Krogdahl et al., 2023) were reported in 2 recent studies with Atlantic salmon (60–190 g). Krogdahl et al. (2020) also tested the effects of components associated with lipid and sterol metabolism on the intestinal health of post-smolt Atlantic salmon (330 g) fed plant-based diets (ca. 1,190 mg choline/kg diet). The study showed that supplementation of either cholesterol, taurocholate, methionine, taurine, or cysteine to the basal diet could not replicate the effect of choline and

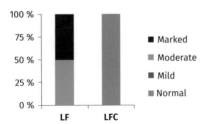

Figure 5.2 Effect of dietary choline (C) at 0.4% inclusion of CC 70% on the severity of steatosis in pyloric caeca of post-smolt Atlantic salmon fed low fishmeal (LF) diets (adapted from Hansen *et al.*, 2020a)

PC in reducing enterocyte vacuolation. The benefits of choline on post-smolt Atlantic salmon somewhat supported the findings of Poston (1990), which showed that supplemental choline was needed to improve the growth of 0.2 g Atlantic salmon fry, but not 1.0, 1.7, and 7.0 g fry. However, the author did not provide information on the measured choline content of the diets and quantification methodology, a common problem with early studies, which makes data comparison and interpretation more challenging. Furthermore, the basal diet contained 35% soy protein isolate [≈2,753 mg/kg; (NRC, 2011)] and 0.5% of choline chloride (CC; 70% CC; free choline is 74.6% of CC), which would contain approximately 5,364 mg choline/kg diet, presumably meeting the dietary choline requirement of young Atlantic salmon. Consequently, the necessity for additional choline supplementation to the Atlantic salmon diet could be misinterpreted as unwarranted. Thus, there is a need to revise the current salmonids feed formulation, at least for Atlantic salmon.

Recent findings also provided more evidence to support the role of choline in lipid metabolism in salmonids. For example, Espe *et al.* (2016) reported choline supplementation to low methionine diets increased phospholipids content in Atlantic salmon liver and muscle. Moreover, Atlantic salmon fed choline supplemented diet showed upregulation of intestinal genes involved in lipoprotein synthesis (*apoA1* and *apoAIV*), cholesterol transport (*abcg5* and *npc1l1*); but unclear as to why genes associated with lipid transport-related (*mtp* and *fatp*) and PC synthesis (*chk* and *pcyt1a*) were downregulated (Hansen *et al.*, 2020a,b).

Choline can affect the stability of other vitamins in premixes and, for this reason, should be added separately to meal mix (Tavčar-Kalcher and Vengušt, 2007; Yang *et al.*, 2021).

Taken together, the literature and industry practices suggest moving the choline content in salmonid feeds from the NRC recommendation (800 mg/kg diet) to values ranging from 500 to 4,000 mg/kg diet) to avoid deficiency symptoms and improve feed utilization efficiency (Table 5.1 and 5.2).

Vitamin C

Unlike many animals, salmonids have limited ability to synthesize vitamin C *de novo* due to the absence of L-gulono-γ-lactone oxidase. Thus, dietary vitamin C is needed to prevent slow growth and deficiency symptoms (Halver *et al.*, 1969; Dabrowski *et al.*, 1990; Krasnov *et al.*, 1998; Drouin *et al.*, 2011; Dawood and Koshio, 2018). The inclusion level of ascorbic acid equivalent recommended by NRC (2011) was 20 mg/kg diet for Atlantic salmon and rainbow trout. However, depending on the form of ascorbic acid derivatives used (NRC, 2011; Dawood and Koshio, 2018), the dietary ascorbic acid required in salmonids can range from 5 to 500 mg/kg diet (McLaren *et al.*, 1947; Dabrowski *et al.*, 1990; Sandnes *et al.*, 1992; NRC, 2011). Broodstock rainbow trout diet should contain at least 100 mg/kg diet to provide 20 mg/kg of ascorbic acid

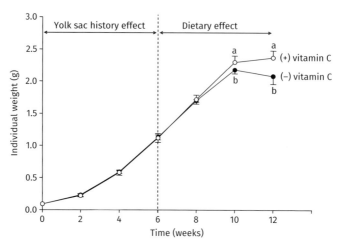

Figure 5.3 Effect of dietary vitamin C (Mg L-ascorbyl-2-phosphate) at 0 (-) and 500 (+) mg/kg diet on growth of alevins of rainbow trout at first feeding (source: adapted from Falahatkar *et al.*, 2011)

in eggs for normal development (Sandnes *et al.*, 1984). Falahatkar *et al.* (2011) also demonstrated that dietary vitamin C had more impact on production performance than vitamin C enrichment of rainbow trout eggs (Figure 5.3).

Ascorbic acid supplementation of >190 mg/kg diet to a plant-based diet was recommended for parr and post-smolt Atlantic salmon (Hamre *et al.*, 2016).

Higher dietary vitamin C inclusion is likely needed to alleviate potential stress and promote wound healing under commercial farming conditions. For example, supplementing 300 mg vitamin C/kg diet to a commercial diet significantly reduced malondialdehyde (MDA; an indicator of cell damage during oxidative stress) and increased SOD and total antioxidant capacity in rainbow trout (180 g) exposed to a sub-acute dosage of diazinon (pesticide) (Mirvaghefi *et al.*, 2015). In coho salmon (*Oncorhynchus kisutch*), the vitamin C requirement was estimated at ~200 mg/kg diet when using growth and antioxidant capacity as response variables (Xu C.M. *et al.*, 2022a). Moradi *et al.* (2022) reported that 400–800 mg vitamin C/kg diet were sufficient to restore hematologic parameters (e.g., red blood cells, white blood cells, hematocrits), as well as immunological and antioxidative responses of rainbow trout (43 g) fed with antibiotics (oxytetracycline, 100 mg/kg fish biomass). More dose-response research should investigate the stress resistance effects of vitamin C (or with other micronutrients) in salmonids exposed to environmental contaminants and pathogenic agents.

Based on the existing literature and industry practices, elevating the vitamin C content in salmonid feeds from 20 mg/kg diet (NRC, 2011) to values ranging from 200 to 1,000 mg/kg diet would be necessary to enhance stress resistance and reproduction performance (Table 5.1 and 5.2). When providing supplemental ascorbic acid in heat-treated feeds, it is strongly advisable to use a stabilized form like EC-coated or phosphorylated forms to minimize losses of activity.

A PROPOSAL FOR OPTIMIZING VITAMIN NUTRITION IN SALMONIDS AND FUTURE RESEARCH

From a sustainability perspective, health, welfare and optimum performance, the need to provide formulations tailored to different life stages of the target species is now more than ever.

Here, vitamins stand as a strategic tool to improve and support the sustainable growth of aquaculture by increasing feed efficiency and survival. Table 5.1 summarizes the recommended vitamin inclusion levels based on updated literature.

Studies in salmonid nutrition have shown clearly how single vitamins can prevent and treat deficiencies. The complexity of vitamin nutrition, along with current formulation challenges in salmonid aquaculture, is conducive to expanding our focus, transitioning from an isolated nutrient approach to emerging concepts such as feed synergy and dietary patterns that are being increasingly adopted and proven effective in human nutrition (Jacobs and Tapsell, 2013; Mozaffarian et al., 2018). These concepts consider how multiple nutrients and bioactive molecules interact, sometimes synergistically, to impact overall health and welfare (Jacobs et al., 2009) (Figure 5.4).

In this chapter, we identified the untapped potential that interactions among vitamins and other nutrients could contribute to sustainability by an overall improvement in production efficiency, by improving disease resistance, and nutrient utilization in salmonid aquaculture. We also evaluated existing literature and identified critical knowledge gaps to address in vitamin nutrition. For example, the immunopotentiation of vitamin D and astaxanthin on resistance to stress and pathogens, the role of vitamin A on broodstock's progeny, the optimum dietary level of vitamin B_6 to increase DHA retention in the fillet, and the lack of recent studies on vitamin K. There is a need to quantify the vitamin supply (e.g., biotin, folate) from gut microbiota and formulate dietary premixes accordingly. Finally, we highlighted how certain experimental designs and lack of clear description and information (e.g., analyzed vitamin content of practical experimental diets, the chemical form of the vitamin used) could limit the robustness of conclusions in vitamin nutrition studies and their applications. Thus, there is a need to adopt and/or develop standardized vitamin nutrition methods to validate results and facilitate knowledge transfer.

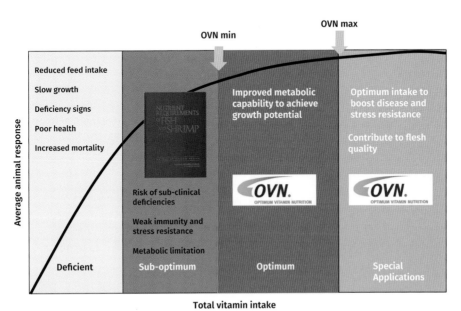

Figure 5.4 The OVN Optimum Vitamin Nutrition® concept (source: dsm-firmenich Animal Nutrition and Health, 2022b adapted by Liu A. et al., 2022).

Table 5.1 A summary of the recommended dietary vitamin content for farmed salmonids since NRC (2011) with recommendations for optimal vitamin nutrition

Vitamins (IU or mg/kg diet)	NRC (2011)*	Optimum Vitamin Nutrition dsm-firmenich 2022	Species	Initial weight (g)	Vitamin dietary content in published research	Benefits	References
A (IU)	2,500	4,000–8,000	AS	195	8,721	Partially alleviated PAH toxicity	Berntssen et al. (2016)
			RT	BS	58,600–193,800	Increased fecundity and swim-up fry growth	Fontagné-Dicharry et al. (2010)†
			CHS	186	367–22,935	Increased growth and antioxidant capacity	Xu C.M. et al. (2022b)
D (IU)	1,600	2,500–8,000	AS	228	2,400–3,600	Increased tissue vitamin D_3 content	Prabhu et al. (2019)
E	50–60	200–600	AT	228	150	Increased body vitamin E content and retention	Hamre et al. (2016)
			AT	6.5	730 vitamin E (+ 0.28 g ARA)	Improved survival post SRS challenge	Dantagnan et al. (2017)
			CT	9.7	78.7–82.2	Improved some immune responses	Saheli et al. (2021)
			RT	2	200 and 1,000	Increased lymphocyte counts	Rahimi et al. (2015)
			RT	42.6	500 vitamin E (+ 1.0 Se)	Improved growth and maintain immune status under high stocking density	Naderi et al. (2017a,b,c, 2019);
			RT	108	100 vitamin E (+ 0.5 Se)	Improved antioxidant capacity against sub-acute diazinon from pesticide	Mirvaghefi et al. (2015)
			RT	11.9	220 vitamin E + (420 vitamin C + 0.4 Se)	Improved production performance, antioxidant capacity and immune responses against ammonia stress	Harsij et al. (2020)

K	<10	8–12	–	–	–	–	–
Astaxanthin		50–100	RT	0.6	100	Improved survival post IHNV challenge; higher serum and tissue carotenoid contents	Amar et al. (2012)
			RT	BS	12.5–92.9	Improved fry growth	Bazyar Lakeh et al. (2010)[†]
			RT	196	50 and 80	Asta better coloring source	Ghotbi et al. (2011)
			RT	18.5	100	Increased muscle astaxanthin content	Rahman et al. (2016)
			RT	20.1	200	Improved muscle and skin pigmentation	Noori and Razi (2018)
	47–100	47–100	RT	28–300	47–100	Improved growth, antioxidant capacity, antiinflammatory and nonspecific immune responses	Zhao W. et al. (2022; 2023) Kheirabadi et al. (2022)
Thiamine (B_1)	1–10	15–30	LT	0.1	20	Improved growth and survival	Lee B.J. et al. (2012)
Riboflavin (B_2)	4–7	25–40	AS	18 and 228	3.4	–	Hemre et al. (2016)
			AS	38 and 68	17	Increased tissue B_2 content	Vera et al. (2020)
Niacin (B_3)	10–150	150–200	AS	18 and 228	65	Increased tissue B_3 content	Hemre et al. (2016)
				38 and 68	148	Increased tissue B_3 content	Vera et al. (2020)
Pantothenic acid (B_5)	20	40–60	AS	18 and 228	22	Increased tissue B_5 content	Hemre et al. (2016)
				38 and 68	58	Increased tissue B_5 content	Vera et al. (2020)

Table 5.1 (continued)

Vitamins (IU or mg/kg diet)	NRC (2011)*	Optimum Vitamin Nutrition dsm-firmenich 2022	Species	Initial weight (g)	Vitamin dietary content in published research	Benefits	References
Pyridoxin (B$_6$)	3–6	20–35	AS	18 and 228 38 and 68	10 17	Increased tissue B$_6$ content Increased tissue B$_6$ content	Hemre et al. (2016) Vera et al. (2020)
			AS	30	19	Likely to stimulate fatty acid elongase and desaturase enzymes	Senadheera et al. (2012)
Biotin (B$_7$)	0.15–1	1.0–1.2	AS	18 and 228	0.3	–	Hemre et al. (2016)
Folate (B$_9$)	1–2	10–15	AS	18 and 228 38 and 68	3–6 6.5	Increased tissue B$_9$ content Increased tissue B$_9$ content	Hemre et al. (2016) Vera et al. (2020)
			RT	7.6	>6	Improved lysozyme activity and IgM production	Soheil et al. (2013)
Cobalamin (B$_{12}$)	0.02	0.17–0.25	AS	18 and 228	0.17	Increased tissue B$_{12}$ content	Hemre et al. (2016)
C	20	200–1,000	AS	18 and 228 38 and 68	190 and 89 251	Increased body vitamin C content Increased body vitamin C content	Hamre et al. (2016) Vera et al. (2020)
			RT	0.1	500	Improved growth and survival	Falahatkar et al. (2011)
			RT	2	200 and 1,000	Increased lymphocyte counts	Rahimi et al. (2015)
			RT	108	300	Improved antioxidant capacity against sub-acute diazinon from pesticide	Mirvaghefi et al. (2015)
			CHS	183	93–225	Improved growth and antioxidant capacity	Xu C.M. et al. (2022a)

Choline	800	500–4,000	AS	2	>4,000	Increased tissues PL content when fed low methionine diet	Espe et al. (2016)
			AS	330	2,980	Reduced enterocyte vacuolation	Krogdahl et al. (2020)
			AS	362	4,250	Improved growth and prevented LMS symptoms	Hansen et al. (2020a)
			AS	456	3,350	Prevented LMS symptoms	Hansen et al. (2020b)
			AS	60–190	≥2,600	Prevented LMS symptoms	Kaur et al. (2023) Krogdahl et al. (2023)

Notes: ARA, arachidonic acid; AS, Atlantic salmon (*Salmo salar*); BS, broodstock; CS, Chinook salmon (*Oncorhynchus tshawytscha*); CT, Caspian trout (*S. caspius*); CHS, coho salmon (*O. kisutch*); COM, commercial diet; IgM, immunoglobin M; IHNV, infectious hematopoietic necrosis virus; LMS, lipid malabsorption syndrome; LT, lake trout (*Salvelinus namaycush*); RT, rainbow trout (*Oncorhynchus mykiss*); SRS, salmonid rickettsial syndrome.

* Dietary requirement values covered Atlantic salmon, rainbow trout, or Oncorhynchus spp. Requirements were listed as not tested, required, or not required for some salmonid species.

† Not covered by NRC (2011).

‡ Supplemented with graded levels of premixes containing the target vitamin.

1,000 IU Vitamin A (retinol) = 0.344 mg vitamin A acetate

1,000 IU Vitamin D_3 (cholecalciferol) = 0.025 mg vitamin D_3

Table 5.2 An overview of metabolic functions, symptoms of deficiency and sensitivity of vitamins in salmonid nutrition. For more details, see Mai et al. (2022), NRC (2011), and Bureau and Cho (1999)

Vitamins	Primary functions	Other functions	Deficiency signs	Main sensitivity
A	Maintaining epithelial integrity	Detoxification Fecundity	Anorexia Ascites Erosion of fins Exophthalmos Hemorrhage (eye, skin) Lesions (eye) Skin depigmentation	Light Oxidizing agents
D	Calcium-phosphorus metabolism	Antimicrobial Antiviral	Hepatic steatosis Scoliosis	Light Oxidizing agents
E	Biological antioxidant	Immunity (lysosome activity, immunoglobulin M)	Anemia Ascites Exophthalmos Skin depigmentation	–
K	Blood coagulation	–	Anemia Hemorrhage (eye, gills) Prolonged clotting time	Light Alkalis
Astaxanthin	Biological antioxidant Pigmentation	Reproduction Stress tolerance Immunity	Not yet demonstrated	Light Heat
Thiamine (B$_1$)	Carbohydrate metabolism	–	Anorexia Cataracts Loss of equilibrium	Heat Alkalis
Riboflavin (B$_2$)	Energy metabolism	–	Poor feed efficiency Cataracts Dark skin coloration Erosion of fins Hemorrhage (eye, skin, fins) Lesions (eye) Photophobia	Light Alkalis

Vitamin	Function		Deficiency signs	Sensitivity
Niacin (B$_3$)	Energy metabolism	–	Poor feed efficiency Abdominal edema Anorexia Hemorrhage (skin) Lesions (skin) Photophobia	–
Pantothenic acid (B$_5$)	Energy and lipid metabolism	–	Anorexia Distended operculum Hemorrhage (skin) Lesions (skin)	Acids Alkalis
B$_6$ (Pyridoxin)	Protein and amino acid metabolism	Fatty acid metabolism (DHA retention in fillet)	Anemia Anorexia Dark skin coloration Erratic spiral swimming Exophthalmos	–
Biotin (B$_7$)	Carbohydrate and lipid metabolism	–	Poor feed efficiency Anorexia Dark skin coloration Lesions (skin)	–
Folate (B$_9$)	Protein and nucleic acid metabolism	Immunity (lysosome activity, immunoglobulin M)	Poor feed efficiency Anemia Anorexia Dark skin coloration	Oxidizing agents Reducing agents
B$_{12}$ (Cobalamin)	Blood cell function and protein metabolism	–	Anorexia Fragmented erythrocytes	Reducing agents Acids Alkalis
Choline	Energy metabolism	Prevention of lipid malabsorption syndrome	Poor feed efficiency	–

Table 5.2 (continued)

Vitamins	Primary functions	Other functions	Deficiency signs	Main sensitivity
Inositol	Neurological osmolytes and signaling molecules	–	Poor feed efficiency Anorexia Lesions (skin)	–
C	Antioxidant and detoxification	Wound healing Stress/disease resistance Reproduction	Anorexia Abnormal cartilage Ascites Exophthalmos Hemorrhage (gill, skin, intramuscular) Lesions (eye) Lordosis Scoliosis	Oxidizing agents Alkalis

Table 5.3 Comparison of vitamin contents of salmon diets not supplemented with vitamin premix between 1990 and 2022 (vitamin C excluded). Dietary vitamins supplied by raw materials of typical salmon feeds formulated between 1990 and 2022 (vitamin C excluded). (Liu A., *et al.*, 2022)

Composition	1990	2022
Vitamin (mg/kg)		
Thiamine (B_1)	1.077	4.180
Riboflavin (B_2)	4.294	1.912
Niacin (B_3)	46.451	31.625
Pantothenic acid (B_5)	10.533	10.430
Pyridoxine (B_6)	3.616	3.885
Biotin (B_7)	0.160	0.174
Folate (B_9)	0.291	0.407
Cobalamin (B_{12})	105.608	17.605
Choline	2303.524	1379.468
Myo-inositol	316.589	209.248
A	89.761	32.772
D	1.059	0.013
E	60.722	80.251
K	1.190	0.549
Typical salmon feed formula (% as-fed)		
Fishmeal, anchovy	30.0	5.0
Poultry by-product meal	20.0	0.0
Corn gluten meal	7.5	10.0
Soy protein concentrate	0.0	30.0
Guar meal	0.0	3.5
Wheat gluten meal	0.0	10.0
Wheat flour	15.8	10.2
Fish oil, anchovy	21.5	8.1
Canola/Rapeseed oil	0.0	16.8
Supplement*	5.2	6.4
Proximate composition (% as-fed)		
Dry matter	94	95
Crude protein	43	41
Crude lipid	28	28
Ash	8	6

* Supplement: crystalline amino acids, lecithin, vitamin C (0.3% inclusion in both diets), CC 60% (0.4% inclusion in both formulas), monocalcium phosphate, astaxanthin.

Chapter 6

Optimum vitamin nutrition for warm water fish

INTRODUCTION

Regarding biomass, warm water species continue to dominate global finfish (hereafter fish) aquaculture production and are expected to increase in the coming years (FAO, 2022). Tacon (2020) projected formulated feed usage to rise to 18,236 tonnes for fed carps, 13,248 tonnes for tilapias, 8,588 tonnes for catfishes, and 469 tonnes for eels by 2050. While these key aquaculture species contribute significantly to global food security, nutrition, and the blue economy, the industry still faces emerging or amplified challenges associated with sustainability, diseases, anthropogenic pollution, and climate change (Naylor *et al.*, 2021).

To address these challenges, the aquaculture industry has made substantial progress through vaccines, selective breeding, and optimizing nutrition through vitamin, prebiotic, and probiotic supplements to maximize growth and boost animal immune systems and resistance to pathogens (Dawood *et al.*, 2019; Gjedrem *et al.*, 2012; Mondal and Thomas, 2022). Vitamins are critical in optimizing fish performance, feed efficiency, and health. The key biological roles of all vitamins and vitamin-like compounds are summarized in Table 6.1. Recent studies on fish such as grass carp (*Ctenopharyngodon idella*), darkbarbel catfish (*Pelteobagrus vachelli*), and Wuchang bream (*Megalobrama amblycephala*) further demonstrated that vitamins could be 'natural' immune boosters against pathogens and environmental stressors (e.g., *Aeromonas hydrophila*, ammonia, temperature) (Cheng K. *et al.*, 2020a,b; Li M. *et al.*, 2013; Ming *et al.*, 2012; Pan *et al.*, 2017; 2018). Accordingly, optimizing vitamin nutrition for fish can contribute to sustainability by promoting animal welfare and nutrient retention, decreasing feed and food waste, and reducing antibiotic use.

Like salmonids, the dietary vitamin requirements remain unreported for several warm water fish species (Liu A. *et al.*, 2022; Mai *et al.*, 2022; NRC, 2011). Moreover, a considerable portion of carp, tilapia, and catfish species is farmed in earthen ponds that contain natural food sources (e.g., phytoplankton, benthic organisms, insects) irrespective of the farming system intensity (Boyd *et al.*, 2020; Rahman, 2015), which presents another layer of challenge to refining vitamin and other nutrient requirements. Furthermore, raw materials in formulated feeds are subject to market change and regulations. Knowledge of species-specific vitamin nutrition is critical to gaining flexibility in feed formulation and optimizing resource use. Consequently, an update on dietary vitamin inclusion levels is warranted (Cabello, 2006; Shepherd *et al.*, 2017; Tocher, 2015). Lastly, increasing public awareness of sustainable food production and transparency reinforces the need to regularly revise vitamin supplementations in modern formulations containing novel ingredients (Koehn *et al.*, 2022; Naylor *et al.*, 2021).

This chapter focuses on the Optimum Vitamin Nutrition® (OVN™) of tilapia, carp, catfish, bass, bream, eel, and other emerging warm water fish species in published papers after 2011, but we also considered older studies when relevant. In this chapter, temperate and warm water species refer to those for which water temperature optima is >20°C. The biochemistry of individual vitamins is available in Chapter 4 of this book as well as in other papers (Cui *et al.*, 2022; Dawood and Koshio, 2018; El-Sayed and Izquierdo, 2022; Hansen *et al.*, 2015; Hernandez and Hardy, 2020; Krossøy *et al.*, 2011; Lim *et al.*, 2018; Lock *et al.*, 2010; Mai *et al.*, 2022; Waagbø, 2010; Yossa *et al.*, 2015) and would not be discussed in depth in this chapter.

TILAPIA (*OREOCHROMIS* SPP.)

Fat-soluble vitamins

Since 2011, there has been limited information on the nutrition of lipid soluble vitamins for tilapia species, except for vitamin E. Presumably, because tilapia and several major freshwater species are primarily farmed in fertilized pond systems that host diverse sources of foods (e.g., phytoplankton, zooplankton, bacteria) containing various vitamins (FAO, 2022). Newly established vitamin A (retinol) inclusion levels for Nile tilapia (*Oreochromis niloticus*; initial BW 5.3–7.5 g) did not exceed the levels recommended for hybrid tilapia (initial BW 1.6 g) (Hu *et al.*, 2006; NRC, 2011) (Table 6.2). To the best of our knowledge, no revised inclusion level and new information on vitamin D_3 (cholecalciferol; hereafter vitamin D) (374.8 IU/kg diet) (Shiau and Hwang, 1993) and vitamin K (5.2 mg/kg diet) (Lee, 2003) for tilapias were generated in recent years.

A comprehensive review of vitamin E by El-Sayed and Izquierdo (2022) became available recently. Based on studies carried out over the last decade, the revised dietary vitamin E inclusion levels (43.2–752.5 mg/kg diet) for tilapias (Jiang M. *et al.*, 2020; Wu F. *et al.*, 2017) were generally higher than the values recommended by NRC (2011) (60 mg/kg diet) and past research (25–100 mg/kg diet) (Table 6.2). Key findings from these studies showed that tilapias fed adequate levels of vitamin E improved production performance and antioxidant capacity, maintained normal organ morphology (e.g., liver, intestine, gill), lipid metabolism and inflammation-related transcriptional responses, increased white blood cell counts, and increased survival from *Streptococcus iniae* infection (Table 6.2) (Jiang, M., *et al.*, 2020; Lim, C., *et al.*, 2010b; Qiang *et al.*, 2019; Wu, F., *et al.*, 2017) (Figure 6.1).

Water-soluble vitamins and vitamin-like nutrients

A dietary thiamine (vitamin B_1) level of ca. 3.5 mg/kg diet was reported to improve growth and prevent deficiency symptoms (e.g., neurological disorders, anorexia, mortality) in Nile tilapia fingerlings (3.6 g); and the requirement for dietary thiamine decreased in 5.3 g fish (Table 6.2) (Lim, C., *et al.*, 2011). Because the authors did not chemically analyze the thiamine content in the final diet and had seemingly estimated the requirement based on "thiamine hydrochloride" but stated "thiamine," the dietary thiamine requirements were likely to be approximately 21% lower than the reported values (thiamine is 78.67% of thiamine hydrochloride).

According to growth performance, the dietary niacin (vitamin B_3) requirement of 87.2 g genetically improved farmed tilapia (GIFT; *O. niloticus*) (20.4 mg/kg diet) (Jiang, M., *et al.*, 2014a) was similar to that of smaller (2.1 g) hybrid tilapia (*O. niloticus* × *O. aureus*) (26 mg/kg diet) (Shiau and Suen, 1992). The requirement for greater niacin storage in the liver of GIFT was 84.6 mg niacin/kg diet. Furthermore, Liu W. *et al.* (2020) recently reported that GIFT fed diets containing high niacin content (>1,000 mg/kg diet) significantly reduced the glycogen and lipid deposition in the liver of GIFT. There were no signs of niacin toxicity in GIFT fed dietary niacin levels up to 1,123 mg/kg diet.

Over the last decade, for various tilapia species, newly estimated dietary inclusion for folate (0.4–0.7 mg/kg diet) and vitamin B_6 (10 mg pyridoxine hydrochloride/kg diet) did not increase

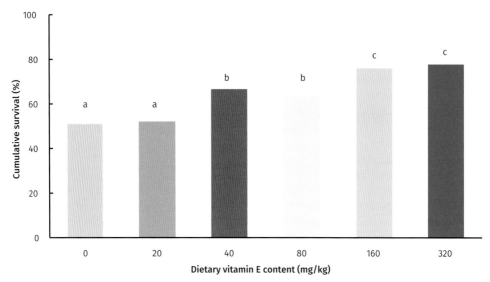

Figure 6.1 Effects of dietary vitamin E levels on the cumulative survival of Nile tilapia fingerlings (*Oreochromis niloticus*) after 96 h of *Streptococcus iniae* infection (bars with different letters (a, b, c) indicate significant differences ($p<0.05$)) (source: adapted from Qiang *et al.*, 2019).

beyond the NRC (2011) recommendation (Table 6.2). The dietary requirements for riboflavin (B$_2$), pantothenic acid (B$_5$), biotin (B$_7$), cobalamin (B$_{12}$), and inositol have been quantified for a few tilapia species (Table 6.2). However, research on these vitamins beyond identifying the requirements and deficiency symptoms remained relatively stagnant, likely because farmed tilapias could partly satisfy their need for these vitamins by preying on food sources from fertilized ponds.

With the increasing use of plant ingredients in modern aquafeed, the high occurrence of mycotoxins presents a risk to the growth and health of farmed fish (Gonçalves *et al.*, 2018). A recent study on Nile tilapia fingerlings showed that supplementing herbal extracted phosphatidylcholine at 800 mg/kg diet could improve growth and alleviate aflatoxin B$_1$-induced oxidative effects (Baldissera *et al.*, 2019; Souza *et al.*, 2020). Increased supplementation to 1,200 mg phosphatidylcholine/kg diet further enhanced the antioxidative status. However, it was unclear how much choline was present in the diet and the herbal extracts considering that phosphatidylcholine only contains, on average, a third of free choline molecules. Therefore, the dietary choline requirement of Nile tilapia (7.1 g) is unlikely to be higher than the NRC (2011) recommendation, which was based on smaller hybrid tilapia (0.6 g; 1,000 mg/kg diet) (Shiau and Lo, 2000). It would be worth investigating if choline was solely responsible for the antioxidative effects or if other components from the herbs (i.e., *Trachyspermum ammi*, *Azadichara indica*, *Achyranthes aspera*) offered extra advantageous properties. Nevertheless, the herbal extract showed promising results as a feed additive against aflatoxin for farmed Nile tilapia.

Among the vitamin nutrition studies in farmed tilapias, vitamin C has received substantial research attention recently (Table 6.2). Key findings showed that tilapias fed vitamin C (≥45 mg/kg diet) increased BW, catalase (CAT) activity, lysozyme, and reduced mortality from *Aeromonas hydrophila* infection (Huang, F., *et al.*, 2016; Ibrahem *et al.*, 2010; Ibrahim *et al.*, 2020; Lim, C., *et al.*, 2010b; Wu, F., *et al.*, 2015). Vitamin C supplementation at higher levels (599–942 mg/kg diet) to broodstock tilapias was recommended to improve egg production, sperm motility, hatching rate, and larval growth and survival (Sarmento *et al.*, 2017; 2018). According to these new findings, the dietary vitamin C requirement of farmed tilapias should be revised. Dietary vitamin C is also known to mitigate stress induced by crowding and bacteria (Dawood and Koshio, 2018; Ibrahim et al., 2020; Mustafa et al., 2013). For instance, tilapias fed dietary vitamin C supplementation (500–1,000 mg/kg diet) above the required levels mitigated stress induced by crowding, environmental contaminants (e.g., cadmium, engine oil), and *Aeromonas hydrophila* by upregulating gene expression of glutathione peroxidase (GPx) and glutathione reductase (GSR), increasing antioxidant enzyme activities, and reducing cortisol and glucose levels (El-Sayed et al., 2016; Ibrahem et al., 2010; Mohamed et al., 2019; Mustafa et al., 2013; Vieira et al., 2018).

In summary, to optimize production performance, health, and resistance to stressors of tilapias under common farming conditions, existing literature and current industry practices (commercial-in-confidence) suggest revising OVN™ values (IU/kg or mg/kg diet) to the following levels: vitamin A 4,000–8,000 IU/kg; vitamin D 1,500–2,000 IU/kg; vitamin E 100–750 mg/kg; vitamin K 5–10 mg/kg; vitamin B$_1$ (thiamine) 10–20 mg/kg; vitamin B$_2$ (riboflavin,) 15–20 mg/kg; niacin 80–120 mg/kg; pantothenic acid 40–50 mg/kg; biotin 0.5–1.0 mg/kg; folate 10–20 mg/kg; vitamin B$_{12}$ (cobalamin) 0.02–0.05 mg/kg; vitamin C 150–1,500 mg/kg; and choline 600–1,200 mg/kg (Table 6.2).

CARP

Fat-soluble vitamins

Aoe et al. (1968) first determined the dietary vitamin A requirement of common carp (*Cyprinus carpio*) (initial weight 2.2 g) to vary widely from 4,000 to 20,000 IU/kg diet. Almost half a century after this study, the dietary vitamin A requirement was determined for Gibel carp (*Carassius auratus gibelio* var. CAS III) (Shao et al., 2016) and grass carp (*Ctenopharyngodon idella*) (Jiang, W.D., et al., 2019, 2020; Wu, F., et al., 2016; Zhang, L., et al., 2017) (Table 6.3), showing that dietary vitamin A requirement appears to decrease with increasing fish size.

Optimal dietary requirements of carp for vitamin D have yet to be established. A dietary vitamin D requirement (412–1,480 IU/kg diet, measured response-dependent) was determined recently for black carp (*Mylopharyngodon piceus*) to sustain optimal growth, antioxidant capacity, and innate immune responses (Figure 6.2) (Wu C. et al., 2020).

The quantification of dietary vitamin D requirements of other carp species is warranted. On the other hand, the recommended dietary vitamin E (α-tocopherol) inclusion levels (100.4–139.6 mg/kg diet) from the literature over the last decade were generally within the range of past research recommendations (62.9–200 mg/kg diet) for farmed carps (Table 6.3).

There is a paucity of information on vitamin K in carp nutrition. Jiang, M., et al. (2007) recommended a dietary requirement of 1.9 mg/kg diet for grass carp fingerlings (3.5 g) based on the blood coagulation time, given the prominent role of vitamin K as a blood coagulator (Krossøy et al., 2011). However, there were no histological changes and no significant effect of dietary vitamin K on the production performance of grass carp. On the other hand, increased vitamin K supplementation led to better production performance in Jian carp (*Cyprinus carpio* var. Jian) fingerlings (8.9 g), and the dietary vitamin K requirement (3.1 mg/kg diet) was higher than that of grass carp (Yuan et al., 2016). Yuan et al. (2016) later demonstrated that vitamin K influenced the abundance of intestinal bacteria (i.e., *Lactobacilli*, *Escherichia coli*, *Aeromonas*) and increased the activities of digestive (e.g., lipase, trypsin, amylase) and antioxidant enzymes (e.g., superoxide dismutase (SOD), glutathione S-transferase) in Jian carp. Based on the studies mentioned above, it is likely that the dietary vitamin K requirement would be different among other carp species and sizes. Thus, there is a need to fill the knowledge gaps in this study area.

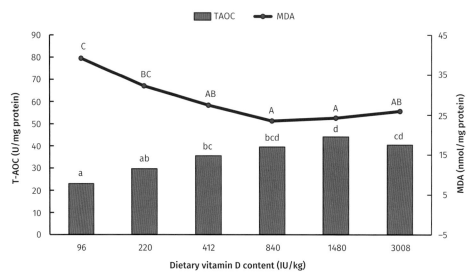

Figure 6.2 Effects of dietary vitamin D on the total antioxidant capacity (T-AOC) and malondialdehyde (MDA) in the liver of black carp (*Mylopharyngodon piceus*) fed graded levels of vitamin D (bars and line with different letters (a, b and A, B respectively) indicate significant differences ($p<0.05$)) (source: Wu C. *et al.*, 2020)

Water-soluble vitamins and vitamin-like nutrients

The dietary thiamine requirement (0.5 mg/kg diet) was first quantified in common carp by Aoe *et al.* (1969). Since then, studies on other carp species have reported thiamine requirement to be approximately double the amount required for common carp (Table 6.3). For example, Jian carp fingerlings (8.2 g) required 1.0 mg thiamine/kg diet for good growth (Huang, H. H., *et al.*, 2011) and 1.6 mg to reduce mortality when challenged by the pathogen *Aeromonas hydrophila* (Feng *et al.*, 2011). The dietary thiamine requirement of grass carp fingerlings can increase to 5.0 mg/kg diet when using thiamine liver content as a biomarker, which appeared to be sensitive to dietary thiamine changes (Jiang, M., *et al.*, 2014b).

Similarly, newly recommended dietary niacin inclusion levels for rohu and mrigal carp (30–42.1 mg/kg diet) were slightly higher than those reported by past research for common carp and grass carp (25.5–28.0 mg/kg diet; Table 6.3). In addition to growth, juvenile grass carp (255.6 g) fed beyond 34.0 mg niacin/kg diet was reported to stimulate digestive enzyme activities that were likely attributed to the upregulation of some associated mRNA expression (e.g., trypsinogen 1, chymotrypsinogen, target of rapamycin, S6 kinase 1) (Li, S. Q., *et al.*, 2016). More dietary niacin is also required (39.8 mg/kg diet) to improve the intestinal immune status (e.g., lysozyme activity) (Feng *et al.*, 2016). The studies on grass carp might indicate that other carp species would require higher dietary niacin to optimize the immune system, especially if challenged by stressors that occur under commercial farming conditions.

Recently, Cui *et al.* (2022) provided a review of myo-inositol in aquatic animals. Newly quantified dietary inositol requirements of carp species (50–415 mg/kg diet) did not exceed those recommended a decade ago (116–518 mg/kg diet) (Table 6.3). However, smaller Jian carp demand more dietary inositol (660–770 mg/kg diet) when exposed to *Aeromonas hydrophila* (Jiang, W.D., *et al.*, 2016).

Over the last decade, for various carp species, new estimated dietary inclusions for vitamin C (35.7–156.6 mg/kg diet), riboflavin (5.0–7.4 mg/kg diet), pantothenic acid (37.0 mg/kg diet), pyridoxine (3.2–8.6 mg/kg diet), biotin (0.2–0.9 mg/kg diet), folate (1.0–2.1 mg/kg diet), and choline (566–2,667 mg/kg diet) did not increase beyond the previous literature recommendations covered by NRC (2011) (Table 6.3). No new research on cobalamin nutrition in carps was published in recent years.

The vitamin-like functions of astaxanthin in cold-water species are relatively well understood. By comparison, there are a paucity of studies on the utilization and benefits of astaxanthin in warm water species. While NRC (2011) has not established a dietary astaxanthin requirement for farmed carps, research efforts on astaxanthin as a feed additive for carp (as ornamental fish and food fish) did not fall short over the last decade.

The primary role of astaxanthin in a segment of carp feed is pigmentation because color is a key criterium that dictates the ornamental carp market value (Sun et al., 2012). However, in recent years the literature indicated that higher dietary astaxanthin inclusion levels (50 to 400 mg/kg diet) may confer physiological benefits (Table 6.3). For example, improved sperm quality (Tizkar et al., 2015), production performance and activities of digestive enzymes (e.g., protease, amylase) and antioxidant enzymes (e.g., SOD, catalase), and less cumulative mortality (Wu, S., and Xu, 2021). However, the astaxanthin inclusion level in feed might be influenced by the astaxanthin-containing microorganisms (e.g., algae, fungi, bacteria) (Lim, K.C., et al., 2018) naturally present in the farming system.

In summary, based on existing literature and current industry practices for carps, we suggest revising the OVN™ values (IU/kg or mg/kg diet) to the following: vitamin A 8,000–11,000 IU/kg; vitamin D 1,500–2,000 IU/kg; vitamin E 100–900 mg/kg; vitamin K 5–10 mg/kg; vitamin B_1 (thiamine) 10–20 mg/kg; vitamin, B_2 (riboflavin) 15–20 mg/kg; niacin 80–120 mg/kg; pantothenic acid 40–50 mg/kg; biotin 0.5–1.0 mg/kg; folate 4.0–7.0 mg/kg; vitamin B_{12} (cobalamin) 0.02–0.05 mg/kg; vitamin C 150–400 mg/kg and choline 600–1,800 mg/kg (Table 6.3).

CATFISH

Fat-soluble vitamins

Of the lipid-soluble vitamins, recent studies on catfish indicated that the dietary vitamin A, vitamin D, and vitamin E inclusion levels should be revised. Battisti et al. (2017) recommended that 23 g silver catfish (*Rhamdia quelen*) fed 2,610 IU retinol/kg diet could enhance growth (Table 6.4). The newly quantified dietary vitamin A inclusion level for silver catfish was higher than that of 50 g channel catfish (*Ictalurus punctatus*) (1,000 IU/kg diet) (Dupree, 1970). The discrepancy might be due to size and species differences. The role of vitamin A as an antioxidant in catfish is scarce. However, Battisti et al. (2017) showed that 2,610 IU retinol/kg diet could improve the antioxidative status (lower activities of catalase, glutathione-S-transferase (GST), thiobarbituric acid (TBARS)) of silver catfish presumably to spare the energy needed for antioxidant enzyme production.

Early studies showed that dietary vitamin D requirements of channel catfish fingerlings (initial BW 0.5–5.0 g) ranged widely from 250 to 1,000 IU/kg diet (Brown and Robinson, 1992; Lovell and Li, 1978; NRC, 2011). Higher dietary vitamin D requirement reported when similar-sized channel catfish (6.0 g) were fed 2,000 IU/kg diet (Andrews et al., 1980). Like channel catfish, a recent study on yellow catfish (*Pelteobagrus fluvidraco*) (5.0 g) reported that 2,260 IU vitamin D/kg diet may enhance the resistance to the pathogen, *Edwardsiella ictaluri*, by increasing CAT activity, lysozyme, and regulating several type I interferon and immunity-related gene expression (Cheng et al., 2020a,b). However, some limitations in these studies need to be considered: firstly, several fish died within the first 24 h of *E. ictaluri* exposure; and secondly, the extent of increasing dietary vitamin D in preventing fish mortality was unclear because there was no indication of statistical analysis being carried out.

When cultured under non-stressful conditions, studies over this past decade on various farmed catfish species showed that the recommended dietary vitamin E inclusion levels (33.0–45.7 mg/kg diet) (Lim, C., *et al.*, 2010a; Lu, Y., *et al.*, 2016) remained similar to previously reported NRC levels (25–50 mg/kg diet) (Table 6.4). Additional benefits, such as higher vitamin E in fillets of striped catfish (*Pangasianodon hypophthalmus*) were reported when fish were fed 192.5 mg vitamin E/kg diet (Al-Noor *et al.*, 2012). Higher dietary vitamin E inclusion can significantly help mitigate responses induced by external stressors (Table 6.4). For example, darkbarbel catfish (*Pelteobagrus vachelli*) fed approximately 400 mg vitamin E/kg diet showed higher growth, lysozyme activity, phagocytic index, and total immunoglobulin level when exposed to high ammonia concentrations (5.7 mg/l total ammonia nitrogen) for 60 days (Li M. *et al.*, 2013). However, this level of vitamin E could not prevent mortalities from another 14 days of exposure to *E. ictaluri*. Moreover, an additional supplementation of vitamin E (240 mg/kg diet) to a commercial diet (unknown nutrient compositions) protected African catfish (*Clarias gariepinus*) against anthropogenic pollutants such as potassium dichromate (Azeez and Braimah, 2020a) and copper sulfate (Azeez and Braimah, 2020b). These studies have important implications as key aquaculture species continue to encounter undesirable culturing conditions. Much evidence pointed towards an increasing reliance of vitamin E in formulated feeds.

Water-soluble vitamins

The requirements for several water-soluble vitamins appeared to be species dependent. For instance, the dietary thiamine requirement of yellow catfish fingerlings (Figure 6.3) (Zhao H. *et al.*, 2020) was 6–7 times higher than that of channel catfish fingerlings (1.0 mg/kg diet) (Murai and Andrews, 1978). Intestinal trypsin and amylase activities were also higher at 6.0 and 7.9 mg thiamine/kg diet in yellow catfish fingerlings (Zhao H. *et al.*, 2020).

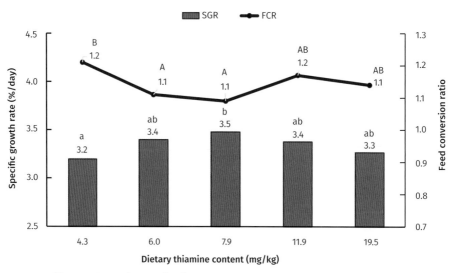

Figure 6.3 Specific growth rate (SGR, %/day) and feed conversion ratio of juvenile yellow catfish (*Pelteobagrus fulvidraco*) fed diets containing increasing thiamine for 63 days (bars and line with different letters (a, b and A, B respectively) indicate significant differences (p<0.05)) (source: Zhao, H., *et al.*, 2020)

The dietary pantothenic acid requirement of green catfish (*Mystus nemurus*) fingerlings (~2.3 mg/kg diet) (Hien and Doolgindachbaporn, 2011) was lower than that of channel catfish fingerlings (10–15 mg/kg diet) (Murai and Andrews, 1979; NRC, 2011; Wilson *et al.*, 1983) (Table 6.4). In general, the recommended dietary vitamin C inclusion levels (46.4–114.5 mg/kg diet) from recent studies (Daniel *et al.*, 2018; Liang *et al.*, 2017) on farmed catfish did not deviate from the range recommended by NRC (2011) (11–150 mg/kg diet) (Table 6.4). However, Pitaksong *et al.* (2013) showed that hybrid catfish (*Clarias macrocephalus* × *C. gariepinus*) fed 988.4 mg vitamin C/kg diet had beneficial effects on growth and could ameliorate adverse changes in the immune responses (e.g., low hemoglobin, white blood cells, lysozyme) of hybrid catfish under acute thermal (19°C) and acid (pH 5.5) stress. A combination of 502.8 mg vitamin C with 59.3 mg vitamin E/kg diet could achieve similar results (Pitaksong *et al.*, 2013). A higher dietary vitamin C inclusion level (1,200 mg/kg diet) was reported to improve weight gain and feed efficiency of stinging catfish (*Heteropneustes fossilis*) (Pal and Chakrabarty, 2012).

To the best of our knowledge, no new research on riboflavin, niacin, pyridoxine, folate, and cobalamin nutrition for farmed catfish species was published in recent years.

Vitamin-like nutrients

Recent study showed that yellow catfish fingerlings require higher dietary choline (1,156.4 mg/kg diet) (Luo *et al.*, 2016) to obtain better growth, feed efficiency, and reduce hepatocyte vacuolization than that of channel catfish (400 mg/kg diet) (Wilson and Poe, 1988; Zhang Z. and Wilson, 1999). Furthermore, Luo *et al.* (2016), for the first time, demonstrated that dietary choline strongly influences some mRNA expression of genes (e.g., microsomal triglyceride transfer protein, apolipoprotein b, lipoprotein lipase) that are involved in lipid metabolism in muscle and liver of yellow catfish. Thus, reaffirming the underlying mechanism of choline in lipid metabolism.

Like tilapias, the essentiality for "vitamin K" has yet to be demonstrated for different species of catfish. Murai and Andrews (1977) concluded that channel catfish fingerlings either do not require dietary "vitamin K" (as menadione) or have an extremely low requirement when a series of experiments failed to produce a deficiency response (i.e., production performance, blood coagulation time, hematologic indices). Since this study, there appeared to be no further publication on "vitamin K" nutrition for catfish. Similarly, inositol is also not required for channel catfish fingerlings (Burtle and Lovell, 1989). However, there is no evidence indicating that other catfish species possess the same ability to synthesize inositol *de novo*. Thus, it would be prudent to ensure that other farmed catfish species are not deficient in vitamin K and inositol, as they might serve as functional additives when the fish are exposed to external stressors.

Astaxanthin is a good example of a functional additive even though it is not an essential micronutrient for farmed catfish. Supplementing 80 mg astaxanthin/kg diet to a diet using practical raw materials affected some antioxidant indices (e.g., lysozyme, SOD, hepatic heat shock proteins mRNA expressions) and reduced mortality after exposing yellow catfish fingerlings to crowding and pathogen stresses (Liu F. *et al.*, 2016).

Based on existing literature and industry practices, we recommend in farmed catfish feeds (IU/kg or mg/kg det) including vitamin A 8,000–12,000 IU/kg; vitamin D 1,500–2,000 IU/kg; vitamin E 100–300 mg/kg; vitamin K 5–10 mg/kg; vitamin B$_1$ (thiamine) 10–20 mg/kg; vitamin B$_2$ (riboflavin) 15–20 mg/kg; niacin 80–120 mg/kg; pantothenic acid 40–50 mg/kg; vitamin B$_6$ (pyridoxine) 15–25 mg/kg; biotin 0.5–1.0 mg/kg; folate 4.0–7.0 mg/kg; vitamin B$_{12}$ (cobalamin) 0.02–0.05 mg/kg; vitamin C 150–1,000 mg/kg; and choline 800–1,100 mg/kg (Table 6.4).

BASS AND BREAM

Fat-soluble vitamins

Newly quantified dietary vitamin A requirement (3,914 IU retinol/kg diet) of Wuchang bream (*Megalobrama amblycephala*) (Liu, B., *et al.*, 2016) fell within the range (1,700–135,000 IU retinol/kg diet) of those determined for European sea bass (*Dicentrachus labrax*) (Villeneuve *et al.*, 2005) and hybrid sea bass (*Morone chrysops × Morone saxatilis*) (Hemre *et al.*, 2004). Whereas no vitamin K nutrition studies on bass and bream were available since Kaushik *et al.* (1998) (Krossøy *et al.*, 2011).

Considering that species of bass and bream are very diverse, the dietary vitamin D requirements are expected to vary widely; however, there is relatively little information on bass and bream species before Darias *et al.* (2010). While juveniles generally require less dietary vitamin D than at the larval stage, species variation played a bigger role than size and cultured environments. Regardless of freshwater or marine bass and bream species, the reported dietary vitamin D requirements (4,970–27,600 IU/kg diet) were higher than that of farmed tilapia, carp, and catfish species (Tables 6.2, 6.3, 6.4). European sea bass larvae (9-day post-hatch) was recommended to receive 27,600 IU/kg diet to maintain weight gain, reduce skeletal deformities (i.e., pugheadness, caudal fin, curvatures of the skull), and induce maturation of digestive function (i.e., higher amylase and trypsin activities, TRPV6 and osteocalcin gene expressions) (Darias *et al.*, 2010). In contrast, Wuchang bream fingerlings (17.7 g) required much less vitamin D (4,970–5,430 IU/kg diet) to enhance growth and feed efficiency (Miao *et al.*, 2015a).

Dioguardi *et al.* (2017) also reported that some immunity-related gene expressions (e.g., lectins, hepcidin) were upregulated when European sea bass was fed increasing levels of dietary vitamin D (0 to 37,500 IU/kg diet). However, several parameters (e.g., lower total antioxidant capacity and hepatic heat shock protein expression, distorted hepatocytes, and intestinal hair cells) were markedly compromised when Wuchang bream was fed excess vitamin D (highest test level 200,0000 IU/kg diet) (Miao *et al.*, 2015b). Consequently, the fish survival (10%) dropped significantly after 96 h of *A. hydrophila* exposure. Thus, the vitamin D inclusion levels for young bass and bream species should not go beyond the current recommended dietary requirement levels for the respective species until a tolerance threshold is determined.

Since the quantification of dietary vitamin E requirement (28 mg/kg diet) of hybrid striped bass fingerlings (1.8 g) (Kocabas, 1999; NRC, 2011), studies over the past decade on farmed sea bass and bream have provided recommendations on vitamin E inclusion levels that varied widely (55.5–2,900 mg/kg diet) (Table 6.5). An evaluation of these studies was given by El-Sayed and Izquierdo (2022). In brief, fish (e.g., bream, bass, salmonids) fed aquafeeds that are susceptible to peroxidation (e.g., PUFA) would typically need higher dietary antioxidants to protect the fish against oxidative damage. For example, largemouth bass (*Micropterus salmoide*) fingerlings fed diets with practical raw materials would require 73–108 mg vitamin E/kg diet to improve growth and feed efficiency (Li, S., *et al.*, 2018a). However, additional dietary vitamin E (160–400 mg/kg diet) significantly reduced the muscle malondialdehyde (MDA) concentration in largemouth bass fingerlings when fed oxidized fish oil (Chen, Y.J., *et al.*, 2013). Additional dietary vitamin E (2,900 mg/kg diet) showed more drastic benefits (e.g., lower mortality, muscular lesions) on European sea bass larvae when fed elevated DHA (5%) (Betancor *et al.*, 2011).

Co-supplementation with vitamin C or Se could further reduce the incidences of muscular lesions and skull deformities in European sea bass larvae (Betancor *et al.*, 2012a,b). Moreover, gilthead sea bream larvae fed a microdiet containing 1,783 vitamin E mg/kg diet, 1,921 vitamin C mg/kg diet, and 7,000 taurine mg/kg diet could significantly reduce incidences of skeletal deformities (e.g., lordosis, kyphosis, abnormal branchiostegal rays and maxillary bones)

(Izquierdo *et al.*, 2019). However, excess dietary vitamin E (3,000 mg/ kg diet) and vitamin C (3,600 mg/kg diet) should be avoided for gilthead sea bream larvae as pro-oxidative effects can occur (Izquierdo *et al.*, 2019). Likewise, the dietary vitamin E inclusion levels should be maintained at, and not exceed, the recommended values for the respective species of farmed bream and bass (Table 6.5).

Water-soluble vitamins and vitamin-like nutrients

Over the last decade, the dietary requirements of several B complex vitamins had been determined for some farmed bass and bream species. These include niacin (30.6–31.3 mg /kg diet), pantothenic acid (18.8–25.7 mg/kg diet), biotin (0.05 mg/kg diet), folate (0.7–5.4 mg/kg diet), choline (1,198–2,232 mg/kg diet), and vitamin B_{12} (cobalamin) (0.06–0.12 mg/kg diet) (Figure 6.4, Table 6.5). On the other hand, there is no new information regarding riboflavin nutrition in recent years and the requirements of vitamin B_1 (thiamine) and vitamin B_6 (pyridoxin) are still absent for this group of farmed fish.

In general, newly recommended dietary vitamin C inclusion range (50–251.5 mg/kg diet) for Asian sea bass/barramundi (*Lates calcarifer*), blunt snout bream, and largemouth bass did not deviate much from the NRC (2011) recommendation (5.0–207.2 mg/kg diet; Table 6.5). Although Ming *et al.* (2012) showed that Wuchang bream (*Megalobrama amblycephala*) fed much higher vitamin C supplementation (700 mg/kg diet) could increase production performance, lysozyme, catalase, and HSP70 mRNA expressions to mitigate heat stress (34°C). At a lower tested vitamin C range (0.2–500 mg/kg diet), Wan *et al.* (2014) reported similar responses in HSP mRNA expressions and improved nonspecific immunity indices (e.g., SOD, MDA, cortisol) when Wuchang bream were under pH (9.5) stress. Thus, a dietary vitamin C level between 150–200 mg/kg diet seemed adequate, and more research is required to evaluate the necessity of supplementing higher levels of vitamin C to farmed bass and bream feeds.

Choline is typically supplemented to maintain weight gain and liver health of farmed fish because some fish species exhibit similar histological changes to mammals when fed high-lipid diets (Jia *et al.*, 2006), imbalanced fatty acids diets (Shimada *et al.*, 2014; Spisni *et al.*, 1998) or elevated levels of non-marine lipid (i.e., vegetable oil). For example, gilthead sea bream fed diets containing 60% soybean oil had signs of hepatic steatosis, with swollen hepatocytes containing lipid vacuoles and displaced nuclei (Caballero *et al.*, 2004). Blunt snout bream fingerlings exhibited classical responses (i.e., higher weight gain and reduced liver lipid content) when fed with the required dietary choline (1,198–2,232 mg/kg diet) (Jiang, G.Z., *et al.*, 2013). Within this dietary choline level, increased intestinal lipase activity, and upregulation of lipoprotein synthesis-related mRNA expression (i.e., apolipoprotein B-100 and triacylglycerides transfer protein) were reported when blunt snout bream was given high-lipid diets (Li, J.Y., *et al.*, 2014; 2016).

Moreover, supplementing choline to a high-lipid diet was reported to reduce liver lipid content, cell damage, and inflammatory response (i.e., down-regulate NFκB expression) in juvenile black seabream (*Acanthopagrus schlegelii*) (Jin *et al.*, 2019). Thus, the revised dietary choline inclusion level for farmed bream was more than double the level of hybrid striped bass (500 mg/kg diet).

Astaxanthin for farmed bass has attracted some attention in recent years due to its functionality in enhancing growth and immunity status (Table 6.5). For example, a series of studies on juvenile Asian sea bass (~28 g) showed that the dietary astaxanthin content at ≥50 mg/kg diet increased serum growth hormone, growth rate, feed efficiency (Lim, K.C., *et al.*, 2019a), lysozyme, phagocytic activity, and reduced serum cortisol, alanine aminotransferase (ALT), and aspartate aminotransferase (AST) levels (Lim, K.C., *et al.*, 2019b). At higher dietary astaxanthin

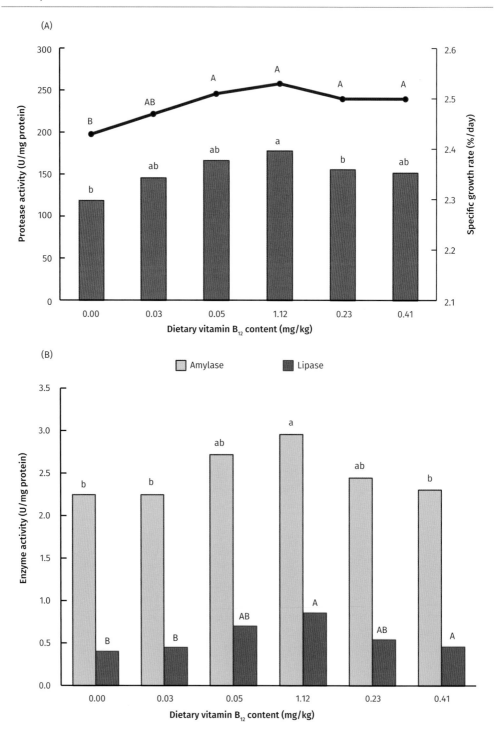

Figure 6.4 Specific growth rate (%/day) (A) and intestinal digestive enzyme activities (B) of fingerling blunt snout bream (*Megalobrama amblycephala*) fed diets containing graded levels of dietary vitamin B$_{12}$ (bars and line with different letters (a, b and A, B) indicate significant differences (p<0.05)) (source: Li, X.F., *et al.*, 2016)

(122–150 mg/kg diet) level, the beneficial effect on the innate immune responses and survival was more pronounced when Asian sea bass was infected by virulent, *Vibrio alginolyticus,* for 14 days (Lim, K.C., *et al.,* 2021). Similarly, Yu, W., *et al.* (2021) reported spotted sea bass fed 58.4–91.5 mg astaxanthin/kg diet (as *Haematococcus pluvialis*) significantly improved antioxidative/immune responses (e.g., T-AOC, SOD, MDA, lysozyme) (Figure 6.5) and survival (>60%) after 7-day of *V. harveyi* challenge. Furthermore, recent evidence in largemouth bass indicated that the astaxanthin could attenuate peroxidation-induced apoptosis by down-regulating the mRNA expressions of cell death mediators (Caspase 3 and Caspase 9) (Xie *et al.,* 2020). Taken together, astaxanthin at 50–150 mg/kg diet can be a functional additive in bream and bass feeds to enhance pigmentation, production performance, and to prepare the fish for potential diseases.

Based on existing literature and current industry practices for bass and bream farming, we recommend revising the OVN™ values (IU/kg or mg/kg diet) to the following: vitamin A 8,000–12,000 IU/kg; vitamin D 1,700–2,200 IU/kg; vitamin E 200–400 mg/kg; vitamin K 8–12 mg/kg; vitamin B$_1$ (thiamine) 20–30 mg/kg; vitamin B$_2$ (riboflavin) 20–30 mg/kg; niacin 100–140 mg/kg; pantothenic acid 50–100 mg/kg; biotin 0.8–1.0 mg/kg; folate 4.0–6.0 mg/kg; vitamin B$_{12}$ (cobalamin) 0.1–0.2 mg/kg; vitamin C 150–200 mg/kg; choline 500–1,000 mg/kg (Table 6.5).

EEL

To date, the eel aquaculture industry still relies heavily on wild-caught glass eels despite decades of arduous research and development efforts in artificial breeding and rearing of egg to glass eel stage economically (Tanaka, 2015). Considering that Japanese eel (*Anguilla japonica*; Endangered) and European eel (*Anguilla anguilla*; Critically Endangered) are both on the IUCN

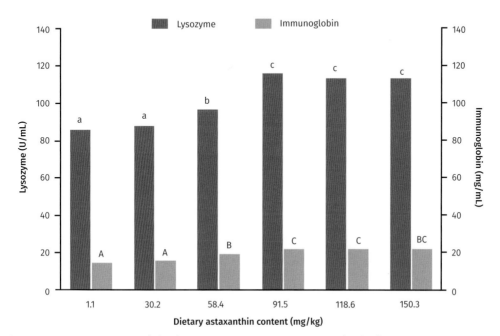

Figure 6.5 Lysozyme activities (U/mL) and immunoglobin concentration (mg/mL) of spotted seabass (*Lateolabrax maculatus*) fed diets containing graded levels of astaxanthin (as *Haematococcus pluvialia*) (bars with different letters (a, b) indicate significant differences (p<0.05)) (source: Yu W. *et al.,* 2021)

Red List of Threatened species (Pike *et al.*, 2020), the limited eel supply and different research priority could have contributed to a lack of progress in vitamin nutrition research in farmed eels.

While symptoms (e.g., fin and skin hemorrhages, abnormal swimming) were described in Japanese eel deficient in several vitamins (i.e., vitamin C, vitamin E, thiamine, riboflavin, niacin, pantothenic acid, pyridoxine, biotin, inositol) (Arai *et al.*, 1972), no vitamin deficiency data on other farmed eels such as European eel and American eel (*Anguilla rostrata*) are available. Furthermore, the quantitative dietary requirement of both lipid-soluble and water-soluble vitamins is absent for farmed eel species, except for vitamin E and vitamin C. As such, NRC (2011) did not provide formulation recommendation for eels.

Vitamin nutrition for eels was recently reviewed by Hamidoghli *et al.* (2019) and Mai *et al.* (2022). In brief, vitamin E and vitamin C have received more attention than other vitamins. The dietary vitamin E requirement of Japanese eel ranged from 21.2 to 213 mg/kg diet with broodstock eels (360 g) having higher dietary needs than juveniles (15 g) (Bae *et al.*, 2013; Shahkar *et al.*, 2018). However, broodstock requirement was determined based on tissue vitamin E content which often have higher saturation points than growth rate and feed efficiency. The dietary vitamin C requirement of Japanese eel ranged from 27 to 912 mg/kg diet (Bae *et al.*, 2012; Ren *et al.*, 2007; Shahkar *et al.*, 2015). Japanese eel fed vitamin C up to the required levels had better growth performance, higher vitamin C content in testes, and improved certain immune responses such as higher SOD and lower TBARS levels.

In summary, there is an urgent need to quantify the dietary vitamin requirements for farmed eel at various life stages. Based on the existing literature and industry practices, we recommend including the following amounts in farmed eel feeds (IU/kg or mg/kg diet): vitamin A 8,000–12,000 IU/kg; vitamin D 1,500–2,000 IU/kg; vitamin E 150–300 mg/kg; vitamin K 3–6 mg/ kg; vitamin B$_1$ (thiamine) 15–25 mg/kg; vitamin B$_2$ (riboflavin) 20–30 mg/kg; niacin 80–120 mg/ kg; pantothenic acid 50–60 mg/kg; vitamin B$_6$ (pyridoxine) 10–15 mg/kg; biotin 0.3–0.5 mg/ kg; folate 4–6 mg/kg; vitamin B$_{12}$ (cobalamin) 0.1–0.2 mg/kg; vitamin C 150–300 mg/kg; choline 800–1,200 mg/kg (Table 6.6).

OTHER WARM WATER SPECIES

Fat-soluble vitamins

The functionalities of some vitamins and their requirements have been identified for other key warm water aquaculture species over the last decade, such as yellowtail kingfish (*Seriola lalandi*), grouper (*Epinephelus spp.*), sturgeon (*Acipenseridae*), snakehead (*Channa punctatus*), and olive flounder (*Paralichthys olivaceus*) (Table 6.7). The dietary vitamin A requirement (2,242–4,102 IU/kg diet) of orange-spotted grouper (*Epinephelus coioides*) (Yang, Q., *et al.*, 2017) was similar to that of greasy grouper (*Epinephelus tauvina*) (3,101 IU/ kg diet) (Shaik Mohamed *et al.*, 2003). Orange-spotted grouper fed vitamin A outside the requirement range showed adverse effects, including poorer growth, feed efficiency, lysozyme, lipolytic enzyme activities, and lower mRNA expressions of hepatic lipase (Yang, Q., *et al.*, 2017).

Limited information was available on both vitamin D and vitamin K nutrition in other warm water species. Under stress-free conditions, the dietary vitamin D requirement (1,403–1,683 IU/ kg diet) of Siberian sturgeon (Wang, L., *et al.*, 2017) was higher than carp, tilapia, and catfish species but lower than bass and bream species (Table 6.7).

Like other major aquaculture species, to promote stress resistance, vitamin E has received more research attention in minor warm water species such as snakehead, cobia (*Rachycentrol canadum*), meager (*Argyrosomus regius*), and yellow drum (*Nibea albiflora*) in recent years. The dietary vitamin E requirements and their effects (e.g., growth, antioxidant capacity, immune responses) on these minor warm water species were comparable to other major aquaculture species (Table 6.7). For instance, meager fed high dietary vitamin E showed improved growth, muscle TBARS level, and lower incidence of granulomas (Lozano et al., 2017; Ruiz et al., 2019). Juvenile yellow drum fed a diet containing 36.2–69.4 mg vitamin E/kg diet had lower MDA level and higher survival when challenged by *Vibrio alginolyticus* (Wang, L., et al., 2019). However, yellow drum exhibited adverse effects when fed vitamin E levels outside this range. Snakehead also showed a similar response when fed excess vitamin E (220 mg/kg diet) (Abdel-Hameid et al., 2012). Thus, it would be prudent to maintain the dietary vitamin E levels within the recommended range to avoid potential pro-oxidative effects.

The dietary requirements of all vitamins were quantified in Japanese yellowtail (*Seriola quinqueradiata*) in the early 1990s (Table 6.7) (Masumoto, 2002). Since then, no new information on vitamin nutrition in *Seriola spp.* was available until recent years. Le et al. (2014a; 2014b) studied the interaction between vitamin E (40 and 180 mg/kg diet) and Se (0–2 mg/kg diet) in yellowtail kingfish (*Seriola lalandi*). However, with only 2 tested vitamin E levels, it is difficult to conclude whether the dietary vitamin E inclusion level for yellowtail kingfish should be revised.

Water-soluble vitamins and vitamin-like nutrients

Zehra and Khan (2018a,b,c,d) recently quantified the dietary requirement of thiamine, riboflavin, niacin, pyridoxine for snakehead fingerlings. The authors observed better growth, feed efficiency, and higher liver storage of the respective vitamin in snakehead when fed increasing dietary content of these water-soluble vitamins (Table 6.7). Moreover, Xun et al. (2019) reported that adequate amount of dietary thiamine increased the gut microbiome diversity and antioxidant enzyme activities (e.g., SOD, CAT, GPX) in golden pompano (*Trachinotus ovatus*). While more research is needed to elucidate the implication of gut microbiome diversity in fish, this study indicated that dietary thiamine could promote good gut health in golden pompano.

Recently, Liu, A., et al. (2019, 2020, 2021a,b) conducted a series of studies on choline nutrition in yellowtail kingfish (~150 g). The study showed that yellowtail kingfish had comparable dietary choline requirement (1,940 mg/kg diet; digestible basis) compared with Japanese yellowtail (1,570–2,160 mg/kg diet; recalculated) when accounting for the differences between the 2 experiments and diet digestibility. Furthermore, juvenile yellowtail kingfish fed adequate choline improved lipid digestibility and appeared to protect liver health. The authors confirmed that the current industry practice of supplementing 3,000 mg choline/kg diet to a fish-meal-based diet would be sufficient to maintain good growth and liver health of the fish at 16 and 24°C. Orange-spotted grouper showed typical responses (e.g., improved growth and liver choline storage) when fed adequate choline (1,094–1,580 mg/kg diet) (Qin, D.G., et al., 2017). Higher dietary choline (>2,940 mg/kg diet) was found to improve survival (100%) of giant grouper (*Epinephelus lanceolatus*) after a 10 h ammonia challenge (Yeh et al., 2015). The authors suggested that choline possibly played a role in regulating the hypothalamic-pituitary-interrenal axis (mediate stress response) in giant grouper (Jiang, X., et al., 2012; Pankhurst, 2011). However, more evidence is required to demonstrate choline's stress resistance role in fish. Nevertheless, in an intensive system where ammonia level can be high after feeding, additional dietary choline might be beneficial to other *Epinephelus spp.*

Like vitamin E, improving the antioxidant capacity and immune response had been the focus of vitamin C nutrition in several warm water species over the last decade. Moving away from avoiding deficiency, newly determined vitamin C inclusion levels had increased when exposed to external stressors. For example, previously determined dietary vitamin C requirement of cobia was 44.7–55.9 mg/kg diet; however, up to 96.6 mg/kg diet was necessary to minimize mortality after a 7 d *Vibrio harveyi* challenge (Zhou, Q., *et al.*, 2012). In addition to higher vitamin C storage in the liver and muscle, silver pomfret (*Pampus argenteus*) required more vitamin C (455–800 mg/kg diet) when experiencing transportation-induced stress (Peng, S.M., *et al.*, 2013). Pacific bluefin tuna also appeared to have high dietary vitamin C requirement (454 mg/kg diet) for good growth, feed efficiency, and higher vitamin C storage in liver and brain (Biswas *et al.*, 2013).

CONCLUSION AND FUTURE RESEARCH

This chapter highlighted that vitamins could serve as a strategic tool to facilitate aquaculture's sustainable growth by improving fish growth, feed efficiency, stress resistance, health, and survival. Organized by major groups of species, Tables 6.2 to 6.6 consist of recommended dietary vitamin inclusion levels and their benefits based on recent scientific studies.

In general, recent studies demonstrated that most vitamin supplementation affected standard antioxidant indices (e.g., SOD, CAT, T-AOC, MDA), gene transcription, immune responses, and blood indices (e.g., lysozyme activity, white blood cells, immunoglobulin, ALT, AST etc.). However, the responses were sometimes inconsistent and could depend on various factors including tissue type, experimental design, diet formulation and raw materials, and fish life stage.

As feed formulation and fish farming conditions continue to evolve, future studies could establish reference intervals for common antioxidant and immune markers to aid clinical assessment and interpretation of dietary effects on the target species.

Furthermore, many warm water species consume plant-based feed, and some farms might be exposed to contaminants such as pesticides, herbicides, and mycotoxins. Environmental flux in an earthen pond can be different to sea cages and recirculating aquaculture systems, thus, presenting another layer of the challenge but highlighting the necessity to update optimal vitamin inclusion levels tailored for the target species and their growing conditions.

We recommend exploring the potential benefits of vitamins in response to contaminants in feed or system water. We also encourage more studies to investigate the effect of vitamins on intestinal microflora and intestinal health as there is some evidence of increasing the amount of intestinal *Lactobacillus* and reducing *Aeromonas* in Jian carp when fed biotin, inositol, and vitamin C.

In addition, providing enough dietary information (e.g., analyzed vitamin content of final diets, the chemical form of the vitamin used) in a study is integral to producing meaningful and reliable conclusions and recommendations. When reporting dietary requirements, the inconsistent use of terminologies has been a common issue in aquaculture nutrition studies. An approach to remedy this issue is to chemically analyze the vitamin content (free form) in the final diet. Finally, inappropriate use of statistical analysis or presentation of results becomes a missed opportunity to provide reliable information for readers.

Table 6.1 An overview of metabolic functions of vitamin/vitamin-like nutrients in animal nutrition and their deficiency symptoms in warm water fish nutrition. For more details, see Mai *et al.* (2022), NRC (2011), and Bureau and Cho (1999). Adapted from Liu *et al.* (2022).

Vitamins/vitamin-like nutrients	Classical functions	Novel functions	Deficiency symptoms in warm water fish
A	Vision process Cell metabolism	Detoxification Fecundity	Anorexia Deformity (bone, operculum) Change of skin color Hemorrhage
D	Calcium-phosphorus metabolism	Antimicrobial Antiviral	Deformity (bone, fin)
E	Biological antioxidant	Immunity	Hepatocyte necrosis Muscle dystrophy Lordosis
K	Blood coagulation	–	Anorexia Hemorrhage (skin)
B$_1$ (Thiamine)	Carbohydrate metabolism – coenzyme for oxidative decarboxylation	–	Abnormal swimming Anorexia Change of skin color Hemorrhage (skin, fins) Loss of equilibrium
B$_2$ (Riboflavin)	Energy metabolism – coenzyme in cellular redox reactions	–	Abnormal swimming Anemia Anorexia Cataract Change of skin color Erosion of fins Hemorrhage (skin, fins) Lethargy Photophobia
Niacin (B$_3$)	Energy metabolism – coenzyme in cellular redox reactions	–	Anorexia Anemia Exophthalmia Hemorrhage Lesion Lethargy
Pantothenic acid (B$_5$)	Energy metabolism – coenzyme for acetylation and acylation reactions	–	Abnormal swimming Anemia Anorexia Clubbed gills Hemorrhage Lethargy

Table 6.1 (continued)

Vitamins/vitamin-like nutrients	Classical functions	Novel functions	Deficiency symptoms in warm water fish
B₆ (Pyridoxin)	Coenzyme for amino acid and lipid metabolism	Fatty acid metabolism (DHA retention in fillet)	Abnormal swimming Anemia Anorexia Exophthalmos Lesion Lethargy Neurological disorder
Biotin (B₇)	Carbohydrate and lipid metabolism – coenzyme for carboxylation/decarboxylation reactions	–	Abnormal swimming Anorexia Change of skin color Lethargy Reduced hematocrit Tetany
Folate (B₉)	Nucleic acid metabolism – coenzyme for one-carbon metabolism Blood cell function	Immunity Antioxidation	Anemia Anorexia Lethargy
B₁₂ (Cobalamin)	Protein metabolism – coenzyme for one-carbon metabolism Blood cell function	–	Anorexia Anemia
C	Antioxidant Detoxification Collagen synthesis	Wound healing Stress/disease resistance Reproduction	Abnormal swimming Anemia Anorexia Bone deformities Change of skin color Exophthalmos Hemorrhage Lesions Lethargy
Choline	Lipid metabolism Cell structural integrity	Prevention of lipid malabsorption syndrome	Anorexia Change of skin color Change in liver lipid content Hemorrhage Increased mortality Inflammation infiltration (liver, intestine)
Inositol	Neurological osmolytes and signaling molecules Cell structural integrity	Antioxidation Immunity	Increased mortality Change of skin and intestine color Hemorrhage Abnormal lipid accumulation in liver
Astaxanthin	Biological antioxidant Pigmentation	Reproduction Stress tolerance Immunity	–

Table 6.2 A summary of the recommended dietary vitamin content for farmed tilapias since NRC (2011) with recommendations for optimal vitamin nutrition

Vitamins (IU or mg/kg diet)	NRC (2011)	Literature covered by NRC (2011)	Optimum Vitamin Nutrition® dsm-firmenich 2022	Species*	Initial weight (g)	Vitamin dietary content in published research	Benefits	References
A (IU)	5,850	5,850–6,970	4,000–8,000	Nile tilapia	5.3	3,910	Improved feed intake, weight gain, feed efficiency and blood cell count compared to the control (0 IU/kg diet) and prevent gross deficiency signs	Guimarães et al. (2014)
				Nile tilapia	7.5	4,000	Improved survival, feed intake, weight gain, feed efficiency and blood cell count compared to the control (0 IU/kg diet)	Guimarães et al. (2016)
D (IU)	374.8	374.8	1,500–2,000	–	–	–	–	–
E	60	25–100	100–750	Nile tilapia	6.7	62.9	Improved survival post Streptococcus iniae challenge	Lim, C., et al. (2010b)
				Nile tilapia	1.9–2.0	<1,000 (supplemented)	Improved feed intake, weight gain, feed efficiency, hematologic indices, immune response, and histopathological changes in liver and gill post copper oxychloride challenge compared to the control (no supplement)	Hassaan et al. (2014)
				GIFT	79.4	45.7–752.5	Improved survival, weight gain and feed efficiency compared to the control (6.4 mg/kg diet), fillet texture, and antioxidant enzyme activities	Wu, F., et al. (2017)

Table 6.2 (continued)

Vitamins (IU or mg/kg diet)	NRC (2011)	Literature covered by NRC (2011)	Optimum Vitamin Nutrition® dsm-firmenich 2022	Species*	Initial weight (g)	Vitamin dietary content in published research	Benefits	References
				GIFT	0.7	80–160	Improved weight gain, feed efficiency, antioxidant enzyme activities, and higher survival post *Streptococcus iniae* challenge compared to treatments with <80 mg/kg diet	Qiang et al. (2019)
				Nile tilapia	14.0	101.6 (+1.1 Se)	Improved growth, antioxidant enzyme activities, immune responses, and intestinal morphometry	Dawood et al. (2020a)
				GIFT	80.3	43.2–76.1	Improved weight gain, feed efficiency and antioxidant enzyme activities at ≥40 mg/kg diet	Jiang, M., et al. (2020)
				Nile tilapia	14.7	300 (supplemented)	Improved growth, antioxidant enzyme activities, immune response, and maintained intestine and liver morphology/health at high stocking density and post *Aeromonas sobria* challenge	Ahmed et al. (2021)
				Nile tilapia	19.8 (f), 18.6 (m)	120 (f), 160 (m) (supplemented)	Improved weight gain, feed efficiency at ≥80 mg/kg diet and reproductive hormones at ≥40 mg/kg diet	Zhang, X., et al. (2021)

								Reference
K	–	–	5–10	Nile tilapia	12.8	400 (supplemented)	Improved weight gain and feed efficiency, reduced histopathological changes in gill and intestine, and TAN and UA in water	Elkaradawy et al. (2021)
				Nile tilapia	3.6	50 (supplemented, +80 Zn)	Improved production performance	Rohani et al. (2021)
				–	–	–		–
B$_1$ (Thiamine)	–	–	10–20	Nile tilapia	3.6	3.3–3.4 (supplemented)	Improved survival and maintained blood parameters compared to non-supplemented diet	Lim et al. (2011)
					5.3	0.9 (supplemented)		–
B$_2$ (Riboflavin)	6	5.0–6.0	15–20	–	–	–		–
Niacin (B$_3$)	26	26–121	80–120	GIFT	87.2	20.4–84.6	Improved growth and liver niacin storage	Jiang et al. (2014a)
Pantothenic acid (B$_5$)	10	10	40–50	–	–	–		–
B$_6$ (Pyridoxin)	15	3.0–16.5	15–25	Nile tilapia	8.4	10.0 of pyridoxine hydrochloride	Improved weight gain, feed intake, feed efficiency and alleviated some hematologic indices induced by heat stress (32°C) compared to non-supplemented diet	Teixeira et al. (2012)
Biotin (B$_7$)	0.06	0.06	0.5–1.0	–	–	–		–
Folate (B$_9$)	1	0.8	10–20	GIFT	60.2	0.4–0.7	Improved growth and liver folic acid storage	Wu, J.P., et al. (2016)
B$_{12}$ (Cobalamin)	NR	NR	0.02–0.05	–	–	–		–

Table 6.2 (continued)

Vitamins (IU or mg/kg diet)	NRC (2011)	Literature covered by NRC (2011)	Optimum Vitamin Nutrition® dsm-firmenich 2022	Species*	Initial weight (g)	Vitamin dietary content in published research	Benefits	References
C	20	16.0–420	150–1,500	Nile tilapia	6.7	77.3	Improved weight gain, feed intake, feed efficiency and hematologic indices compared to 10 mg/kg diet.	Lim et al. (2010b)
				Nile tilapia	11.0	500	Improved growth, immune response, and survival post Aeromonas hydrophila challenge	Ibrahem et al. (2010)
				Nile tilapia	30.0	1,000 (supplemented)	Improved stress response induced by crowding	Mustafa et al. (2013)
				GIFT	220.1	53.1–185.9	Improved growth and vitamin C storage in muscle and liver	Wu, F., et al. (2015)
				GIFT	70.0	45.0–118.6	Improved growth (at 45 mg/kg diet), vitamin C storage in muscle and liver, and hematologic indices at >110 mg/kg diet)	Huang, F., et al. (2016)
				Nile tilapia	45 (m)	599–942	Improved body weight, sperm concentration and motility	Sarmento et al. (2017)
				Nile tilapia	70 (f), 153 (m)	599	Improved growth, fecundity, larval weight, larval survival, and larval resistance to air exposure	Sarmento et al. (2018)

Nutrient				Species		Dose	Effects	Reference
				Nile tilapia	14.1	516.3 (+1.2 Se)	Improved feed efficiency, antioxidant enzyme activities, immune response, and intestinal morphometry compared to non-supplemented diet	Dawood et al. (2020b)
				Nile tilapia	14.7	473.5	Improved growth, antioxidant enzyme activities, immune response, and resistance to Aeromonas sobria challenge	Ibrahim et al. (2020)
				Nile tilapia	21.2	500 mg (supplemented)	Improved growth, antioxidant enzyme activities, and immune responses when exposed to imidacloprid toxicity. Reduced imidacloprid concentration in tissue	Naiel et al. (2020)
				Nile tilapia	12.4	400 (supplemented)	Improved growth, digestive and antioxidant enzyme activities, immune response, and resistance to Streptococcus agalactiae challenge	El Basuini et al. (2021)
Choline	1,000	1,000	600–1,200	Nile tilapia	148.0	1,000	Alleviated some size-sorting stress hematologic indices	Fernandes Jr et al. (2016)
				Nile tilapia	7.1	731–1,200 of herbal extracts (supplemented)	Improved growth and antioxidant status, post aflatoxin B$_1$ challenge	Baldissera et al. (2019), Souza et al. (2020)
Inositol	400	400	-	-	-	-	-	-

Notes: GIFT, genetically improved farmed tilapia (*Oreochromis niloticus*); f, female; m, male; NR, not required; TAN, total ammonia nitrogen; UA, unionized ammonia; QS, *Quillaja Saponaria*.

* The scientific names of the species were Nile tilapia (*Oreochromis niloticus*).

1,000 IU Vitamin A (retinol) = 0.3 mg retinol = 0.344 mg retinyl acetate = 0.55 mg retinyl palmitate
1,000 IU Vitamin D$_3$ (cholecalciferol) = 0.025 mg vitamin D$_3$

Table 6.3 A summary of the recommended dietary vitamin content for farmed carps since NRC (2011) with recommendations for optimal vitamin nutrition

Vitamins (IU or mg/kg diet)	NRC (2011)	Literature covered by NRC (2011)	Optimum Vitamin Nutrition® dsm-firmenich 2022	Species*	Initial weight (g)	Vitamin dietary content in published research	Benefits	References
A (IU)	4,000	4,000–20,000	8,000–11,000	Gibel carp	69.4	1,716–2,698	Higher plasma vitamin A concentration	Shao et al. (2016)
				Grass carp	5.0	4,769	Improved growth	Wu, F., et al. (2016)
				Grass carp	262.0	2,213–2,622	Improved growth, antioxidant enzyme activities, protection against enteritis, and histopathological changes in the head kidney and spleen post *Aeromonas hydrophila* challenge	Zhang, L., et al. (2017), Jiang, W.D., et al. (2019), Jiang, W.D., et al. (2020)
D (IU)	–	–	1,500–2,000	Black carp	4.7	412–1,480	Improved production performance, antioxidant enzyme activities, and immune responses compared to <412 IU/kg diet	Wu, C., et al. (2020)
				Rohu	–	150 (supplemented)	Reverse nitrite-induced inhibition of serum testosterone and estradiol production	Ciji et al. (2013)
E	100–132	62.9–200	100–900	Grass carp	11.2	100.4	Improved growth and some antioxidant enzyme activities	Li, J., et al. (2014)
				Common carp	60	90.9 (supplemented)	Prevent deficiency symptoms and maintain antioxidant enzyme activities	Wang, K., et al. (2016)

K				Species			Effect	Reference
				Grass carp	266.4	116.2–136.4	Improved weight gain, feed intake, feed efficiency, antioxidant enzyme activities, immune responses, survival, and minimized skin lesions post *Aeromonas hydrophila* challenge compared to <116 mg/kg diet	Pan *et al.* (2017)
				Grass carp	266.4	139.6	Minimized gill rot morbidity post *Flavobacterium columnare* challenge	Pan *et al.* (2018)
				Rohu	225	500 (supplemented, +500 vitamin C +10 Zn)	Improved weight gain and survival	Rahman *et al.* (2018)
				Rohu	4.3	100–914	Improved weight gain, feed efficiency, liver vitamin E storage, and antioxidant enzyme activities against oxidized fish oil compared to non-supplemented diet	Fatima *et al.* (2019)
–	1.9	5–10		Jian carp	8.9	3.13	Improved weight gain, feed intake, feed efficiency, digestive and antioxidant enzyme activities, and intestinal microbial composition compared to ≤1.5 mg/kg diet	Yuan *et al.* (2016)

Table 6.3 (continued)

Vitamins (IU or mg/kg diet)	NRC (2011)	Literature covered by NRC (2011)	Optimum Vitamin Nutrition® dsm-firmenich 2022	Species*	Initial weight (g)	Vitamin dietary content in published research	Benefits	References
				Jian carp	8.2	1.0–1.6	Improved weight gain, feed intake, feed efficiency, survival, digestive and antioxidation enzyme activities, immune response, and survival post *Aeromonas hydrophila* challenge	Huang, H.H., et al. (2011), Feng et al. (2011), Li, X.Y., et al. (2014)
B_1 (Thiamine)	0.5	0.5	10–20	Grass carp	10.7	1.3–5.0	Improved weight gain, feed efficiency and liver thiamine storage	Jiang, M., et al. (2014b)
				Grass carp	243	0.8–1.2	Improved weight gain, feed intake, feed efficiency and the immune responses of intestine and gill	Wen, L.M., et al. (2015; 2016)
				Mrigal carp	6.2	1.8–3.3	Improved weight gain, feed efficiency and liver thiamine storage	Zehra and Khan (2017)
				Catla	3.5	0.7–1.6	Improved production performance, liver thiamine storage, and antioxidant enzyme activities	Khan and Khan (2021a)
				Jian carp	23.4	5.0	Improved weight gain, feed efficiency and digestive enzyme activities	Li, W., et al. (2010)
B_2 (Riboflavin)	7	4.0–7.0	15–20	Grass carp	275.8	5.9–7.4	Improved weight gain, feed intake, feed efficiency, antioxidant enzyme activities, and gill immune response	Chen, L., et al. (2015a)

Vitamin				Species			Effects	References
Niacin (B₃)	28	25.5–28.0	80–120	Rohu	4.2	33.0	Improved weight gain and feed efficiency	Ahmed (2011)
				Mrigal carp	4.5	30.0	Improved weight gain and feed efficiency	Ahmed (2011)
Pantothenic acid (B₅)	30	23–50	40–50	Grass carp	255.6	34.0–42.1	Improved weight gain, feed intake, feed efficiency, digestive enzyme activities, intestinal immune response	Li, S.Q., et al. (2016), Feng et al. (2016)
				Grass carp	253.4	37.0	Improved gill antioxidant enzyme activities and immune response	Li et al. (2015a)
B₆ (Pyridoxin)	6	5.0–11.4	15–25	Mrigal carp	6.3	5.6–8.6	Improved weight gain, feed efficiency and liver pyridoxine storage	Zehra and Khan (2017)
				Catla	3.7	3.2–6.9	Improved production performance, liver pyridoxine storage, and immune response	Khan and Khan (2021b)
Biotin (B₇)	1	1.0	0.5–1.0	Jian carp	7.7	0.2	Improved weight gain, feed intake, feed efficiency, digestive enzyme activities, and increased the counts of intestinal Lactobacillus and Bacillus	Zhao, S., et al. (2012)
				Catla	7.9	0.4–0.9	Improved weight gain, feed efficiency, survival, liver biotin storage, and PC activity	Khan and Khan (2019)
Folate (B₉)	NR	3.6–4.3	4.0–7.0	Grass carp	267.7	1.0–2.1	Improved weight gain, feed intake, feed efficiency, gill and intestinal antioxidant enzyme activities and immune response	Shi, L., et al. (2015; 2016)
B₁₂ (Cobalamin)	NR	0.094	0.02–0.05	–	–	–	–	–

Table 6.3 (continued)

Vitamins (IU or mg/kg diet)	NRC (2011)	Literature covered by NRC (2011)	Optimum Vitamin Nutrition® dsm-firmenich 2022	Species*	Initial weight (g)	Vitamin dietary content in published research	Benefits	References
C	45	45–750	150–400	Jian carp	12.6	40.9–45.1	Improved growth, digestive enzyme activities, and increased the counts of intestinal *Lactobacillus* and *Bacillus*	Liu, Y., et al. (2011)
				Catla	14.0	500–1,000 (supplemented)	Ameliorate changes in hematologic response and enzyme activities post deltamethrin challenge	Vani et al. (2011)
				Mrigal carp	0.8	35.7–42.0	Improved growth, feed efficiency and liver vitamin C storage	Zehra and Khan (2012)
				Grass carp	264.4	156.6	Improved gill immune response and minimized gill rot morbidity post *Flavobacterium columnare* challenge	Xu, H.J., et al. (2016a)
				Grass carp	264.4	92.8–122.9	Improved growth, feed intake, feed efficiency, head kidney and spleen immune response, and minimized skin lesion morbidity post *Aeromonas hydrophila* challenge	Xu, H.J., et al. (2016b)
				Grass carp	8.5	67.2	Improved growth, feed intake, feed efficiency, and liver and muscle antioxidant enzyme activities	Nasar et al. (2021)

Nutrient			Species			Observed effects	Reference
Choline	1,500	600–1,800	Jian carp	7.9	566–607	Improved growth, digestive enzyme activities, intestinal morphology, antioxidant enzyme activities, immune response, higher survival post *Aeromonas hydrophila* challenge	Wu, P., et al. (2011, 2013, 2014, 2017)
			Gibel carp	5.5	2,500–2,667	Improved growth	Duan et al. (2012)
			Grass carp	266.9	1136.5–1,555.0	Improved weight gain, feed intake, feed efficiency, fillet quality (shear force, cooking loss), gill antioxidant enzyme activities, and gill immune response	Zhao, H.F., et al. (2015, 2016)
			Grass carp	9.3	1,330.7–1548.1	Improved weight gain, feed intake, feed efficiency, maintained intestine and gill morphology/health, and minimized gill rot morbidity post *Flavobacterium columnare* challenge	Yuan, Z.H., et al. (2020a,b, 2021)
			Grass carp	221.3	296.9–415.1	Improved intestinal immune response post *Aeromonas hydrophila* challenge	Li, S.A., et al. (2018)
Inositol	440	116–440	—				
			Jian carp	22.3	518	Improved weight gain, digestive enzyme activities, increased the amount of intestinal *Lactobacillus*	Jiang, W.D., et al. (2009)
			Jian carp	22.3	660–770.5	Improved antioxidant enzyme activities and immune response post *Aeromonas hydrophila* challenge	Jiang, W.D., et al. (2016)

Table 6.3 (continued)

Vitamins (IU or mg/kg diet)	NRC (2011)	Literature covered by NRC (2011)	Optimum Vitamin Nutrition® dsm-firmenich 2022	Species*	Initial weight (g)	Vitamin dietary content in published research	Benefits	References
				Common carp	26.3	50 and 100 (supplemented)	Increased immune response and survival post *Aeromonas hydrophila* challenge	Jagruthi *et al.* (2014)
				Common carp	5.0	100 and 200	Improved antioxidant status against ammonia stress	Sowmya and Sachindra (2014)
				Koi	26.3	80	Stimulated skin coloration and improved feed efficiency	Nguyen *et al.* (2014)
				Koi	76.2 mm	50	Stimulated skin coloration	Boonyapakdee *et al.* (2015)
Astaxanthin	–	–	–	Common carp	6.2	190.4	Improved growth, immune response, and survival post *Aeromonas hydrophila* challenge	Sowmya and Sachindra (2015)
				Goldfish	47.2 (female), 54.2 (male)	149.6	Improved hatching rate, egg survival rate (incubation period), and sperm quality	Tizkar *et al.* (2013; 2015; 2016)
				Crucian carp	10.0	400	Improved weight gain, digestive and antioxidant enzyme activities, and higher survival post *Aeromonas hydrophila* challenge	Wu and Xu (2021)

Notes: * Dietary requirement values covered common carp or rohu. Requirements were listed as not tested, required, or not required for some carp species.
† The scientific names of the species were black carp (*Mylopharyngodon piceus*); catla (*Catla catla*); koi and common carp (*Cyprinus carpio*), goldfish and crucian carp (*Carassius auratus*); grass carp (*Ctenopharyngodon idella*); Gibel carp (*Carassius auratus gibelio*); Jian carp (*Cyprinus carpio var. Jian*); mrigal carp (*Cirrhinus mrigala*); rohu (*Labeo rohita*); tiger prawn (*Penaeus monodon*); golden apple snail (*Pomacea canaliculata*).

1,000 IU Vitamin A (retinol) = 0.3 mg retinol = 0.344 mg retinyl acetate = 0.55 mg retinyl palmitate
1,000 IU Vitamin D (cholecalciferol) = 0.025 mg vitamin D

Table 6.4 A summary of the recommended dietary vitamin content for farmed catfish since NRC (2011) with recommendations for optimal vitamin nutrition

Vitamins (mg kg⁻¹ diet)	NRC (2011)	Literature covered by NRC (2011)	Optimum Vitamin Nutrition® dsm-firmenich 2022	Species*	Initial weight (g)	Vitamin dietary content in published research	Benefits	References
A (IU)	2,000	1,000–2,000 of RA	8,000–11,000	Silver catfish	23.0	2,610	Improved growth and antioxidant enzyme activities	Battisti et al. (2017)
D_3 (IU)	500	250–1,000	1,500–2,000	Yellow catfish	5.0	2,260	Increased resistance to oxidative stress and inflammation responses post Edwardsiella ictaluri challenge	Cheng, K., et al. (2020a,b)
E	50	25–50	100–300	Channel catfish	3.2	30.8–43.0	Maintained production performance and liver lipid content	Lim, C., et al. (2010a)
				Striped catfish	125.1	192.5	Improved weight gain and vitamin E storage in fillet	Al-Noor et al. (2012)
				Darkbarbel catfish	1.0	403.1	Improved weight gain, feed efficiency, antioxidant enzyme activities, and immune response under high ammonia and post Edwardsiella ictaluri challenge	Li, M., et al. (2013)
				Hybrid catfish	21–27	59.3 (+502.8 vitamin C)	Improved growth, hematologic indices, and immune response under acid (pH 5.5) and cold stress (19°C)	Pitaksong et al. (2013)
				Yellow catfish	2.0	33.0–45.7	Improved weight gain, feed efficiency, antioxidant enzyme activities, immune response, and resistance to Aeromonas hydrophila	Lu, Y., et al. (2016)
				African catfish	120	240 (supplemented)	Toxicity from potassium dichromate	Azeez and Braimah (2020b)

Table 6.4 (continued)

Vitamins (mg kg⁻¹ diet)	NRC (2011)	Literature covered by NRC (2011)	Optimum Vitamin Nutrition® dsm-firmenich 2022	Species*	Initial weight (g)	Vitamin dietary content in published research	Benefits	References
				African catfish	120	240 (supplemented)	Toxicity copper sulfate	Azeez and Braimah (2020a)
K	R	NR–R	5–10	–	–	–	–	–
B_1 (Thiamine)	1	1.0 of THC	10–20	Yellow catfish	3.8	6.0–7.4	Improved weight gain, feed efficiency and digestive enzyme activities	Zhao, H., et al. (2020)
B_2 (Riboflavin)	9	6.0–9.0	15–20	–	–	–	–	–
Niacin (B_3)		7.4–14	80–120	–	–	–	–	–
Pantothenic acid (B_5)	15	10.0–15	40–50	Green catfish	10.6	5–40 of calcium d-pantothenate	Improved weight gain and survival	Hien and Doolgindachbaporn (2011)
B_6 (Pyridoxin)	3	3.0	15–25	–	–	–	–	–
Biotin (B_7)	R	0.25–2.5	0.5–1.0	–	–	–	–	–
Folate (B_9)	1.5	1.0–1.5	4.0–7.0	–	–	–	–	–
B_{12} (Cobalamin)	R	–	0.02–0.05	–	–	–	–	–

				Species			Effect	Reference
C	15		150–1,000	Stinging catfish	–	1,200	Improved weight gain, feed efficiency	Pal and Chakrabarty (2012)
				Hybrid catfish	21–27	988.4	Improved growth, hematologic indices, and immune response under acid (pH 5.5) and cold stress (19°C)	Pitaksong et al. (2013)
				Yellow catfish	2.0	114.5	Improved growth and lysozyme, and higher survival post Aeromonas hydrophila challenge	Liang et al. (2017)
				Striped catfish	3.2–3.4	46.4–76.4	Improved weight gain, feed efficiency, survival and liver vitamin C storage	Daniel et al. (2018)
				Hybrid sorubim catfish	14.6	ND	Seemed to increase the integrity of the intestinal mucosa and stimulate erythropoiesis	Rodrigues et al. (2018)
Choline	400	400	600–1,100	Yellow catfish	3.5	1,156.4	Improved weight gain, feed intake, feed efficiency, liver lipid content and vacuoles	Luo et al. (2016)
Inositol	NR	NR	–	–	–	–	–	–
Astaxanthin	–	–	–	Yellow catfish	2.1	80	Improved survival post crowding stress (150 g/L) and Proteus mirabilis challenge	Liu, F., et al. (2016)

Notes: ND, not determined; NR, not required; R, required; RA, retinyl acetate; THC, thiamine hydrochloride.

* The scientific names of the species were African catfish (Clarias gariepinus), channel catfish (Ictalurus punctatus), darkbarbel catfish (Pelteobagrus vachelli), green catfish (Mystus nemurus), hybrid catfish (Clarias macrocephalus × C. gariepinus), hybrid sorubim catfish (Pseudoplatystoma reticulatus × P. corruscans), stinging catfish (Heteropneustes fossilis), silver catfish (Rhamdia quelen), striped catfish (Pangasianodon hypophthalmus), vundu (Heterobranchus longifilis), yellow catfish (Pelteobagrus fluvidraco).

1,000 IU Vitamin A (retinol) = 0.3 mg retinol = 0.344 mg retinyl acetate = 0.55 mg retinyl palmitate

1,000 IU Vitamin D_3 (cholecalciferol) = 0.025 mg vitamin D_3

Table 6.5 A summary of the recommended dietary vitamin content for major farmed bream and bass since NRC (2011) with recommendations for optimal vitamin nutrition

Vitamins (IU or mg/kg diet)	NRC (2011)	Literature covered by NRC (2011)	Optimum Vitamin Nutrition® dsm-firmenich 2022	Species*	Initial weight (g)	Vitamin dietary content in published research	Benefits	References
A (IU)	31	1,700–135,000	8,000–12,000	European sea bass	9 dph	9,000	Maintained skeleton and fin structures	Georga et al. (2011)
				Wuchang bream	2.4	3,914	Improved weight gain, feed efficiency and survival post Aeromonas hydrophila challenge	Liu, B., et al. (2016)
				Wuchang bream	17.7	4,970–5,430	Improved weight gain and feed efficiency	Miao et al. (2015a)
D (IU)	–	–	1,700–2,200	European sea bass	100	3,750–37,500 (supplemented)	Improved feed efficiency, increased serum peroxidase activity and phagocytic activity of peritoneal cavity leucocytes	Dioguardi et al. (2017)
				Gilthead seabream	0.86 mg	1,000–1,200	Improved survival and decreased skeletal deformities	Sivagurunathan et al. (2022)
				European sea bass	1.9 mg	2,900	Reduced larval mortality and lesions when fed elevated DHA (3 and 5%)	Betancor et al. (2011)
E	–	28	200–400	Largemouth bass	7.5	73–108	Improved weight gain, feed intake, feed efficiency, liver vitamin E storage, and enhanced some antioxidant and immunity indices	Li, S., et al. (2018)
				Red sea bream	1.8	>71.0	Improved weight gain, feed efficiency and antioxidant indices when fed oxidized fish oil	Gao et al. (2012a)

Vitamin			Species				Reference
K	–	8–12	Red sea bream	28.9	171.7 (+843.8 vitamin C)	Improved antioxidant indices when fed oxidized fish oil	Gao et al. (2012b)
			Blunt snout bream	0.6	55.5	Improved growth and liver n-3 PUFA content	Zhang, Y., et al. (2017)
			Gilthead sea bream	0.1 mg (16 dph)	1,783 (+1,921 vitamin C, +7,000 taurine)	Improved length, survival, and reduced skeletal deformities	Izquierdo et al. (2019)
B_1 (Thiamine)	–	20–30	–	–	–	–	–
B_2 (Riboflavin)	4.1–5.0	20–30	–	–	–	–	–
			–	–	–	–	–
Niacin (B_3)	–	100–140	Blunt snout bream	3.6	30.6–31.3	Improved weight gain, feed intake, feed efficiency, survival, and liver niacin storage	Li, X.F., et al. (2017)
	–	50–100	Hybrid striped bass	1.6	18.8	Improved weight gain, feed efficiency, survival	Raggi et al. (2016)
Pantothenic acid (B_5)	–		Blunt snout bream	6.0	24.1	Improved weight gain, feed efficiency, survival, higher liver pantothenic acid storage, digestive enzyme activities, and fatty acid synthesis-related gene expressions	Qian et al. (2015)
B_6 (Pyridoxin)	–	20–25	–	–	–	–	–
Biotin (B_7)	0.046	0.8–1.0	Japanese sea bass	–	–	–	–
Folate (B_9)	–	4–6	Blunt snout bream	27.0	0.7–1.0	Improved weight gain, feed efficiency, digestive and antioxidant enzyme activities, and immune response post heat stress	Sesay et al. (2016, 2017)

Table 6.5 (continued)

Vitamins (IU or mg/kg diet)	NRC (2011)	Literature covered by NRC (2011)	Optimum Vitamin Nutrition® dsm-firmenich 2022	Species*	Initial weight (g)	Vitamin dietary content in published research	Benefits	References
B$_{12}$ (Cobalamin)	–	–	0.1–0.2	Blunt snout bream	0.7	0.1	Improved growth and digestive enzyme activity	Li, X.F., et al. (2016)
C	20–30	5.0–207.2	150–200	Asian sea bass	0.01	170	Prevent spinal deformities	Fraser and De Nys (2011)
				European sea bass	45 dph	50	Improved growth and reduced skeletal deformities	Darias et al. (2011)
				Asian sea bass	78.9	43.2 (supplemented)	Improved weight gain, feed efficiency and promoted wound healing process	Catacutan et al. (2012)
				European sea bass	0.4	2,998 (+3,179 vitamin E)	Reduced TBARS content and incidence of muscular lesions when fed high DHA diets (5%)	Betancor et al. (2012b)
				Largemouth bass	6.7	102.6–147.8	Improved weight gain, feed efficiency, survival, and antioxidant capacities	Chen, Y.J., et al. (2015)
				Largemouth bass	10.9	120.5–145.4	Improved weight gain, feed efficiency, digestive and antioxidant enzyme activities, immune responses, and liver morphology. Improved survival post Aeromonas hydrophila challenge	Yusuf et al. (2020, 2021)
				Wuchang bream	133.4	750.3	Improved weight gain, feed efficiency, antioxidant enzyme activities, and immune response, and resistance to heat stress test (34°C)	Ming et al. (2012)

Nutrient			Species			Effect	Reference
			Blunt snout bream	6.4	133.7–251.5	Improved antioxidant enzyme activities and immune responses post pH challenge (pH 9.5)	Wan et al. (2014)
			Red sea bream	2.0	573.7	Improved weight gain, feed efficiency, survival, immune responses, and resistance to low salinity challenge	Dawood et al. (2016)
			Red sea bream	2.0	403.2	Improved weight gain, feed efficiency, immune responses, and resistance to low salinity challenge	Dawood et al. (2017)
			Blunt snout bream	1.8	1,198–1,525 (supplemented)	Improved growth and liver choline storage, and reduced liver lipid content	Jiang et al. (2013)
	500	600–1,000	Blunt snout bream	9.8	1,782	Improved feed efficiency when fed high-lipid diets (11% lipid)	Li, J.Y., et al. (2014)
			Blunt snout bream	42.2	2,232	Improved lipase activity when fed high-lipid diets (15% lipid)	Li, J.Y., et al. (2016)
			Black sea bream	8.2	3,000 of CC (supplemented)	Reduced lipid content in whole-body and muscle when fed high-lipid diet (16% lipid)	Jin et al. (2019)
Choline			Blunt snout bream	43.7	2,000–4,000 of CC (supplemented +12,000–8,000 betaine)	Improved some antioxidant enzyme activities when fed high-lipid diet (11%)	Adjoumani et al. (2019)
Inositol	NR	–	–	–	–	–	–
	–		Red porgy	226.9	92.6	Improved skin pigmentation 120-d prior to harvest	Kalinowski et al. (2011)
Astaxanthin	–	50–150	European sea bass	0.4	100	Improved immune responses and higher survival post osmotic stress test	Saleh et al. (2018)

Table 6.5 (continued)

Vitamins (IU or mg/kg diet)	NRC (2011)	Literature covered by NRC (2011)	Optimum Vitamin Nutrition® dsm-firmenich 2022	Species*	Initial weight (g)	Vitamin dietary content in published research	Benefits	References
				Asian sea bass	28.1	49.0	Improved weight gain, feed efficiency	Lim, K.C., et al. (2019a)
				Asian sea bass	28.2	49.0	Improved hematology, serum biochemistry, and immune responses	Lim, K.C., et al. (2019b)
				Largemouth bass	15.3	75–150	Improved weight gain and some antioxidative responses induced by high-lipid diet (18.1%)	Xie et al. (2020)
				Asian sea bass	28.3	122–150	Improved survival, serum biochemistry, and immunity indices post Vibrio alginolyticus challenge	Lim, K.C., et al. (2021)
				Spotted sea bass	30.6	58.4–91.5	Higher weight gain and improved survival and immune responses post Vibrio harveyi challenge	Yu et al. (2021)

Notes: CC, choline chloride; DHA, docosahexaenoic acid; dph, days post-hatch; TBARS; thiobarbituric acid-reactive substances.

* Dietary requirement values covered Asian sea bass/barramundi (*Lates calcarifer*) and European sea bass (*Dicentrarchus labrax*). Requirements were listed as not tested, required, or not required for some species.

† The scientific names of the species were European sea bass (*Dicentrarchus labrax*), hybrid striped bass (*Morone chrysops × Morone saxatilis*), Japanese sea bass (*Lateolabrax japonicus*), largemouth bass (*Micropterus salmoide*), spotted sea bass (*Lateolabrax maculatus*), striped bass (*Morone saxatilis*), black sea bream (*Acanthopagrus schlegelii*), red sea bream (*Pagrus major*), Wuchang/blunt snout bream (*Megalobrama amblycephala*), gilthead sea bream (*Sparus aurata*).

1,000 IU Vitamin A (retinol) = 0.3 mg retinol = 0.344 mg retinyl acetate = 0.55 mg retinyl palmitate

1,000 IU Vitamin D_3 (cholecalciferol) = 0.025 mg vitamin D_3

Vitamins (IU or mg/kg diet)*	NRC (2011)	Optimum Vitamin Nutrition® dsm-firmenich 2022	Species†	Initial weight (g)	Dietary vitamin content in published research	Benefits	References
A (IU)	–	8,000–12,000	–	–	–	–	–
D (IU)	–	1,500–2,000	–	–	–	–	–
E	–	150–300	Japanese eel	15.0	<21.6 of TA	Improved weight gain, feed efficiency and increased whole-body vitamin E storage	Bae et al. (2013)
			Japanese eel	360	212.9	Increased liver vitamin E storage, improved some serum chemistry and immunity indices	Shahkar et al. (2018)
K	–	3.0–6.0	–	–	–	–	–
B₁ (thiamine)	–	15–25	–	–	–	–	–
B₂ (Riboflavin)	–	20–30	–	–	–	–	–
Niacin (B₃)	–	80–120	–	–	–	–	–
Pantothenic acid (B₅)	–	50–60	–	–	–	–	–
B₆ (Pyridoxin)	–	10–15	–	–	–	–	–
Biotin (B₇)	–	0.3–0.5	–	–	–	–	–
Folate (B₉)	–	4.0–6.0	–	–	–	–	–
B₁₂ (Cobalamin)	–	0.1–0.2	–	–	–	–	–
C	–	150–900	Japanese eel	15.0	41.1–43.9	Improved weight gain, feed efficiency, survival	Bae et al. (2012)
			Japanese eel	360	410.7 and 911.7	Increased liver and testes vitamin C contents	Shahkar et al. (2015)
Choline	–	800–1,200	–	–	–	–	–
Inositol	–	–	–	–	–	–	–

Notes: TA, DL-α-tocopheryl acetate.

* No recommendation from NRC (2011) or literature covered by NRC (2011).

† The scientific name of the species was Japanese eel (*Anguilla japonica*).

1,000 IU Vitamin A (retinol) = 0.3 mg retinol = 0.344 mg retinyl acetate = 0.55 mg retinyl palmitate

1,000 IU Vitamin D₃ (cholecalciferol) = 0.025 mg vitamin D₃

Table 6.7 A summary of the recommended dietary vitamin content for other farmed warm water species (excluding ornamental fish) since NRC (2011).

Vitamins (IU or mg/kg diet)	Literature covered by NRC (2011)	Species*	Initial weight (g)	Dietary vitamin content in published research	Benefits	References
A (IU)	18,933	Japanese yellowtail	–	–	–	–
	9,000	Olive flounder	–	–	–	–
	3,101	Orange-spotted grouper	7.4	2,242–4,102	Improved weight gain, feed efficiency, lipolytic enzyme activities, and immune response	Yang, Q., et al. (2017)
	–	Dourado	18	8,818	Improved weight gain	Koch et al. (2018)
D (IU)	NR	Japanese yellowtail	–	–	–	–
	–	Siberian sturgeon	3.5	1,403–1,683	Improved growth and increased serum osteocalcin concentration and vitamin D metabolites	Wang, L., et al. (2017)
	119	Yellowtail kingfish	47.5	180	Increased lysozyme activity and fillet vitamin content	Le et al. (2014a,b)
	104–115	Malabar grouper	–	–	–	–
	31	Red drum	–	–	–	–
E	–	Yellow drum	32.3	68.8	Improved growth and some antioxidative enzyme activities and higher survival post *Vibrio alginolyticus* challenge	Wang L. et al. (2019)
	–	Snakehead	5.0	140–169 of DL-α-tocopheryl acetate	Improved weight gain, feed efficiency and hematologic parameters	Abdel-Hameid et al. (2012)
	–	Cobia	6.1	78–111 of DL-α-tocopheryl acetate	Improved weight gain, feed efficiency, survival, liver vitamin E storage, and immune response	Zhou, Q.C., et al. (2013)

Vitamin		Species			Performance	Reference
–	–	Olive flounder	1.1	73.2 (+431.9 vitamin C)	Maintained production performance	Gao et al. (2014)
–	–	Silver pomfret	29.6	196.3	Improved weight gain, feed efficiency	Hossain et al. (2016)
–	–	Meager	62.9	285–451 of DL-α-tocopheryl acetate	Maintained normal liver morphology and improved muscle TBARS level	Lozano et al. (2017)
–	–	Meager	79.3	450 (+230 vitamin C, +23 vitamin K)	Improved growth, tended to reduce granulomas, and upregulated some gene expressions of antioxidant enzyme ($tnf\alpha$ and cat)	Ruiz et al. (2019)
K	–	–	–	–	–	–
B_1 (Thiamine)	11.2	Japanese yellowtail	–	–	–	–
	–	Golden pompano	11.2	12.6–12.9	Increased gut microbiome diversity and improved production performance, liver vitamin B_1 storage, and some digestive enzyme activities and immune responses	Xun et al. (2019)
	–	Snakehead	4.9	2.3–2.6	Improved growth, liver vitamin B_1, storage, and some antioxidant enzyme activities	Zehra and Khan (2018d)
B_2 (Riboflavin)	11	Japanese yellowtail	–	–	–	–
	–	Snakehead	4.8	5.7–7.7	Improved weight gain, feed intake, feed efficiency, liver vitamin B_2, storage, and some antioxidant enzyme activities	Zehra and Khan (2018c)
Niacin (B_3)	12	Japanese yellowtail	–	–	–	–
	–	Snakehead	4.7	37.1–52.3	Improved weight gain, feed efficiency, survival and liver vitamin B_3 storage, and hematologic parameters	Zehra and Khan (2018a)

Table 6.7 (continued)

Vitamins (IU or mg/kg diet)	Literature covered by NRC (2011)	Species*	Initial weight (g)	Dietary vitamin content in published research	Benefits	References
Pantothenic acid (B$_5$)	35.59	Japanese yellowtail	–	–	–	–
	–	Malabar grouper	15.3	10.4–11.9	Improved weight gain, feed efficiency, survival and liver coenzyme A and vitamin B$_5$ storage	Lin et al. (2012)
B$_6$ (Pyridoxin)	11.7	Japanese yellowtail	–	–	–	–
	–	Snakehead	4.7	7.6–10.4	Improved weight gain, feed intake, feed efficiency, survival, and liver vitamin B$_6$ storage	Zehra and Khan (2018b)
Biotin (B$_7$)	0.67	Japanese yellowtail	–	–	–	–
Folate (B$_9$)	1.2	Japanese yellowtail	–	–	–	–
	–	Malabar grouper	7.3	0.7–0.9	Improved weight gain, feed efficiency and some antioxidant enzyme activities and immune responses	Lin et al. (2011)
	–	Siberian sturgeon	4.4	1.8–4.4	Improved weight gain, feed efficiency and liver vitamin B$_9$ storage	Falah et al. (2020)
	–	Flathead gray mullet	39.1	30–50 (supplemented)	Improved growth and maintained hematologic parameters	Badran and Ali (2021)
B$_{12}$ (Cobalamin)	0.053	Japanese yellowtail	–	–	–	–

		Species				
C	14–122	Japanese yellowtail	–	–	–	–
	8.3–46.2	Malabar grouper	–	–	–	–
	28–93	Olive flounder	–	–	–	–
	15	Red drum	–	–	–	–
	44.7–53.9	Cobia	5.5	13.6–96.6	Improved weight gain, feed efficiency, survival, liver vitamin C storage, some antioxidant enzyme activities and immune responses, and higher survival post *Vibrio harveyi* challenge	Zhou, Q., et al. (2012)
	28.2–87	Chu's croaker	14.2	71.5–176.2	Improved weight gain, feed efficiency, survival, vitamin C storage, and some antioxidant enzyme activities	Zou et al. (2020)
	–	Chinese sucker	7.1	84.6–126.1	Improved growth, vitamin C storage, and antioxidant enzyme activities	Huang, F., et al. (2017)
	–	Meager	79.3	230 (+450 vitamin E, +23 vitamin K)	Improved growth, tended to reduce granulomas, and upregulated some gene expressions of antioxidant enzyme (*tnfα* and *cat*)	Ruiz et al. (2019)
	–	Pacific bluefin tuna	0.27	453.8	Improved vitamin C storage in liver and brain	Biswas et al. (2013)
	–	Silver pomfret	6.2	455.3–800.5	Improved vitamin C storage in liver and muscle, and higher survival post transportation stress test	Peng et al. (2013)
Choline	696	Cobia	–	–	–	–
	588	Red drum	–	–	–	–
	598–634	Yellow perch	–	–	–	–
	2,920 of CC	Yellowtail kingfish	153.3–157.3	1,930–1,940 (digestible basis)	Improved weight gain, feed intake, feed efficiency, lipid digestibility, and maintained normal liver morphology	Liu, A., et al. (2019; 2020; 2021b)
	–	Giant grouper	9.3	>2,940	Improved survival post ammonia challenge	Yeh et al. (2015)

Table 6.7 (continued)

Vitamins (IU or mg/kg diet)	Literature covered by NRC (2011)	Species*	Initial weight (g)	Dietary vitamin content in published research	Benefits	References
	–	Orange-spotted grouper	87.9	1,093.7–1,579.7	Improved growth and liver choline storage	Qin, D.G., et al. (2017)
	1,700–3,200	Siberian sturgeon	37.7	1,500	Improved weight gain, feed efficiency	Yazdani Sadati et al. (2014)
	–	Olive flounder	5.9	847–1,047	Improved weight gain, feed efficiency, survival, and liver choline storage (reared at 15°C)	Won et al. (2019)
	–	Chinese sucker	5.5	493.8–549.7	Improved weight gain, feed efficiency and choline storage	Lu et al. (2020)
	–	Snakehead	14.6	50–100 (supplemented)	Reduced Se accumulation in tissues, improved some antioxidant enzyme activities, and reduced hepatic gene expression of HSP70 and HSP90 when exposed to high Se concentration	Li, M.Y., et al. (2020)
	–	Large yellow croaker	33.3	37.5 (supplemented)	Improved skin color	Yi et al. (2014)
Astaxanthin	–	Golden pompano	23.7	200 (supplemented)	Improved weight gain, feed efficiency, survival, and some liver antioxidant enzyme activities (in vitro and in vivo)	Xie, J.J., et al. (2017)
	–	Loach	3.0	100–151 (supplemented)	Improved weight gain, feed efficiency, antioxidant enzyme activities, and immune responses	Chen, X.M., et al. (2020)

Notes: CC, choline chloride; HSP, heat shock protein; HIF-1α, hypoxia-inducible factor-1α; NR, not required; TBARS, thiobarbituric acid-reactive substances.

* Beluga sturgeon (*Huso huso*), Chinese sucker (*Myxocyprinus asiaticus*), Chu's croaker (*Nibea coibor*), cobia (*Rachycentrol canadum*), dourado (*Salminus brasiliensis*), flathead gray mullet (*Mugil cephalus*), giant grouper (*Epinephelus lanceolatus*), golden pompano (*Trachinotus ovatus*), Japanese yellowtail (*Seriola quinqueradiata*), large yellow croaker (*Larimichthys crocea*), loach (*Paramisgurnus dabryanus*), Malabar grouper (*Epinephelus malabaricus*), meager (*Argyrosomus regius*), olive flounder (*Paralichthys olivaceus*), orange-spotted grouper (*Epinephelus coioides*), Pacific bluefin tuna (*Thunnus orientalis*), red drum (*Sciaenops ocellatus*), Siberian sturgeon (*Acipenser baerii*), silver pomfret (*Pampus argenteus*), snakehead (*Channa punctatus*), yellowtail kingfish (*Seriola lalandi*), yellow drum (*Nibea albiflora*), yellow perch (*Perca flavescens*).

1000 IU Vitamin A (retinol) = 0.344 mg vitamin A acetate

1000 IU Vitamin D₃ (cholecalciferol) = 0.025 mg vitamin D₃

Chapter 7

Optimum vitamin nutrition for shrimps

INTRODUCTION

The terms 'shrimp' and 'prawn' are often used interchangeably in aquaculture albeit their anatomical differences. To be consistent with the existing literature, we will refer to 'shrimp' as an umbrella term for species of the genus *Litopenaeus, Penaeus, Marsupenaeus, Fenneropenaeus,* and *Macrobrachium* in this chapter.

The metabolic functions and deficiency symptoms from vitamin and vitamin-like compounds are not well described in shrimp (Table 7.1). Studies on several vitamins remain absent with research focus skewed toward selected vitamins. For instance, no new study has been published on vitamins A, B_5 (pantothenic acid), B_7 (biotin), and B_{12} (cobalamin) in shrimp over the last 2 decades. In contrast, 9 studies on the vitamin E in shrimp have been published since 2012 and 6 on vitamin C since 2016 (Table 7.2).

It is challenging to ascertain the vitamin requirements of shrimp. The slow feeding rate and bottom-feeding nature of shrimp are conducive to vitamin leaching, although significant improvements concerning chemical and physical processing technologies have been developed over the years to protect vitamins from leaching (Gadient and Schai, 1994; Marchetti et al., 1999). Cannibalism makes the determination of precise vitamin requirements even more difficult (Romano and Zeng, 2017). Furthermore, shrimp are commonly farmed in ponds and can obtain vitamins from eating microorganisms and other invertebrates (Brown et al., 1999; Moss et al., 2006; Olsen et al., 2000; Phillips, 1984). Finally, a lack of understanding of the interactions among vitamins and other factors (e.g., nutrients, temperature, salinity, crowding) in shrimp limits the opportunity to attain optimal production performance under suboptimal culturing conditions. Because of the reasons listed above, it is common practice to over-formulate shrimp – and even fish – feeds with vitamins, especially the water-soluble ones (Aldrich, 2007; Cahu, 1999; Conklin, 1989; Hardy and Brezas, 2022; Rokey, 2007).

In this chapter, the Optimum Vitamin Nutrition® (OVN™) levels provided for shrimp farming are intended for applications under typical industry practices because the farming condition in ponds is not always optimal and can fluctuate (e.g., water temperature, salinity, density, animal size, dissolved oxygen). Compared to the recommended levels in the literature, industry generally supplement levels exceeding the levels needed to prevent clinical deficiency signs. The OVN™ ranges intend to compensate for factors which can influence the animal's requirements and corresponding husbandry conditions, including feed intake and feeding frequency. These ranges are based on university and industry research (some confidential), published requirements, and on-farm practices. Higher inclusion levels should be added to the product to account for processing and shelf life storage losses to achieve the targeted intake. The listed OVN™ levels are general guidelines, and in all cases, national feed legislation must be followed.

LIPID-SOLUBLE VITAMINS: PHYSIOLOGICAL ROLES, DEFICIENCY SIGNS AND PROPOSED SUPPLEMENTATION LEVELS

Vitamin A

Vitamin A (retinoids) is an essential nutrient for several aquaculture species, including salmonids (Kitamura *et al.*, 1967a, b), tilapias (Guimarães *et al.*, 2014; Hu *et al.*, 2006), carps (Yang, Q.H., *et al.*, 2008; Zhang, L., *et al.*, 2017), and shrimp (He, H., *et al.*, 1992; Hernandez *et al.*, 2009; Reddy *et al.*, 1999). Vitamin A plays a prominent role in several developmental processes, including the vision process, embryonic development, and differentiation of cell types (Hernandez and Hardy, 2020; Liñán-Cabello *et al.*, 2002). These functions are widely accepted in fish (Mai *et al.*, 2022; NRC, 2011), but limited studies are available on shrimp.

Early studies have identified provitamin A carotenoids (e.g., β-carotene) in Penaeid shrimp and proposed a potential metabolic pathway to biosynthesize vitamin A (Liñán-Cabello *et al.*, 2002; Schiedt *et al.*, 1993). However, little evidence indicates that shrimp can synthesize enough vitamin A to meet their metabolic requirement. Therefore, vitamin A is typically supplemented in today's shrimp feeds. The need to supplement Pacific white shrimp (*Litopenaeus vannamei*) feeds with vitamin A was shown when 40 mg post-larvae (PL) fed a vitamin A-free diet exhibited lower growth than those fed a diet containing 4,800 IU vitamin A/kg diet (He, H., *et al.*, 1992). However, the study did not have comparable data or graded inclusion level of vitamin A indicating 4,800 IU vitamin A/kg diet would provide optimal growth or health for the PL. Shiau and Chen (2000) reported that giant tiger shrimp (*Penaeus monodon*) (initial weight = 0.7 g) required 8,362 IU/kg diet of dietary vitamin A to improve growth performance, which was almost double the "requirement" (4,800 IU/kg diet) (He *et al.*, 1992; NRC, 2011) of *L. vannamei* PL. The dietary vitamin A requirement (7,500 IU/kg diet) of 0.1 g Kuruma shrimp (*Marsupenaeus japonicus*) was somewhat similar to those of *P. monodon* (Hernandez *et al.*, 2009). The lower dietary vitamin A requirement of *L. vannamei* could be attributed to size, species, or experimental differences. For broodstock, a higher supplementation of vitamin A (15,000 IU/kg diet) was recommended for the development of the ovaries of *M. japonicus* (20.2 g) (Alava, 1993). In addition, high dietary vitamin A (9,520–1,119,560 IU/kg diet) negatively affected the growth and feed efficiency but not the survival of 1.0 g *P. monodon* (Shiau and Chen, 2000). The results indicated that hypervitaminosis A could occur in growing shrimp, but between 9,520 and 1,119,560 IU/kg diet is not toxic enough to be deadly (Shiau and Chen, 2000). Therefore, in a nutrient-replete and stress-free situation, dietary vitamin A inclusion level exceeding 9,520 IU/kg diet should be avoided for juvenile shrimp.

Research on vitamin A nutrition for farmed shrimp stagnated for more than a decade, and to the best of our knowledge, there has been no published study on vitamin A nutrition for farmed shrimp in recent years. There is also an apparent lack of clinical and pathological assessments, as none were reported in the aforementioned studies. A dependence on using general production metrics (e.g., weight gain, feed efficiency, survival) to identify vitamin A deficiency in farmed shrimp is not completely reliable. Therefore, future studies could incorporate health-related assessments and provide deeper insight into the metabolic functions of vitamin A in farmed shrimp species. Furthermore, the vitamin A inclusion levels should be revised and account for shrimp species, size, physiological status (e.g., molt cycle, reproduction status), and environmental variations that may influence their dietary requirements.

There is a need to assess how vitamin A might interact with vitamin E and zinc in shrimp. The synergistic effect between vitamin A and vitamin E to suppress the peroxidation of phosphatidylcholine has been demonstrated *in vitro* (Tesoriere *et al.*, 1996). This protection against phospholipid oxidation stress is of relevance in shrimp nutrition and deserves further attention. Finally, the pivotal role of zinc in the cellular transport of vitamin A has not been described in shrimp yet. The transport of vitamin A depends on the retinol-binding proteins family (Tanumihardjo et al., 2016; Mai et al., 2022), which can be impaired if the dietary zinc level is below the requirement in vertebrates (Mobarhan et al., 1992).

Based on early studies and existing industry practice, we suggest revising the NRC (2011) requirement levels (4,800–180,000 IU/kg diet) to a dietary inclusion level of 7,000–12,000 IU/kg diet (Table 7.2). We recommend broodstock shrimp feed with higher inclusion levels and juveniles with lower vitamin A inclusions to ensure optimal growth, welfare, and reproduction performance of farmed shrimp.

Vitamin D

The need for vitamin D_3 (cholecalciferol) supplementation has been demonstrated in several farmed shrimp species, including L. vannamei (He, H., et al., 1992) and P. monodon (Reddy et al., 1999) by removing ≈4,800 and 4,000 IU vitamin D_3/kg diet from the supplemented diet, respectively. However, lower growth was used as the primary indicator to infer vitamin D_3 deficiency, and no pathological assessment was carried out (He, H., et al., 1992; Reddy et al., 1999). While the essentiality of vitamin D_3 and its potential role in shrimp nutrition was recognized, there is a paucity of information on this vitamin (e.g., requirements, metabolism functions) in farmed shrimp.

In vertebrates, vitamin D_3 is a precursor to 1,25-dihydroxy-cholecalciferol (calcitriol), the active form of vitamin D_3, which stimulates calcium absorption from the intestine and regulates calcium–phosphorus movement between different tissues (Lock et al., 2010). The metabolic role of vitamin D_3 may also be necessary for mineralizing the exoskeleton of shrimp and other crustaceans. Although there is evidence that M. japonicus could absorb enough dissolved calcium from seawater to exclude dietary calcium (Deshimaru and Yone, 1978), marine shrimp may require a higher dietary vitamin D_3 to compensate for the lack of external calcium when reared in low salinity water. As such, quantifying the dietary vitamin D_3 requirement of other key shrimp species is warranted, considering that shrimp such as L. vannamei can be exposed to or cultured in low salinity water.

Shiau and Hwang (1994) reported that young P. monodon (0.2 g) reared in seawater required approximately 4,000 IU vitamin D_3/kg diet to significantly improve growth, whereas the whole-body calcium and phosporus concentrations were unaffected by increasing dietary vitamin D_3. More recently, Wen M. et al. (2015) showed that similar-sized L. vannamei (0.4 g) required 6,366 IU vitamin D_3/kg diet when reared at low salinity water. While growth was not affected, L. vannamei showed an increased whole-body deposition of ash, calcium, phosphorus, and magnesium when fed an increasing dietary vitamin D_3 under a low salinity environment. Higher deposition in minerals and lower growth might indicate that shrimp prioritized available energy to form exoskeleton over muscle growth in low salinity conditions. Interestingly, Dai et al. (2021) reported that feeding a much higher dietary vitamin D_3 level (19,200 IU/kg diet) to similar-sized L. vannamei (0.5 g) could improve growth, feed efficiency, total antioxidant capacity (T-AOC), catalase activity, and increased carapace calcium and phosphorus concentrations and relative mRNA expressions of lysozyme in hepatopancreas. However, the experimental designs of Wen M. et al. (2015) and Dai et al. (2021) were different (e.g., rearing system, highest tested vitamin D_3 level, salinity). Further research is needed to confirm the functions of vitamin D_3 in shrimp, especially regarding its potential metabolic roles during molting (Lemos and Weissman, 2021). Vitamin D_3 is the predominant chemical form of vitamin D supplement in aquaculture diets. The intermediate metabolite of vitamin D_3, $25OHD_3$, has never been tested in shrimp and may prove beneficial compared to vitamin D_3.

In summary, available information on vitamin D$_3$ in shrimp is lacking, and the dietary requirement should be revised. Based on the updated literature and industry practices, we recommend adjusting the dietary vitamin D$_3$ inclusion level for major farmed shrimp species from the NRC (2011) value to 4,000–6,500 IU/kg diet (Table 7.2).

Vitamin E

Vitamin E (α-tocopherol) nutrition in shrimp has received substantial research attention in the last decade compared with other vitamins. Dietary vitamin E is required to improve the growth of shrimp (He, H., et al., 1992; Lee and Shiau, 2004; Reddy et al., 1999). However, unlike fish, there is no published record of pathological changes in vitamin E-deficient shrimp. Therefore, it is necessary to determine the clinical/pathological symptoms of vitamin E deficiency in shrimp and to establish biomarker(s) to aid with on-farm diagnosis of their nutritional and health status.

Vitamin E is well known for its role as an antioxidant for cell membranes that are prone to peroxidation (Combs and McClung, 2017; El-Sayed and Izquierdo, 2022). Thus, in fish, a higher dietary vitamin E level was needed to mitigate lipid peroxidation when they were fed high-lipid diets (El-Sayed and Izquierdo, 2022). While lipid peroxidation-induced pathological changes are less likely to occur in shrimp because shrimp feeds do not contain much PUFA, environmental changes (e.g., variable salinity, low oxygen events, contaminant) can induce oxidative stress. There is growing evidence that dietary vitamin E is needed to enhance the antioxidant capacity and immune responses of *P. monodon*, *L. vannamei*, and *M. rosenbergii* (Dandapat et al., 2000; Lee and Shiau, 2004; Liu, Y., et al., 2007). For example, Liu et al. (2007) reported that *L. vannamei* exposed to rapid salinity change (from 30‰ to 5–50‰ within 24 h) after being fed a practical diet containing 100 mg vitamin E/kg diet had increased muscle superoxide dismutase (SOD), catalase, glutathione peroxidase (GPX), and gill Na$^+$/K$^+$-ATPase activities than those fed low dietary vitamin E. The authors suggested that vitamin E could be an effective antioxidant against acute osmotic disturbance in shrimp. Oriental river prawn (*Macrobrachium nipponense*) fed 94 mg vitamin E/kg diet showed improved growth and changes in hepatopancreas SOD and catalase activities compared with those fed vitamin E-deficient diets (Zhao et al., 2016). Male *M. nipponense* fed a diet supplemented with 80–160 mg vitamin E/kg diet showed improved growth, lysozyme activity, MDA, and T-AOC in the hepatopancreas (Li, Y., et al., 2019). *M. nipponense* also had higher survival when challenged by *Aeromonas hydrophila* for 12-hours (Li, Y., et al., 2019). Higher dietary vitamin E levels of 100.8 to 184.4 mg/kg diet for female *M. nipponense* was suggested by Li, Y., et al. (2018) to improve the immune response (e.g., lysozyme activity) and antioxidant capacity (e.g., MDA, catalase activity). These results further indicated that dietary vitamin E is needed to improve the immunity and antioxidant status of farmed shrimp. Future studies should investigate the efficacy and interactive effect of this vitamin with other nutrients against external stressors.

The role of vitamin E at reproduction and larval stages has received considerable attention in fish nutrition and is growing in shrimp nutrition research. Because the metabolic rates in rapidly growing tissues are high, free-radical production might increase in fast-growing shrimp (El-Sayed and Izquierdo, 2022). Consequently, higher dietary vitamin E content is expected in broodstock or larval feeds. For example, Nguyen et al. (2012) reported that pond-reared female *M. japonicus* broodstock fed with increasing dietary vitamin E and vitamin C significantly improved the hatching rate (>80%) and per cent of larval reaching zoea I (>90%) (Figure 7.1). The authors demonstrated that vitamins E and C interact synergistically and recommended broodstock feeds to contain 600 mg vitamin E/kg diet and 1,000 mg vitamin C/kg diet to obtain better reproduction performance of *M. japonicus*.

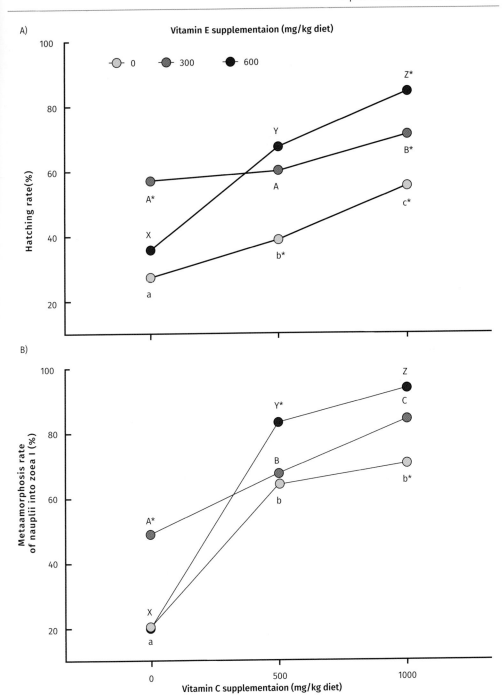

Figure 7.1 Effects of vitamin E and vitamin C supplementations on the reproductive performance parameters of *Marsupenaeus japonicus* broodstock: (A) hatching rate and (B) metamorphosis rate of nauplii into zoea I. For each vitamin E supplemented level, means with different letters are significantly different at *p*<0.05. For each dietary vitamin C level, means with an asterisk are significantly different from other groups (source: Nguyen *et al.*, 2012)

Vitamin E requirement is affected by interactions with other dietary components, particularly PUFA and selenium in mammals and fish (Brody, 1999; Hamre, 2011; Lall and Dumas, 2022). In shrimp, these interactions are presumed but have yet to be demonstrated experimentally (He and Lawrence, 1993a; Mai et al., 2022).

Early studies on the dietary vitamin E inclusion levels in shrimp and crustacean feeds ranged from 85 to 200 mg/kg diet (Alday-Sanz, 2010; Conklin, 1989; Lee and Shiau, 2004) (Table 7.2). By incorporating new information generated over the last decade and current industry practices, we suggest increasing the dietary levels of vitamin E to above the NRC (2011) recommendation (85–99 mg/kg diet) to values ranging between 150 to 300 mg/kg diet to avoid deficiencies, improve production performance, immune status, and stress tolerance.

Vitamin K

There are very limited studies on vitamin K in shrimp nutrition, and research in this area of study has not progressed for more than 2 decades until a recent study conducted by Dai et al. (2022b). As such, there is a clear need to revise the dietary vitamin K requirement of farmed shrimp and expand our understanding of the functions of this vitamin in shrimp. Naturally occurring vitamin K vitamers (i.e., phylloquinone and menaquinones) are lipid-soluble, while the synthetic vitamin K_3 (menadione) is water-soluble and typically supplemented in commercial shrimp feeds as salts (e.g., menadione sodium bisulphite, menadione nicotinamide bisulphite) to maintain growth and health of the animal (Krossøy et al., 2011). While marine ingredients (e.g., fishmeal, fish oil) contain relatively low levels of vitamin K (0.01–2.2 mg/kg ingredient), using plant-based ingredients such as soybean oil (2.7 mg/kg ingredient) and canola oil (1.1 mg/kg ingredient) that generally contain more vitamin K_1 (phylloquinone), could change the dietary recommendation of vitamin K for farmed shrimp (Krossøy et al., 2011; NRC, 2011; Ostermeyer and Schmidt, 2001; Woollard et al., 2002). However, currently, there is no study that investigated the role of different vitamin K forms and their bioavailability in shrimp.

Early reports on the dietary vitamin K requirements were inconsistent in farmed shrimp. Growth and survival were not suitable indicators to determine dietary vitamin K requirements because both He H. et al. (1992) and Reddy et al. (1999) were unable to demonstrate a positive response on PL L. vannamei (0.04 g) and P. monodon (0.09 g) fed a vitamin K-free semi-purified diet. On the other hand, using protein efficiency ratio (PER) and VKD protein precursor concentration as indicators, the dietary vitamin K requirement of P. monodon (0.4 g initial weight) was 31.8–40.2 mg/kg diet (Shiau and Liu, 1994a). However, Shiau and Liu (1994b) suggested that fleshy prawn (Fenneropenaeus chinensis) (1.3 g initial weight) required higher vitamin K_3 (183.0–185.4 mg/kg diet) to obtain better growth, feed efficiency, and PER. The authors pointed out several factors, including species, size, oil source, and study duration, that could have resulted in the disparity in vitamin K_3 requirements in shrimps. In addition, the studies tended to estimate the vitamin K requirements based on the menadione content of the experimental diets rather than the total vitamin K content of the diets. Consequently, the estimated requirement could be different.

To optimize vitamin supplementation and reduce feed cost/waste, information is needed on the metabolic role of vitamin K in shrimp (e.g., plasma prothrombin synthesis, calcium transport), synthesis by the gut microbiota and content as well as bioavailability of naturally occurring vitamin K in the feed ingredients (Krossøy et al., 2011). Based on early studies and current industry practices, we recommend supplementing 40–60 mg vitamin K_3/kg diet in shrimp feeds to maintain optimal growth and health of the shrimp (Table 7.2).

WATER-SOLUBLE VITAMINS: PHYSIOLOGICAL ROLES, DEFICIENCY SIGNS AND PROPOSED SUPPLEMENTATION LEVELS

Vitamin B$_1$ (Thiamine)

Studies on thiamine metabolism and its functions in shrimp are in their infancy and warrant more research in this area to optimize commercial feed formulations. Thiamine is a highly active molecule, and TPP (coenzyme) participates in a series of decarboxylation and transketolase reactions critical in energy production in vertebrates as well as crustaceans. In a recent study, Sun et al. (2022) observed an increase in markers of glucose metabolism (e.g., hexokinase, glucose transporter 2) and lipogenesis (e.g., fatty acid synthase) in *M. nipponense* fed graded levels of thiamine (0.0–95.7 mg/kg diet). Because shrimp utilize much of their energy from carbohydrates, their dietary thiamine requirement(s) needs further consideration and would likely depend on the available carbohydrate content in the feed.

The NRC (2011) recommendation for the dietary thiamine requirement to prevent deficiency and promote growth and health of *M. japonicus* (0.5 g) and *P. monodon* (1.4 g) was 60–120 mg thiamine hydrochloride/kg diet (Deshimaru and Kuroki, 1979) and 14 mg thiamine hydrochloride/kg diet (Chen, H.Y., et al., 1991), respectively (Table 7.2). The thiamine requirement of *M. nipponense* was estimated at 66–68 mg/kg diet based on weight gain and hemolymph thiamine content (Sun, M., et al., 2022). However, depending on the purity of the supplement, the thiamine requirements would be lower because the estimates were based on thiamine hydrochloride and not the total thiamine content in the feed.

Apart from relatively lower growth, feed efficiency, and/or survival, the classical signs of thiamine deficiency (e.g., hyperirritability, ataxia) in vertebrates were not observed in existing shrimp studies. As such, identifying a sensitive biomarker for thiamine deficiency in farmed shrimp would be advantageous for diagnosis. Chen, H.Y., et al. (1991; 1994) suggested that thiamine and TPP levels in haemolymph could be used as thiamine biomarkers for *P. monodon* as they were sensitive to thiamine supplementation. A recent study reported that thiamine supplementation above 54 mg/kg diet significantly improved feed efficiency, PER, haemolymph SOD and CAT, and lysozyme (Huang, X.L., et al., 2015). Dietary thiamine requirements were 44.7 and 152.8 mg/kg diet based on weight gain and hepatopancreas thiamine content, respectively.

Based on updated literature and industry practices, we recommend moving the dietary thiamine inclusion level to 50–150 mg/kg diet (Table 7.2).

Vitamin B$_2$ (Riboflavin)

Quantitative information on the riboflavin (vitamin B$_2$) requirement of shrimp is almost non-existent. In vertebrates, riboflavin is a precursor to coenzymes, FAD and flavin mononucleotide (FMN) (Combs and McClung, 2017; Hansen et al., 2015). Like fish, *P. monodon* PL (0.1 g) fed a low riboflavin content diet (0.5 mg/kg diet) for 15 weeks exhibited short-head dwarfism (i.e., low carapace to body length ratio) (Chen and Hwang, 1992). Other signs of riboflavin deficiency in *P. monodon* were light coloration, irritability, protuberant cuticle at the joints of the abdominal somites (Chen and Hwang, 1992) and severe necrosis and detachment of epithelial cells in the mid-gut (Catacutan and De la Cruz, 1989).

The NRC (2011) recommendations for the dietary riboflavin requirement of juvenile *P. monodon* were 23.0 mg/kg diet (Chen and Hwang, 1992) and 80 mg/kg diet for *M. japonicus* (Table 7.2). The riboflavin requirement (40.9 mg/kg diet) was determined for the first time in *L. vannamei* PL recently (Sanjeewani and Lee, 2023). No prior riboflavin requirement was determined when Conklin (1989) recommended 40 mg/kg diet for shrimp. Chen and Hwang (1992) used whole-body riboflavin content as an indicator to estimate the dietary requirement because haemolymph glutathione reductase activity was not sensitive to dietary riboflavin. Growth and feed efficiency were not responsive to dietary riboflavin in *P. monodon* reared in saline (Chen and Hwang, 1992) and low salinity waters (Reddy *et al.*, 1999). For *L. vannamei*, sufficient dietary riboflavin improved their growth, but higher feed intake was the main driver given that FCR and the apparent digestibility of diet dry matter and protein were similar across treatments (1.4 to 55.1 mg riboflavin/kg diet) (Sanjeewani and Lee, 2023). *L. vannamei* fed ≈40 mg riboflavin/kg diet appeared to enhance SOD, GPX, and lysozyme activities compared with those fed 1.4 or 55.1 mg riboflavin/kg diet (Sanjeewani and Lee, 2023). Their results offered preliminary insight into riboflavin's potential participation in the glutathione redox cycle, however, the efficacy as a participant in shrimp's innate immune system should be tested under challenged condition.

Owing to the paucity of information and the fundamental roles of riboflavin in metabolism, the dietary riboflavin requirements of key farmed shrimp species are warranted. The riboflavin content in plant proteins is relatively lower than in fishmeal (Hansen *et al.*, 2015). Accordingly, commercial shrimp feeds should ensure adequate riboflavin supplementation when using plant-based or other novel ingredients. Furthermore, the dietary riboflavin inclusion level should account for leaching and potential loss from sunlight (if farmed in shallow ponds) because shrimp are slow feeders. Based on existing studies and industry practice, we recommend a dietary riboflavin inclusion level of 40–80 mg/kg diet for shrimp (Table 7.2). However, future research could focus on quantifying the dietary riboflavin requirements of different life stages of shrimp, identifying the vitamin's metabolic role, and quantifying potential riboflavin production from the gut microbiome.

Niacin (vitamin B$_3$)

Like riboflavin, information on niacin (vitamin B$_3$) in shrimp nutrition is scarce despite its essentiality in forming coenzymes, NAD and nicotinamide adenine dinucleotide phosphate (NADP) (Combs and McClung, 2017). Both NAD and NADP participate in cellular redox reactions that are fundamental to virtually all aspects of metabolism. In fish, the most pronounced deficiency signs are damage to skin and fins, such as hemorrhages, lesions, erosion, and dermopathies (Waagbø, 2010). However, pathological deficient symptoms were either not examined or observed in *P. monodon* (Shiau and Suen, 1994) and *L. vannamei* (Xia M.H. *et al.*, 2015), whereas darkening of gills was observed in *L. vannamei* reared at low salinity water (Reddy *et al.*, 1999) (Table 7.1).

The existing recommendations of dietary niacin inclusion levels for shrimp varied considerably and were inconsistent (NRC, 2011) (Table 7.2). For instance, the dietary niacin requirement of *P. monodon* was 7.2 mg/kg diet but the data fitted poorly using a broken-line regression, and the ANOVA results were variable (Shiau and Suen, 1994). A dietary niacin inclusion level of 40 mg/kg diet was suggested for crustaceans (Alday-Sanz, 2010) and 150 mg/kg diet for shrimp (Conklin, 1989). NRC (2011) recommended a 400 mg/kg diet for *M. japonicus*, and the source of the empirical data was absent, which makes verification challenging.

Until recently, dietary niacin requirement estimates on shrimp were based on production performance, such as growth, survival, and feed efficiency, with no measurements of enzyme activities of haemolymph. Xia, M.H., *et al.* (2015) showed that a dietary niacin inclusion level of 109.6–121.2 mg/kg diet improved the growth, feed efficiency, and lysozyme and catalase enzyme activity in the haemolymph of juvenile *L. vannamei* (0.8 g). The results indicated that optimal dietary niacin might enhance innate immunity and disease resistance of growing shrimp, and future research could test the efficacy toward external stressors (e.g., low salinity, suboptimal temperature, pathogen, high stocking density). Hasanthi and Lee (2023) estimated the niacin requirement of juvenile *L. vannamei* (0.4 g) at 130 mg/kg diet based on weight gain, feed conversion, immune parameters (e.g., phagocytic activity, lysozyme activity), antioxidant activities (e.g., glutathione peroxidase) and survival of shrimp challenged with *Vibrio parahaemolyticus*.

Dietary niacin supplementation could depend on interactions between niacin, tryptophan, and pyridoxine (vitamin B_6). Evidence suggested that the conversion of tryptophan into niacin was absent or insufficient in *L. vannamei* (Xia, M.H., *et al.*, 2015). However, the conversion of tryptophan into niacin can be enhanced by pyridoxine supplementation in mammals (Shibata *et al.*, 1995). To our knowledge, this positive interaction has not been verified in shrimp.

While the existing studies on dietary niacin inclusion levels in shrimp feeds varied widely from 7.2–400 mg/kg diet, based on the literature and industry practice, we recommend a dietary niacin inclusion level of 100–250 mg/kg diet to ensure the growth and health of farmed shrimp are not compromised (Table 7.2).

Pantothenic acid (vitamin B_5)

There were only 2 published studies on pantothenic acid (vitamin B_5) in shrimp nutrition, and no new information has been available since 1999. Pantothenic acid is required for the synthesis of coenzyme A (CoA); therefore, it plays a critical role in fatty acids, amino acids, and carbohydrate metabolism (Combs and McClung, 2017). In shrimp, reported clinical signs of pantothenic acid deficiency were irritability and light coloration after 4 weeks of feeding a low pantothenic acid diet (0.02 mg/kg diet) (Shiau and Hsu, 1999). The shells of *P. monodon* fed the non-supplemented diet were thinner and softer than those of the pantothenic acid supplemented groups (Shiau and Hsu, 1999). Because shrimp is particularly vulnerable during the post-molt stage, the dietary pantothenic acid inclusion level in shrimp feed under commercial farming conditions needs to be sufficient to optimize the production performance and welfare of the animal. As such, the optimal dietary pantothenic acid inclusion levels in shrimp feeds should be revised and tailored to the life stage and cultured environment of the farmed shrimp species.

An early study on determining the pantothenic acid requirements reported that 100 mg pantothenic acid/kg diet was sufficient for *F. chinensis* to improve survival and growth (Liu, T., *et al.*, 1995); however, only 2 dietary pantothenic acid levels (50 and 100 mg/kg diet) were used in the study. Therefore, the dietary pantothenic acid requirement should be revised and increase the number of inclusion levels tested. Shiau and Hsu (1999) suggested that *P. monodon* (0.9 g) required 101.4–139.9 mg pantothenic acid/kg diet to optimize growth, and hepatopancreatic CoA and pantothenic acid content. Unfortunately, the requirement estimated based on weight gain (139.9 mg/kg diet) had low goodness of fit value which indicated that the dietary pantothenic acid requirement of *P. monodon* was likely lower than 120 mg/kg diet (Shiau and Hsu, 1999).

While there is a lack of recently published information on pantothenic acid in shrimp nutrition (e.g., metabolism, function, requirement), by incorporating industry knowledge and practices, we recommend a dietary pantothenic acid inclusion level of 100–180 mg/kg diet to ensure the production performance and welfare of farmed shrimp are not compromised until more data become available (Table 7.2).

Pyridoxine (vitamin B_6)

Vitamin B_6 in its coenzyme form, pyridoxal phosphate (PLP) and its aminated form, pyridoxamine phosphate (PMP), is pivotal to protein catabolism, particularly in the transamination and deamination of amino acids (Combs and McClung, 2017; Mai *et al.*, 2022). These biochemical functions and the metabolism of vitamin B_6 have yet to be demonstrated in shrimp. Clinical signs of vitamin B_6 deficiencies observed in fish, such as nervous disorders, have not been reported in shrimp. Symptoms associated with vitamin B_6 deficiencies in shrimp were slow growth, poor survival, low feed efficiency and decreased pyridoxine concentration in hepato-pancreas (Deshimaru and Kuroki, 1979; Boonyaratpalin, 1998; Shiau and Wu, 2003).

For shrimp, studies on vitamin B_6 requirements varied from 60 to 200 mg/kg inclusion in the diet. Deshimaru and Kuroki (1979) reported that 120 mg pyridoxine hydrochloride/kg diet was required for *M. japonicus* to improve weight gain and survival. However, depending on the purity of the supplemental vitamin B_6 used in that experiment, the calculated dietary vitamin B_6 requirement would be less than 100 mg/kg diet for *M. japonicus*. Shiau and Wu (2003) reported that a level of 89.5 mg vitamin B_6/kg diet was needed to improve *P. monodon* growth. For the first time, the authors used hepatopancreatic glutamate–oxaloacetate transaminase (GOT) enzyme activity to quantify the dietary vitamin B_6 requirement in shrimp. Although the estimate (71.4 mg/kg diet) was slightly lower than using weight gain as an indicator, the authors suggested that the difference might be permitted as a biomarker for estimating vitamin B_6 status in farmed shrimp (Shiau and Wu, 2003). Similarly, when GOT and glutamic pyruvic trans-ferase (GPT) enzyme activities were used as biomarkers, Li, E., *et al.* (2010) showed that the dietary vitamin B_6 requirements of *L. vannamei* PL reared under low salinity (3‰) were 107.0 and 108.3 mg/kg diet, respectively. The estimated requirements were lower than using growth (133.3–151.9 mg/kg diet) as the primary response to evaluate vitamin B_6 status. In addition, GOT and GPT activities decreased after reaching the vitamin B_6 requirement level in both studies (Li, E., *et al.*, 2010; Shiau and Wu, 2003). When performing an on-farm evaluation of vitamin B_6 status in shrimp, low GOT and GPT activities could indicate that either the animal is deficient or sufficient in vitamin B_6, which could be misleading if not interpreted with additional data such as growth. Juvenile *L. vannamei* fed a practical diet and reared in higher salinity (26–29‰) water required less dietary vitamin B_6 for growth (110.1–110.4 mg/kg diet) than PL at low salinity water (Figure 7.2) (Cui *et al.*, 2016; Li, E., *et al.*, 2010). The difference in the reported vitamin B_6 requirements could be attributed to the basal diet composition, shrimp size, water salinity, nutrient leaching etc. Other recommended dietary content of vitamin B_6 in shrimp feeds were 60 mg/kg diet (Conklin, 1997), 70 mg/kg diet (Alday-Sanz, 2010), and 100–200 mg/kg diet.

More recently, Cui, P., *et al.* (2016) reported that *L. vannamei* fed dietary vitamin B_6 of 90 and 119 mg/kg diet increased the SOD and lysozyme activities in the haemolymph. The results provided some evidence that vitamin B_6 could enhance the antioxidant enzyme activity and nonspecific immunity in farmed shrimp. Future research could explore the functionality of vitamin B_6 in response to external stressors.

Based on existing literature and industry practice, we recommend a dietary vitamin B_6 inclusion level of 50–175 mg/kg diet to ensure the growth and health of farmed shrimp are not compromised (Table 7.2).

Folate (vitamin B_9)

Folate, as a coenzyme (tetrahydrofolate; THF) and a methyl donor, is a vital participant in the one-carbon metabolism in animals. Vitamin B_6, vitamin B_{12}, methionine, and folate are essential for DNA synthesis, antioxidant generation, and epigenetic regulation in animals (Lyon *et al.*, 2020; Naderi and House, 2018). While there is no publication on folate biochemistry and metabolism in shrimp, the functions of folate are likely to be similar between shrimp and other animals. Reports on folate deficiency symptoms beyond growth and mortalities in farmed shrimp are absent (Reddy *et al.*, 1999; Shiau and Huang, 2001). Thus, future research could describe folate deficiency-related clinical and pathological signs to improve the diagnosis of nutritional pathology in farmed shrimp.

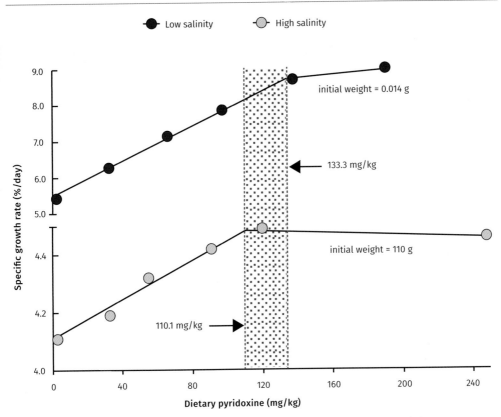

Figure 7.2 Relationship between dietary vitamin B6 content and specific growth rate of *Litopenaeus vannamei* reared at low salinity (3‰) and high salinity (26–29‰). Shaded area represents the range of dietary requirements estimated based on broken-line regression analysis (source: adapted from Li *et al.*, 2010 and Cui *et al.*, 2016)

While some fish, such as channel catfish (*Ictalurus punctatus*) and common carp (*Cyprinus carpio*), can obtain folate from intestinal bacteria (Duncan *et al.*, 1993; Kashiwada, 1971; Aoe *et al.*, 1967), information on the intestinal microbial synthesis of folate in shrimp is lacking. Until more data become available, folic acid supplementation in shrimp feeds is necessary because the folate content in many raw materials (0.1–4.3 mg/kg ingredient) is insufficient to meet the folate requirement of shrimp (NRC, 2011). The NRC (2011) recommended dietary folate requirements were 2.0 mg/kg diet for juvenile *P. monodon* (initial weight of 0.8 g) (Shiau and Huang, 2001) and 5.0 mg/kg diet for juvenile *F. chinensis* (Liu, T., *et al.*, 1995) (Table 7.2). The dietary folic acid requirement (2.2 mg/kg diet) of giant freshwater prawn (*M. rosenbergii*) was quantified in a more recent study (Asaikkutti *et al.*, 2016). The estimate was like that of *P. monodon*. Although present in minute amounts, *M. rosenbergii* PL were fed practical diets containing raw materials with folate. As a result, the total folate content of the diet would be slightly higher than the measured folate content in the feed.

Asaikkutti *et al.* (2016a) reported novel effects of folate in shrimp. For example, increasing dietary folate affected digestive enzyme activities (i.e., protease, amylase, lipase) and haemocytes. High folate intake increased the activities of antioxidant and metabolic enzymes in the muscle and hepatopancreas of *M. rosenbergii*, which, although unproven, could have impacted weight gain, survival, and digestive enzyme activities in this study. Therefore, possible adverse effects of excess folate supplementation in shrimp warrant further studies.

Based on industry practice, we recommend a dietary folic acid inclusion level of 10–20 mg/kg diet to ensure the growth and health of farmed shrimp are not compromised.

Biotin (vitamin B$_7$) and vitamin B$_{12}$ (cobalamin)

Like most other B complex vitamins, publicly available information on biotin and vitamin B$_{12}$ in shrimp nutrition is scarce, and there is a lack of progress in this area of study since Shiau and Chin (1998) and Shiau and Lung (1993), respectively. Because the metabolic roles of these vitamins in crustaceans are almost non-existent, information on biotin and vitamin B$_{12}$ relied heavily on mammalian and fish models. As such, we referred to Hansen *et al.* (2015), Yossa *et al.* (2015), and Mai *et al.* (2022) for a summary of the biochemistry and general deficiency symptoms in mammals and fish.

Biotin

Of the major farmed species, the biotin requirement for improved growth and/or feed efficiency was demonstrated in *P. monodon* (Reddy *et al.*, 1999; Shiau and Chin, 1998). Because no clinical or pathogenic sign was described, it would be difficult for researchers and producers to detect biotin deficiency in farmed shrimp during production. Biotin is a coenzyme for carboxylases such as pyruvate carbo carboxylase and acetyl-CoA carboxylase, and thus heavily involved in gluconeogenesis, fatty acid synthesis and oxidation, and function of the TCA cycle (Table 7.1). However, Shiau and Chin (1998) were unable to establish a dose-response of hepatopancreatic PC and acetyl-CoA carboxylase enzyme activities to changing dietary biotin levels. The authors suggested that these enzymes might not be suitable biomarkers for biotin deficiency detection in shrimp. Thus, weight gain and PER were used to determine the dietary biotin requirement (2.0–2.4 mg/kg diet) of juvenile *P. monodon* (0.3 g) (Shiau and Chin, 1998). Liu, T., *et al.* (1995) suggested a dietary requirement of 4.0 mg biotin/kg diet for *F. chinensis* whereas the recommended dietary biotin content (1.0 mg/kg diet) in crustacean feed was lower in comparison (Alday-Sanz, 2010; Conklin, 1989). Recently, Xia, M., *et al.* (2014) showed that juvenile *L. vannamei* required more than 0.068 mg biotin kg^{-1} diet to improve growth. While 0.4–9.4 mg biotin/kg diet did not yield significantly better production performance, the T-AOC, catalase, and acetyl-CoA carboxylase activity increased. The authors recommended biotin requirements of 0.7 and 2.25 mg/kg diet for juvenile *L. vannamei* based on feed efficiency and serum acetyl-CoA carboxylase activity, respectively (Xia, M., *et al.*, 2014). On this basis, species-specific vitamin requirement estimates should be established to maximize optimal production performance under on-farm rearing conditions.

Vitamin B$_{12}$ (cobalamin)

Vitamin B$_{12}$ (cobalamin) are coenzymes (methylcobalamin and adenosylcobalamin) that participate in the one-carbon cycle along with folate and methionine. Metabolic roles of vitamin B$_{12}$, such as coenzyme functions, DNA methylation, and interrelation with methionine and folate, that are critical to terrestrial animals have yet to be characterized in shrimp. Information on the synthesis of vitamin B$_{12}$ from gut bacteria in shrimp is also absent. Therefore, vitamin B$_{12}$ is typically supplemented as cobalamin in shrimp feeds. While the dietary vitamin B$_{12}$ requirement of juvenile *P. monodon* (0.6 g) was estimated at 0.2 mg/kg diet (Shiau and Lung, 1993), the requirement of this vitamin for *P. monodon* was questionable. These authors did not report the model fit for the second-order regression, and by replotting the data, the model did not appear to be a good fit and lacked an apparent response to dietary vitamin B$_{12}$. When using

ANOVA, the dietary requirement of vitamin B_{12} estimated based on per cent weight gain was significantly higher at 0.05 mg/kg diet in *P. monodon*. Consequently, if dietary vitamin B_{12} is needed, the requirement might be more aligned with giant freshwater prawn (*M. rosenbergii*) (<0.1 mg/kg diet) (Conklin, 1997) and *P. monodon* (Liu, T., et al., 1995).

Based on existing literature and industry practice, we recommend biotin levels of 1.0–2.25 mg/kg diet and vitamin B_{12} content of 0.02–0.05 mg/kg diet to ensure the growth and health of farmed shrimp are not compromised (Table 7.2).

Vitamin C (ascorbic acid)

Vitamin C (ascorbic acid) is the most extensively studied vitamin in shrimp nutrition because of its participation in critical physiological functions (Combs and McClung, 2017; Dawood and Koshio, 2018). Vitamin C is an antioxidant and participates in the synthesis of steroid hormones and collagen, and thus is important for reproduction, wound healing, and immune defense (Table 7.1). Unlike most mammals, birds and some fish, shrimp require an exogenous source of vitamin as they lack the enzyme, L-gulono-γ-lactone oxidase, to synthesize vitamin C from glucose (Lightner et al., 1979). Reported vitamin C-deficient signs in shrimp included blackened lesions in tissues, loss of appetite, poorer growth, higher mortality, less molt frequency, and reduced tolerance to stress (Lightner et al., 1977; Niu et al., 2009). There is a lack of established vitamin C biomarkers because publications containing information on clinical and histopathological responses were limited.

Depending on the form of supplemental vitamin C, early reports of the dietary vitamin C requirements in shrimp varied considerably from 40–2,000 mg/kg diet for *P. monodon* (Shiau and Hsu, 1994; Shiau and Jan, 1992), 43–10,000 mg/kg diet for *M. japonicus* (Guary et al., 1976; Moe et al., 2004), 90–191 mg/kg diet for *L. vannamei* (He, H., and Lawrence, 1993b; Niu et al., 2009), 600–2,000 mg/kg diet for *F. californiensis* (Lightner et al., 1979; Qin et al., 2007), 4,000–8,000 mg/kg diet for *F. indicus* (Boonyaratpalin, 1998), and 104–135 mg/kg diet for *M. rosenbergii* (D'Abramo et al., 1994; Hari and Kurup, 2002). However, modern commercial feeds generally use stabilized forms of vitamin C (e.g., L-ascorbyl-2-polyphosphate, L-ascorbyl-2-monophosphates) to minimize leaching and degradation. Based on the stabilized forms of vitamin C supplement, the inclusion levels of vitamin C recommended by NRC (2011) were 600 mg/kg[1] diet for *F. chinensis* and 50–100 mg/kg diet for *M. japonicus*, *L. vannamei*, and *P. monodon* (Table 7.2).

In the last decade, studies on vitamin C nutrition in shrimp have provided more evidence of its role in boosting the antioxidant and immune status, and disease resistance. For instance, monsoon river prawn (*M. malcolmsonii*) PL fed a practical diet containing 100 mg vitamin C/kg diet had higher growth, feed efficiency, digestive enzyme activities (i.e., protease, amylase, lipase), total haemocyte count, and antioxidant enzyme activities (i.e., SOD, CAT, GOT, GPT) in tissues than those fed low vitamin C diet (Asaikkutti et al., 2016b). *M. rosenbergii* PL fed a practical diet supplemented with 150 mg vitamin C/kg diet showed similar responses as *M. malcolmsonii* PL (Asaikkutti et al., 2018). Wu, Y.S., et al. (2016) also reported that *L. vannamei* fed a fishmeal-based diet supplemented with 200 mg vitamin C/kg diet and 1,3/1,6-glucan (1 g/kg diet) induced nonspecific immune responses, including superoxide anion, phenol oxidase, SOD, GOT, and GPT.

Vitamin C can serve as a chelating agent and interact with copper and iron in shrimp (Hsu, T.S., and Shiau, 1999b), like in mammals (Hungerford and Linder, 1983) and fish (Dabrowski and Köck, 1989; Dabrowski et al., 1990). Dietary supplementation with L-ascorbyl-2-polyphosphate and L-ascorbyl-2-monophosphates at ≥14 mg/kg diet (ascorbic acid equivalent) decreased the concentrations of copper and increased iron in the hepatopancreas and muscle of *P. monodon* (Hsu, T.S., and Shiau, 1999b). Therefore, vitamin C might serve as a functional nutrient to circumvent concerns raised by high levels of copper, such as tissue lipid peroxidation, while sparing iron concomitantly. Further studies are required to verify the interactive relationship between vitamin C and minerals in shrimp.

Based on existing literature and industry practices, we recommend a dietary vitamin C inclusion level of 250–500 mg/kg diet to optimize the growth, health, and stress tolerance of farmed shrimp (Table 7.2).

VITAMIN-LIKE NUTRIENTS: PHYSIOLOGICAL ROLES, DEFICIENCY SIGNS AND PROPOSED SUPPLEMENTATION LEVELS

Choline

Compared with other vitamins, research efforts on choline in shrimp nutrition have been substantial over the last few decades. While choline is ubiquitous in raw materials, choline supplementation remains a common practice to ensure optimal growth and health of shrimp under commercial farming conditions.

Choline is typically classified as "vitamin-like" nutrient in literature because most animals can synthesize this nutrient *de novo* or when the essentiality is conditional for certain species. For shrimp, the dietary essentiality of choline varied among studies and appeared to be conditional in *M. japonicus*. The choline requirement for growth and survival was reported in juvenile *M. japonicus* (0.5–1.0 g) (Kanazawa *et al.*, 1976) but not in another study with a similar study design (Deshimaru and Kuroki, 1979).

While choline has no known coenzyme function, this trimethylated compound is a methyl donor that participates in the one-carbon cycle. As such, changes in the dietary intake of other one-carbon participants (e.g., methionine, folate, vitamin B_6, vitamin B_{12}) can affect the dietary requirement of choline in animals (Zeisel, 1981, 2013). For instance, dietary methionine (>16.5 g/kg diet, uncommonly high level in feeds for aquatic species) spared most of the dietary choline requirement of juvenile and PL *M. japonicus*. The results were based on growth as the response criteria (Michael *et al.*, 2006, 2011). Likewise, dietary choline (1.4–1.5 g/kg diet) appeared to spare the need for methionine when *M. japonicus* were fed a methionine-deficient diet (Michael *et al.*, 2006, 2011). Although the conversion pathways were not demonstrated, the aforementioned studies provided evidence that shrimp can synthesize choline, most likely via the methylation of phosphatidylethanolamine (PE) like vertebrates. Thus, the interactive effect of choline with metabolically related nutrients could be one of the factors contributing to the highly variable dietary choline requirements of shrimp reported in early works (from 0 to 6,200 mg/kg diet) (Deshimaru and Kuroki, 1979; Gong *et al.*, 2003; Kanazawa *et al.*, 1976; Shiau and Lo, 2001). Future studies can attempt to elucidate the biosynthesis of choline and its role in the one-carbon metabolism in shrimp. In addition, future designs on choline nutrition should account for other key participants (e.g., vitamin B_6, folate, vitamin B_{12}) of the one-carbon cycle as their requirements might be influenced by each other.

The role of choline as a precursor to acetylcholine (neurotransmitter) has received relatively more attention in mammalian clinical research but not in aquatic animals. Rather, choline's role as the head group of phospholipids (e.g., phosphatidylcholine (PC), sphingomyelin) had been the primary focus in fish and shrimp nutrition for several decades. Likely because PC is generally the dominant phospholipid of cell membranes and serves as the main lipid moiety of the haemolymph lipoprotein transport system in crustaceans (Lee and Puppione 1978; Teshima and Kanazawa, 1980). Shiau and Cho (2002) provided further evidence in choline's role in lipid transport by demonstrating higher dietary choline (7,840 mg/kg diet) was needed in juvenile

P. monodon when fed a diet with 11% lipid compared to 5% lipid (requirement = 6,440 mg choline/kg diet). Given that commercial shrimp feeds usually contain <10% lipid, higher dietary choline might not be warranted under optimal culturing conditions.

In recent years, novel functions of choline such as osmoregulation and antioxidation capacity have been evaluated in shrimp nutrition studies. Michael and Koshio (2016) showed that PL *M. japonicus* fed 45 days of 0.6 g choline chloride (CC)/kg diet took longer time to reach 50% mortality after a 24 h freshwater challenge test (Figure 7.3). Their results indicated that dietary choline could improve osmotic stress tolerance.

Huang, M., *et al.* (2020) provided new insights into the role of choline and demonstrated that *L. vannamei* fed 1,082 mg choline/kg diet could meet the growth requirement, but 2,822 mg choline/kg diet decreased the hepatopancreas MDA concentration, SOD, and GPX activities in shrimp. The authors suggested that dietary choline could influence the antioxidant capacity of *L. vannamei* and potentially reduce the peroxidation damage in the hepatopancreas; however, its efficacy would need to be tested under challenged conditions. Similarly, Shi, B., *et al.* (2021) reported that smaller *L. vannamei* required 3,254 mg choline/kg diet (broken-line regression; 6,488 mg choline/kg diet using quadratic regression) to optimize growth and feed efficiency. Providing 3,294 choline/kg diet might improve antioxidant capacity by increasing catalase activity and decreasing MDA concentration in hepatopancreas. Recently, Lu, J., *et al.* (2022) explored the potential mechanism of choline on the antioxidation capacity and immune

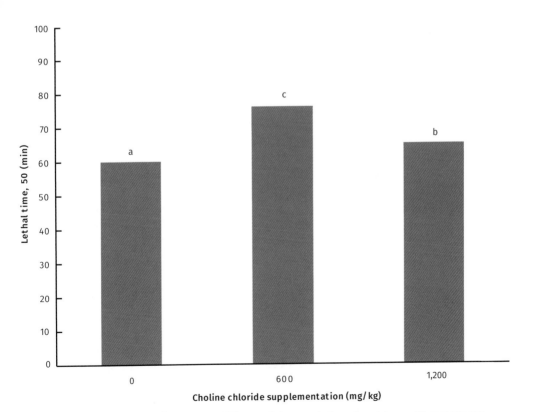

Figure 7.3 The lethal time (min) required to kill 50% of the population of post-larvae *Marsupenaeus japonicus* fed 0, 600, and 1,200 mg choline chloride/kg diet after exposure to freshwater for 2 h (bars with different letters (a,b) indicate significant differences (*p*<0.05)) (source: Michael and Koshio, 2016)

response of juvenile *L. vannamei*. Although the haemolymph MDA SOD, and T-AOC were not highly responsive to dietary choline in this study, apoptosis (%) and ROS in haemocytes dropped significantly in shrimp fed ≥6,550 mg choline/kg diet. Choline appeared to inhibit pro-apoptosis related mRNA expression such as *caspase-1, caspase-3, caspase-8, p53*, and *p38MAPK*. Similar trend in gill and intestinal mRNA expressions were also reported in grass carp (*Ctenopharyngodon idella*) fed increasing levels of choline (Yuan, Z.H., *et al.*, 2020a, 2021). Despite the new evidence mentioned above, a deeper understanding of the mechanism of choline on antioxidation and immune responses in shrimp is required to facilitate its potential functionality in stress mitigation in commercial shrimp feeds.

Based on existing literature and industry practice, we recommend a dietary vitamin choline inclusion level of 600–6,500 mg/kg diet to ensure the growth and health of farmed shrimp are not compromised (Table 7.2).

Inositol

Like choline, *myo*-inositol (active factor of inositol) is categorized as a "vitamin-like" nutrient. For shrimp, *myo*-inositol should be categorized as a vitamin because an exogenous source is required. However, we listed *myo*-inositol as "vitamin-like" to be consistent with the general literature. To date, recommendations for *myo*-inositol inclusion levels in the shrimp nutrition literature ranged from 2,000 to 5,484 mg/kg diet (Table 7.2). The essentiality and dietary requirement of *myo*-inositol were reported in *M. japonicus* (2,000–4,000 mg/kg diet) (Deshimaru and Kuroki 1976; Kanazawa *et al.*, 1976), *P. monodon* (3,400 mg/kg diet) (Reddy *et al.*, 1999; Shiau and Su, 2004), *F. chinesis* (4,000 mg/kg diet) (Liu, T.B., *et al.*, 1993), and *M. nipponense* (2,787–5,484 mg/kg diet) (Liu, A., *et al.*, 2023). The dietary *myo*-inositol requirements varied considerably in shrimp compared with major farmed fish (e.g., 250–500 mg/kg diet for salmonids, 45–770.5 mg/kg diet for carps, 420 mg/kg diet for marine fish) (Mai *et al.*, 2022; NRC, 2011). Differences could be attributed to study variations such as the life stage of shrimp, species, experimental designs, and bioavailability of raw materials.

A more recent study reported that *L. vannamei* (0.6 g) fed 2,705 mg *myo*-inositol/kg diet had higher glutathione peroxidase activity than those fed 449 mg *myo*-inositol/kg diet (Chen, S.J., *et al.*, 2018a). However, growth, feed conversion ratio, survival, and antioxidant enzyme activities (i.e., total antioxidant capacity, SOD) were generally unaffected by the dietary treatment. Chen, S.J., *et al.* (2018b) showed that feeding *L. vannamei* a diet containing 460–1,030 mg *myo*-inositol/kg diet could improve resistance to acute hypoxia stress by increasing total antioxidant capacity, GPX, reducing MDA, and delaying mortality. These results have important practical implications, and studies could further investigate *myo*-inositol's function on stress resistance in shrimp.

Because limited information is available on inositol in shrimp nutrition in general, and Cui, W., *et al.* (2022) recently reviewed the inositol nutrition in aquatic animals, we echo their statement and call for more research on incorporating the interaction of *myo*-inositol with other micronutrients and culturing stress factors when feeding practical diets containing novel ingredients.

Based on existing literature, we suggest a dietary *myo*-inositol content of 2,000–5,500 mg/kg diet to ensure optimal growth and health of the targeted shrimp species (Table 7.2).

Astaxanthin

Astaxanthin has received substantial attention in shrimp nutrition studies in the last 10 years. Astaxanthin is primarily accumulated in the carapace of shrimp and other crustaceans. Crustaceans can synthesize some astaxanthin from β-carotene, zeaxanthin, and canthaxanthin

(Schiedt *et al.*, 1991; 1993). Lim, K.C., *et al.* (2018) recently provided a detailed review of the biochemistry, metabolism, and its role as a functional nutrient in aquatic animals, including farmed shrimp. Astaxanthin is required, particularly at juvenile stages, for optimal survival and growth. The addition of 50 to 200 mg astaxanthin/kg diet improved the growth and survival of *P. monodon* (Niu *et al.*, 2012, 2014), *L. vannamei* (Flores *et al.*, 2007), *M. rosenbergii* (Kumar, V., *et al.*, 2009), *F. indicus* (Kumlu *et al.*, 1998), and *M. japonicus* (Chien and Shiau, 2005; Petit *et al.*, 1997; Yamada *et al.*, 1990). Astaxanthin represents an effective commercial feed additive for enhancing pigmentation (Ju *et al.*, 2011; Wang, W., *et al.*, 2018), reproductive performance (Paibulkichakul *et al.*, 2008; Pangantihon-Kühlmann *et al.*, 1998), antioxidant capacity (Wang, W., *et al.*, 2019; Xie, S., *et al.*, 2018), and stress tolerance (Figure 7.4) (Niu *et al.*, 2014; Zhang, J., *et al.*, 2013).

Recent studies advised that higher addition of astaxanthin was required when shrimp were under stress. As a cautionary note, the levels of astaxanthin tested in several studies were abnormally high and above maximum dietary inclusion allowed in several jurisdictions. For example, Wang, W., *et al.* (2018) reported that supplementing 401–420 mg astaxanthin/kg diet could improve the total haemocyte count and viable cells of juvenile *M. japonicus* (0.4 g) as well as higher median lethal time (LT_{50}) after a freshwater challenge. Similarly, Wang, W., *et al.* (2019) showed that feeding 550 mg astaxanthin/kg diet with 180 mg vitamin E/kg diet could improve the freshwater stress resistance (higher LT_{50}) and digestive enzyme activities (i.e., protease, lipase, amylase) of 2.8 g *M. japonicus*. In *L. vannamei*, dietary astaxanthin supplementation of 100–200 mg/kg diet reduced cell membrane lipid peroxidation and delayed mortality after an acute challenge where salinity decreased from 28 to 5 ppt in a few minutes and shrimp were observed for 45 minutes afterwards (Xie, S., *et al.*, 2018). The lower need for astaxanthin could be because *L. vannamei* has a better salinity tolerance than other marine shrimp

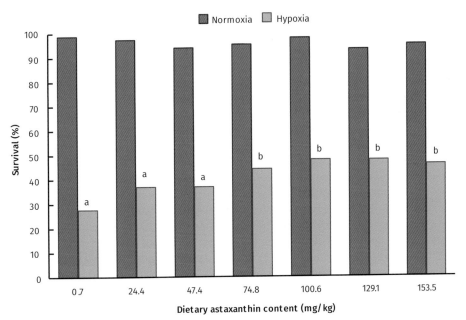

Figure 7.4 Survival (%) of post-larvae *Litopenaeus vannamei* fed graded levels of dietary astaxanthin (mg/kg) followed by hypoxia exposure for 1 h (bars with different letters (a,b) indicate significant differences (*p*<0.05)) (source: Zhang, J., *et al.*, 2013)

species. Furthermore, Yu, Y., *et al.* (2020) demonstrated that 289 mg astaxanthin/kg diet could improve the growth and hepatopancreas morphology of juvenile *L. vannamei* when fed an oxidized fish oil diet. The shrimp also had higher survival after being exposed to an acute salinity change test, i.e., three-fold decrease in salinity for 5 hours. Future studies should investigate the efficacy of astaxanthin against other stressors (e.g., temperature, pathogens, low oxygen and high stocking density).

Based on existing literature and industry practice, we recommend a dietary astaxanthin content of 50–200 mg/kg diet to optimize the growth, survival, health, and pigmentation of farmed shrimp (Table 7.2).

CONCLUSION AND FUTURE RESEARCH

Vitamins and vitamin-like compounds contribute not only to the growth and development but also to the immunity, quality, and sustainability of shrimp. Dietary vitamins improved Ca absorption and shell quality at low salinity (e.g., D_3, B_5), antioxidant capacity (e.g., E, choline, *myo*-inositol, astaxanthin) and lysozyme activity (e.g., B_1, B_3, B_6) among other beneficial responses. Therefore, vitamins and vitamin-like molecules can serve as strategic nutrients to achieve goals other than just weight gain, and the concept of OVN™ fully applies to shrimp as it does to fish.

Nevertheless, scientific knowledge on shrimp vitamin nutrition is scarce compared to fish and terrestrial livestock species. Vitamins C, D and E, as well as the vitamin-like choline and astaxanthin, have garnered attention in recent shrimp vitamin nutrition research. In contrast, requirement studies focusing on other vitamins (e.g., A, K, most B vitamins) have been lacking and, when available, have been often conducted more than 2 decades ago. Therefore, there is a need to determine and/or re-evaluate the recommendations for vitamins that have been overlooked over the past 20 years.

Shrimp vitamin nutrition research unwittingly adopted a reductionist approach. As a result, interactions between vitamins and other nutrients and functionalities under suboptimal rearing conditions have been understudied to date. This knowledge gap must be filled before providing narrower and condition-specific (e.g., temperature, salinity, density, farming system) inclusion levels for individual shrimp species. Until then, the industry would need to rely on closely related species to formulate shrimp feeds. Moreover, providing comprehensive details on the methodological approach used in vitamin requirement studies with shrimp is crucial to help generate practical knowledge. For instance, a lack of information on vitamins used and an incomplete description of dietary treatments (e.g., the distinction between the level of total vs supplemented vitamins was not always clear) in several studies made it difficult to draw reliable and applicable conclusions about vitamin requirements. Graded inclusion levels of vitamins were not always tested, and when they were part of the study design, the range of inclusions was sometimes too wide or too narrow to determine the requirement accurately.

Overall, vitamin nutrition of aquatic species stands substantially as an untapped research focus area, particularly for shrimp species. Recent studies have demonstrated the potential of vitamins and vitamin-like molecules to address current challenges in shrimp aquaculture (e.g., disease and stress resistance), and improve product quality and sustainability of shrimp aquaculture. This chapter has provided trends and implications for future studies that we trust will spur innovative research in the field of shrimp vitamin nutrition.

Table 7.1 An overview of metabolic functions of vitamin/vitamin-like nutrients in animal nutrition and their deficiency symptoms in shrimp beyond growth and survival. For more details, see Mai et al. (2022), NRC (2011), and Bureau and Cho (1999). Adapted from Liu A. et al. (2023)

Vitamins/vitamin-like nutrients	Classical functions	Novel functions	Deficiency symptoms in shrimp
A	Vision process Cell metabolism	Detoxification Fecundity	–
D	Calcium-phosphorus metabolism	Antimicrobial Antiviral	–
E	Biological antioxidant	Immunity Reproduction	–
K	Blood coagulation	–	Decreased calcium deposition Lower prothrombin level
B_1 (Thiamine)	Carbohydrate metabolism – coenzyme for oxidative decarboxylation	Immunity	Anorexia
B_2 (Riboflavin)	Energy metabolism – coenzyme in cellular redox reactions	–	Hyperirritability Light pigmentation Exoskeleton malformation Protruding cuticle at abdominal somites Short-head dwarfism
Niacin (B_3)	Energy metabolism – coenzyme in cellular redox reactions	–	–
Pantothenic acid (B_5)	Energy metabolism – coenzyme for acetylation and acylation reactions	–	Hyperirritability Light pigmentation Thin and soft shell
B_6 (Pyridoxin)	Coenzyme for amino acid and lipid metabolism	Fatty acid metabolism (DHA retention in fillet) Antioxidation	–

Table 7.1 (continued)

Vitamins/vitamin-like nutrients	Classical functions	Novel functions	Deficiency symptoms in shrimp
Biotin (B$_7$)	Carbohydrate and lipid metabolism – coenzyme for carboxylation/decarboxylation reactions	–	–
Folate (B$_9$)	Nucleic acid metabolism – coenzyme for one-carbon metabolism Blood cell function	Immunity Antioxidation	–
B$_{12}$ (Cobalamin)	Protein metabolism – coenzyme for one-carbon metabolism Blood cell function	–	–
C	Antioxidant Detoxification Collagen synthesis Immunity	Wound healing Stress/disease resistance Reproduction	Disturbances in molting Melanized lesions (black death syndrome) Lethargy Abnormal coloration Swollen hepatopancreas
Choline	Lipid metabolism Cell structural integrity	Prevention of lipid malabsorption syndrome Osmoregulation Antioxidation	–
Inositol	Neurological osmolytes and signaling molecules Cell structural integrity	Antioxidation Immunity	–
Astaxanthin	Biological antioxidant Pigmentation	Reproduction Immunity	–

Table 7.2 A summary of the recommended dietary content of vitamin/vitamin-like nutrients for farmed shrimp since NRC (2011) with recommendations for optimal vitamin nutrition.

Vitamins/vitamin-like nutrients (IU or mg/kg diet)[τ]	Literature covered by NRC (2011)	Optimum Vitamin Nutrition dsm-firmenich 2022[¤]	Species	Initial weight (g)	Dietary content in published research	Benefits	References
A (IU)	4,800–180,000	7,000–12,000	–	–	–	–	–
D (IU)	4,000	4,000–6,500[#]	PWS	0.4	6,366	Higher whole-body ash content under low salinity condition	Wen, M., et al. (2015)
			KS	10	4,000 (0.1 mg±)	Improved survival after *Vibrio parahaemolyticus* challenge	Yang, M.C., et al. (2021)
			PWS	0.5	19,200	Improved production performance, antioxidant capacity, immune response, and increased carapace Ca and P concentrations	Dai et al. (2021)
E	85–99	150–300	KS	50 (F)	563–576	Improved hatching rate, larval metamorphosis rate (nauplii to zoea I), and tissues and eggs vitamin E content	Nguyen et al. (2012)
			GS	–	16.4	Improved antioxidant responses post atrazine exposure	Griboff et al. (2014)
			ORP	0.3–0.4	94.1	Improved growth, antioxidant enzyme activities, and vitamin E content in tissues	Zhao, W., et al. (2016)
			ORP	0.4 (F)	100.8–184.4	Improved production performance, spawning, antioxidant enzyme activities, and immune responses	Li, Y., et al. (2018)
			ORP	0.3 (M)	80–160‡	Improved production performance, antioxidant enzyme activities, immune response, and higher survival post *Aeromonas hydrophila* challenge	Li, Y., et al. (2019)

Table 7.2 (continued)

Vitamins/vitamin-like nutrients (IU or mg/kg diet)†	Literature covered by NRC (2011)	Optimum Vitamin Nutrition dsm-firmenich 2022**	Species	Initial weight (g)	Dietary content in published research	Benefits	References
			KS	2.8	180	Improved growth, digestive enzyme activities, pigmentation, astaxanthin content in tissues, and higher LT_{50} when exposed to freshwater	Wang, W., et al. (2019)
			TS	43.1 (F), 41.9 (M)	286.3	Higher PUFA content in hepatopancreas and oocyte	Usman et al. (2019)
			TS	50.5 (M)	200‡	Improved sperm count, spermatophore weight, and percent live sperm	Jiang, S., et al. (2020)
			PWS	5.3	1,000	Improved production performance and antioxidant enzyme activities	Ebadi et al. (2021)
K	30–185	40–60	PWS	0.9	39.1	Improved antioxidant enzyme activities and immune response	Dai et al. (2022b)
B₁ (Thiamine)	14–120	50–150	PWS	0.5	44.6–152.8	Improved production performance, antioxidant enzyme activities, immune response, and thiamine content in hepatopancreas	Huang, X.L., et al. (2015)
			ORP	0.2	0.0–165.0	Increased weight gain and enzyme activities associated with glucose catabolism and fatty acid synthesis	Sun, M., et al. (2022)
B₂ (Riboflavin)	22.5–80	40–80	PWS	0.2	40.9	Improved growth and enhanced antioxidant enzyme activities, immune responses, and intestinal villi height	Sanjeewani and Lee (2023)

Vitamin						Effects	Reference
Niacin (B₃)	7.2–400	100–250	PWS	0.8	109.6–121.2	Improved production performance, antioxidant enzyme activity, and immune response	Xia, M.H., et al. (2015)
			PWS	0.4	0–195	Improved growth, feed conversion, immune parameters, antioxidant enzyme activity and survival	Hasanthi and Lee (2023)
Pantothenic acid (B₅)	100–750	100–180	–	–	–	–	–
B₆ (Pyridoxin)	72–200	50–175#	PWS	0.014	133.3–151.9	Improved growth and antioxidant enzyme activities at low salinity	Li, E., et al. (2010)*
			PWS	1.0	110.1–110.4	Improved production performance, antioxidant enzyme activities, and immune response	Cui, P., et al. (2016)
Biotin (B₇)	0.2–2.4	1.0–2.25	PWS	1.0	0.4–2.25	Improved growth, feed efficiency, and antioxidant enzyme activities	Xia, M., et al (2014)
Folate (B₉)	1.9–5	10–20	GFP	0.3	2.2	Improved production performance and digestive enzyme activities	Asaikkutti et al. (2016a)
B₁₂ (cobalamin)	0.01–0.2	0.02–0.05	–	–	–	–	–
C	40–10,000	250–500	KS	50 (F)	841–880 (source C2MP-Ca)	Improved hatching rate, larval metamorphosis rate (nauplii to zoea I), and tissues and eggs vitamin E content	Nguyen et al. (2012)
			MRP	0.25	105.6 (source C2MP-Ca)	Improved production performance, antioxidant enzyme activities, and digestive enzyme activities	Asaikkutti et al. (2016b)
			PWS	13–15	200 (source not specified)	Improved immune responses	Wu, Y.S., et al. (2016)
			GFP	0.3	150‡ (source C2MP-Mg)	Improved production performance, antioxidant enzyme activities, and digestive enzyme activities	Asaikkutti et al. (2018)
			TS	43.1 (F), 41.9 (M)	406.1 (source C2MP-Ca)	Higher PUFA content in hepatopancreas and oocyte	Usman et al. (2019)

Table 7.2 (continued)

Vitamins/vitamin-like nutrients (IU or mg/kg diet)[†]	Literature covered by NRC (2011)	Optimum Vitamin Nutrition dsm-firmenich 2022[**]	Species	Initial weight (g)	Dietary content in published research	Benefits	References
Choline	600–6,200	600–6,500	PWS	5.3	990 (source C-free)	Improved production performance and antioxidant enzyme activities	Ebadi et al. (2021)
			KS	0.01	600[‡]	Improved production performance and survival. Higher survival after 24-h of freshwater challenge	Michael and Koshio (2016)
			PWS	1.8	1,082–2,822	Improved antioxidant enzyme activities	Huang M. et al. (2020)
			PWS	0.3	3,254–6,488	Improved production performance and antioxidant enzyme activities	Shi, B., et al. (2021)
			PWS	1.4	3,270–6,550	Improved production performance and immune response	Lu, J., et al. (2022)
Myo-inositol	2,000–4,000	2,000–5,500	PWS	0.6	2,705	Improved antioxidant enzyme activities	Chen, S.J., et al. (2018a)
			PWS	0.4	460–1,030	Improved resistance to acute hypoxia stress	Chen, S.J., et al. (2018b)
			ORP	0.2	2,787–5,485	Improved growth and *myo*-inositol content in hepatopancreas	Liu, M.Y., et al. (2023)
Astaxanthin	–	50–200	PWS	0.9–1.0	75–100	Improved astaxanthin retention and pigmentation after cooking	Ju et al. (2011)
			TS	1.2	110	Improved production performance, antioxidant enzyme activities, carotenoid retention efficiency, and higher survival post low DO stress test	Niu et al. (2012)
			PWS	1.0	129.1	Improved production performance, antioxidant enzyme activities, and	Zhang, J., et al. (2013)

			Benefits	Reference
TS	2.1	110	Improved antioxidative status post live transport stress test and improved production performance and carotenoid content in tissues	Niu et al. (2014)
TS	7.3	100[‡]	Improved red scoring of cooked prawn	Wade et al. (2015)
KS	0.4	401–420[‡]	Improved growth, immune responses, pigmentation, and higher LT_{50} when exposed to freshwater	Wang, W., et al. (2018)
KS	2.8	550	Improved growth, digestive enzyme activities, pigmentation, astaxanthin content in tissues, and higher LT_{50} when exposed to freshwater	Wang, W., et al. (2019)
PWS	2.1	100–200[‡]	Improved antioxidant enzyme activities and delayed mortality after acute salinity challenge	Xie, S., et al. (2018)
PWS	0.5	289	Improved growth and hepatopancreatic morphology when fed oxidized fish oil and higher survival after acute salinity challenge	Yu, Y., et al. (2020)

Notes: C2MP-Ca: L-Ascorbic acid-2-monophosphate-calcium salt; C2MP-Mg: L-Ascorbic acid-2-monophosphate-magnesium salt; C-free: L-Ascorbic acid free (not esterified); F: female; KS: Kuruma shrimp (*Marsupenaeus japonicus*); LT50: lethal time to 50% mortality; GFP: giant freshwater prawn (*Macrobrachium rosenbergii*); GS: ghost shrimp (*Palaemonetes argentinus*); M: male; MRP: monsoon river prawn (*Macrobrachium malcolmsonii*), TS: tiger shrimp (*Penaeus monodon*); ORP: oriental river prawn (*Macrobrachium nipponense*); PUFA: polyunsaturated fatty acids; PWS: Pacific white shrimp (*Litopenaeus vannamei*).

* Not covered by NRC (2011).

** Lower inclusion levels are recommended at low stock density (<10 shrimp per m2).

† No recent studies or OVN® recommendations for inositol (refer to section 4.2).

‡ Supplemented level. The vitamin or vitamin-like nutrient content of the final diet was not provided.

Higher inclusion level is recommended at sub-optimal salinity.

Bibliography

Aas, T. S., Åsgård, T., & Ytrestøyl, T. (2022). Utilization of feed resources in the production of Atlantic salmon (*Salmo salar*) in Norway: An update for 2020. *Aquaculture Reports, 26*, 101316. https://doi.org/10.1016/j.aqrep.2022.101316

Abawi, F. G., Sullivan, T. W., & Scheideler, S. E. (1985). Interaction of dietary fat with levels of vitamins A and E in broiler chicks. *Poultry Science, 64*(6), 1192–1198. https://doi.org/10.3382/ps.0641192

Abdel-Hameid, N.-A. H., Abidi, S. F., & Khan, M. A. (2012). Dietary vitamin E requirement for maximizing the growth, conversion efficiency, biochemical composition and haematological status of fingerling *Channa punctatus. Aquaculture Research, 43*(2), 226–238. https://doi.org/10.1111/j.1365-2109.2011.02819.x

Acevedo-Rocha, C. G., Gronenberg, L. S., Mack, M., Commichau, F. M., & Genee, H. J. (2019). Microbial cell factories for the sustainable manufacturing of B vitamins. *Current Opinion in Biotechnology, 56*, 18–29. https://doi.org/10.1016/j.copbio.2018.07.006

Adams, C. R. (1978). Vitamin product forms for animal feeds. In *Vitamin nutrition update-seminar.* Hoffman-La Roche Inc. RCD. Nutley, NJ.

Adjoumani, J. Y., Abasubong, K. P., Phiri, F., Xu, C., Liu, W., & Zhang, D. (2019). Effect of dietary betaine and choline association on lipid metabolism in blunt snout bream fed a high-fat diet. *Aquaculture Nutrition, 25*(5), 1017–1027. https://doi.org/10.1111/anu.12919

Ahmadi, M. R., Bazyar, A. A., Safi, S., Ytrestøyl, T., & Bjerkeng, B. (2006). Effects of dietary astaxanthin supplementation on reproductive characteristics of rainbow trout (*Oncorhynchus mykiss*). *Journal of Applied Ichthyology, 22*(5), 388–394. https://doi.org/10.1111/j.1439-0426.2006.00770.x

Ahmed, I. (2011). Effect of dietary niacin on growth and body composition of two Indian major carps rohu, Labeo rohita, and mrigal, *Cirrhinus mrigala* (Hamilton), fingerlings based on dose–response study. *Aquaculture International, 19*(3), 567–584. https://doi.org/10.1007/s10499-010-9373-0

Ahmmed, M. K., Ahmmed, F., Tian, H. S., Carne, A., & Bekhit, A. E. D. (2020). Marine omega-3 (n-3) phospholipids: A comprehensive review of their properties, sources, bioavailability, and relation to brain health. *Comprehensive Reviews in Food Science and Food Safety, 19*(1), 64–123. https://doi.org/10.1111/1541-4337.12510

Ahmed, S. A. A., Ibrahim, R. E., Farroh, K. Y., Moustafa, A. A., Al-Gabri, N. A., Alkafafy, M., & Amer, S. A. (2021). Chitosan vitamin E nanocomposite ameliorates the growth, redox, and immune status of Nile tilapia (*Oreochromis niloticus*) reared under different stocking densities. *Aquaculture, 541*, 736804. https://doi.org/10.1016/j.aquaculture.2021.736804

Akhtar, S., Pal, M., Sahu, A. K., Alexander, N. P., Gupta, C., Choudhary, S. K., Jha, A. K., & Rajan, M. G. (2010). Stress mitigating and immunomodulatory effect of dietary pyridoxine in *Labeo rohita* (Hamilton) fingerlings. *Aquaculture Research, 41*(7), 991–1002. https://doi.org/10.1111/j.1365-2109.2009.02383.x

Al-Humadi, H., Zarros, A., Kyriakaki, A., Al-Saigh, R., & Liapi, C. (2012). Choline deprivation: An overview of the major hepatic metabolic response pathways. *Scandinavian Journal of Gastroenterology, 47*(8–9), 874–886. https://doi.org/10.3109/00365521.2012.685755

Al-Khalifa, A. S., & Simpson, K. L. (1988). Metabolism of astaxanthin in the rainbow trout (*Salmo gairdneri*). *Comparative Biochemistry and Physiology Part B, 91*(3), 563–568. https://doi.org/10.1016/0305-0491(88)90022-3

Al-Noor, S. M., Hossain, M. D., & Islam, M. A. (2012). The study of fillet proximate composition, growth performance and survival rate of striped catfish (*Pangasius hypophthalmus*) fed with diets containing different amounts of alpha-tocopherol (vitamin-E). *Journal of Bio-Science, 20*, 67–74. https://doi.org/10.3329/jbs.v20i0.17658

Alava, V. R. (1993). Effects of dietary vitamins A, E, and C on the ovarian development of *Penaeus japonicus*. *Bulletin of the Japanese Society of Scientific Fisheries, 59*, 1235–1241. https://doi.org/10.2331/suisan.59.1235

Albrektsen, S., Waagbø, R., & Sandnes, K. (1993). Tissue vitamin B6 concentrations and aspartate aminotransferase (AspAT) activity in Atlantic salmon (*Salmo salar*) fed graded dietary levels of vitamin B6. *Skr. ser. Ernaring* Fik (Ed.), *6*(1) (pp. 21–34).

Albrektsen, S., Hagve, T. A., & Lie, Ø. (1994). The effect of dietary vitamin B6 on tissue fat contents and lipid composition in livers and gills of Atlantic salmon (*Salmo salar*). Comp. *Biochemistry and Physiology, 109a*(2), 403–411. https://doi.org/10.1016/0300-9629(94)90144-9

Albrektsen, S., Kortet, R., Skov, P. V., Ytteborg, E., Gitlesen, S., Kleinegris, D., Mydland, L. T., Hansen, J. Ø., Lock, E. J., Mørkøre, T., James, P., Wang, X., Whitaker, R. D., Vang, B., Hatlen, B., Daneshvar, E., Bhatnagar, A., Jensen, L. B., & Øverland, M. (2022). Future feed resources in sustainable salmonid production: A review. *Reviews in Aquaculture, 14*(4), 1790–1812. https://doi.org/10.1111/raq.12673

Alday-Sanz, V. (2010). *Shrimp book*. Nottingham University Press.

Aldrich, G. (2007). Raw material for feeds (aqua, pet, swine and poultry): Focus on nutrition. In M. N. Riaz (Ed.), *Extruders and expanders in pet food, aquatic and livestock feeds* (pp. 29–53). Agrimedia GmbH.

Allen, P. C. (1988). Physiological basis for carotenoid malabsorption during coccidiosis. In *Proceedings of the 1988 Maryland Nutrition Conference for Feed Manufacturers*. College Park, MD.

Allen, P. C. (1997). Production of free radical species during *Eimeria maxima* infections in chickens. *Poultry Science, 76*(6), 814–821. https://doi.org/10.1093/ps/76.6.814

Allen, P. C., Danforth, H. D., Morris, V. C., & Levander, O. A. (1996). Association of lowered plasma carotenoids with protection against cecal coccidiosis by diets high in n-3 fatty acids. *Poultry Science, 75*(8), 966–972. https://doi.org/10.3382/ps.0750966

Alsop, D., Brown, S., & Van Der Kraak, G. (2007). The effects of copper and benzo[a]pyrene on retinoids and reproduction in zebrafish. *Aquatic Toxicology, 82*(4), 281–295. https://doi.org/10.1016/j.aquatox.2007.03.001

Alsop, D., Matsumoto, J., Brown, S., & Van Der Kraak, G. (2008). Retinoid requirements in the reproduction of zebrafish. *General and Comparative Endocrinology, 156*(1), 51–62. https://doi.org/10.1016/j.ygcen.2007.11.008

Amar, E. C., Kiron, V., Akutsu, T., Satoh, S., & Watanabe, T. (2012). Resistance of rainbow trout *Oncorhynchus mykiss* to infectious hematopoietic necrosis virus (IHNV) experimental infection following ingestion of natural and synthetic carotenoids. *Aquaculture, 330–333*, 148–155. https://doi.org/10.1016/j.aquaculture.2011.12.007

Amezaga, M. R., & Knox, D. (1990). Riboflavin requirements in on-growing rainbow trout, Oncorhynchus mykiss. *Aquaculture, 88*(1), 87–98. https://doi.org/10.1016/0044-8486(90)90322-E

Amlashi, A. S., Falahatkar, B., Sattari, M., & Gilani, M. H. T. (2011). Effect of dietary vitamin E on growth, muscle composition, hematological and immunological parameters of sub-yearling beluga *Huso huso* L *Fish and Shellfish Immunology, 30*(3), 807–814. https://doi.org/10.1016/j.fsi.2011.01.002

Andrews, J. W., Murai, T., & Page, J. W. (1980). Effects of dietary cholecalciferol and ergocalciferol on catfish. *Aquaculture, 19*(1), 49–54. https://doi.org/10.1016/0044-8486(80)90006-X

Aoe, H., Masuda, I., Saito, T., & Takada, T. (1967). Water-Soluble Vitamin Requirements of Carp-V. NIPPON SUISAN GAKKAISHI, 33(11), 1068–1071. https://doi.org/10.2331/suisan.33.1068

Aoe, H., Masuda, I., Mimura, T., Saito, T., & Komo, A. (1968). Requirement of young carp for vitamin A. *Nippon Suisan Gakkaishi, 34*(10), 959–964. https://doi.org/10.2331/suisan.34.959

Aoe, H., Masuda, I., Mimura, T., Saito, T., Komo, A., & Kitamura, S. (1969). Water-Soluble Vitamin Requirements of Carp-VI. NIPPON SUISAN GAKKAISHI, 35(5), 459–465. https://doi.org/10.2331/suisan.35.459

Arai, S., Nose, T., & Haimoto, Y. (1972). Qualitative requirements of young eels, *Anguilla japonika*, for water-soluble vitamins and their deficiency symptoms. *Boll. Freshwater Res. Lab. Tokyo, 22*, 69–83.

Arukwe, A., & Nordbø, B. (2008). Hepatic biotransformation responses in Atlantic salmon exposed to retinoic acids and 3,3',4,4'-tetrachlorobiphenyl (PCB congener 77). *Comparative Biochemistry and Physiology. Toxicology and Pharmacology, 147*(4), 470–482. https://doi.org/10.1016/j.cbpc.2008.02.002

Asaikkutti, A., Bhavan, P. S., & Vimala, K. (2016a). Effects of different levels of dietary folic acid on the growth performance, muscle composition, immune response and antioxidant capacity of freshwater prawn, *Macrobrachium rosenbergii. Aquaculture, 464*, 136–144. https://doi.org/10.1016/j.aquaculture.2016.06.014

Asaikkutti, A., Bhavan, P. S., Vimala, K., Karthik, M., & Cheruparambath, P. (2016b). Effect of different levels dietary vitamin C on growth performance, muscle composition, antioxidant and enzyme activity of freshwater prawn, *Macrobrachium malcolmsonii.* Aquac. Rep. 3, 229–236. https://doi.org/10.1007/s40 011-016-0772-5

Asaikkutti, A., Bhavan, P. S., Vimala, K., & Karthik, M. (2018). Effect of different levels of dietary vitamin C on growth performance, muscle composition, antioxidant and enzyme activity of *Macrobrachium rosenbergii. Proceedings of the National Academy of Sciences, India Section B. Proceedings of the Natl Acad. Sci. India Sec. B—Biol. Sci., 88*(2), 477–486. https://doi.org/10.1007/s40011-016-0772-5

Asaoka, Y., Terai, S., Sakaida, I., & Nishina, H. (2013). The expanding role of fish models in understanding non-alcoholic fatty liver disease. *Disease Models and Mechanisms, 6*(4), 905–914. https://doi.org/10.1242/dmm.011981

Asson-Batres, M. A., Smith, W. B., & Clark, G. (2009). Retinoic acid is present in the postnatal rat olfactory organ and persists in vitamin A–depleted neural tissue. *Journal of Nutrition, 139*(6), 1067–1072. https://doi.org/10.3945/jn.108.096040

Astrup, H. N., & Langebrekke, A. (1985). The effect of a high level of vitamin E and of a ration with hydrogenated free fatty acids upon pork quality. *Meldinger fra Norges Landbrukshogskole, 64*(21), 6.

Atef, S. H. (2018). Vitamin D assays in clinical laboratory: Past, present and future challenges. *Journal of Steroid Biochemistry and Molecular Biology, 175*, 136–137. https://doi.org/10.1016/j.jsbmb.2017.02.011

Athamena, A., Trajkovic-Bodennec, S., Brichon, G., Zwingelstein, G., & Bodennec, J. (2011a). Synthesis of phosphatidylcholine through phosphatidylethanolamine N-Methylation in Tissues of the Mussel *Mytilus galloprovincialis. Lipids, 46*(12), 1141–1154. https://doi.org/10.1007/s11745-011-3590-9

Athamena, A., Brichon, G., Trajkovic-Bodennec, S., Péqueux, A., Chapelle, S., Bodennec, J., & Zwingelstein, G. (2011b). Salinity regulates N-methylation of phosphatidylethanolamine in euryhaline crustaceans hepatopancreas and exchange of newly-formed phosphatidylcholine with hemolymph. *Journal of Comparative Physiology B, 181*(6), 731–740. https://doi.org/10.1007/s00360-011-0562-6

Augustine, P. C., & Ruff, M. D. (1983). Changes in Carotenoid and vitamin A Levels in Young Turkeys Infected with *Eimeria meleagrimitis* or *E. adenoeides. Avian Diseases, 27*(4), 963–971. https://doi.org/10.2307/1590197

Averianova, L. A., Balabanova, L. A., Son, O. M., Podvolotskaya, A. B., & Tekutyeva, L. A. (2020). Production of vitamin B2 (riboflavin) by microorganisms: An overview. *Frontiers in Bioengineering and Biotechnology, 8*, 570828. https://www.frontiersin.org/articles/10.3389/fbioe.2020.570828. https://doi.org/10.3389/fbioe.2020.570828

Azeez, O. I., & Braimah, S. F. (2020a). Mitigating effect of vitamin-E on copper sulphate-induced toxicity in African catfish (*Clarias gariepinus). European Journal of Medical and Health Sciences, 2*(4). https://doi.org/10.24018/ejmed.2020.2.4.411

Azeez, O., & Braimah, S. (2020b). Protective effects of vitamin E on potassium dichromate-induced haemotoxicity and oxidative stress in African catfish (*Clarias gariepinus). Asian Journal of Environment & Ecology, 2020b, 13*(2):18-31. https://doi.org/10.9734/ajee/2020/v13i230177

Babu, S., & Srikantia, S. G. (1976). Availability of folates from some foods. *American Journal of Clinical Nutrition, 29*(4), 376–379. https://doi.org/10.1093/ajcn/29.4.376

Bacou, E., Walk, C., Rider, S., Litta, G., & Perez-Calvo, E. (2021). Dietary oxidative distress: A review of nutritional challenges as models for poultry, swine and fish. *Antioxidants, 10*(4), 525. https://doi.org/10.3390/antiox10040525

Badran, M. F., & Ali, M. A. M. (2021). Effects of folic acid on growth performance and blood parameters of flathead grey mullet, *Mugil cephalus. Aquaculture, 536*, 736459. https://doi.org/10.1016/j.aquaculture.2021.736459

Bae, J. Y., Park, G. H., Yoo, K. Y., Lee, J. Y., Kim, D. J., & Bai, S. C. (2012). Re-evaluation of the Optimum Dietary vitamin C Requirement in juvenile eel, *Anguilla japonica* by using L-ascorbyl-2-monophosphate. *Asian-Australasian Journal of Animal Sciences, 25*(1), 98–103. https://doi.org/10.5713/ajas.2011.11201

Bae, J.-Y., Park, G. H., Yoo, K.-Y., Lee, J.-Y., Kim, D.-J., & Bai, S. C. (2013). Evaluation of optimum dietary vitamin E requirements using DL-α-tocopheryl acetate in the juvenile eel, *Anguilla japonica*. *Journal of Applied Ichthyology*, *29*(1), 213–217. https://doi.org/10.1111/jai.12001

Bailey, L. B., & Gregory, J. F. (2006). Folate. In B. Bowman & R. Russell (Eds.), *Present knowledge in nutrition* (pp. 278–301). International Life Sciences Institute.

Bailey, L. B., Da-Silva, V., West, A. A., & Caudill, M. A. (2013). Folate. In J. Zempleni, J. Suttie, J. Gregory & P. J. Stover (Eds.), *Handbook of vitamins* (5th ed) (pp. 421–446). CRC Press, Taylor & Francis Group, LLC ISBN 9781466515567. https://doi.org/10.1201/b15413

Bajaj, S. R., & Singhal, R. S. (2019). Effect of extrusion processing and hydrocolloids on the stability of added vitamin B12 and physico-functional properties of the fortified puffed extrudates. *LWT*, *101*, 32–39. https://doi.org/10.1016/j.lwt.2018.11.011

Balaghi, M., & Wagner, C. (1992). Methyl group metabolism in the pancreas of folate-deficient rats. *Journal of Nutrition*, *122*(7), 1391–1396. https://doi.org/10.1093/jn/122.7.1391

Balaghi, M., Horne, D. W., Woodward, S. C., & Wagner, C. (1993). Pancreatic one-carbon metabolism in early folate deficiency in rats. *American Journal of Clinical Nutrition*, *58*(2), 198–203. https://doi.org/10.1093/ajcn/58.2.198

Baldissera, M. D., Souza, C. F., Baldisserotto, B., Zimmer, F., Paiano, D., Petrolli, T. G., & Da Silva, A. S. (2019). Vegetable choline improves growth performance, energetic metabolism, and antioxidant capacity of fingerling Nile tilapia (*Oreochromis niloticus*). *Aquaculture*, *501*, 224–229. https://doi.org/10.1016/j.aquaculture.2018.11.021

Barash, P. G. (1978). Nutrient toxicities of vitamin K. In M. Rechcigl, Jr. (Ed.), *CRC handbook series in nutrition and food*. CRC Press.

Barnett, B. J., Cho, C. Y., & Slinger, S. J. (1979). The essentiality of cholecalciferol in the diets of rainbow trout (*Salmo gairdneri*). Comparative Biochemistry and Physiology Part A, 63(2), 291–297. https://doi.org/10.1016/0300-9629(79)90162-2

Barnett, B. J., Cho, C. Y., & Slinger, S. J. (1982a). Relative biopotency of dietary ergocalciferol and cholecalciferol and the role of and requirement for vitamin D in rainbow trout (*Salmo gairdneri*). *Journal of Nutrition*, *112*(11), 2011–2019. https://doi.org/10.1093/jn/112.11.2011

Barnett, B. J., Jones, G., Cho, C. Y., & Slinger, S. J. (1982b). The biological activity of 25-hydroxycholecalciferol and 1,25-dihydroxycholecalciferol for rainbow trout (*Salmo gairdneri*). *Journal of Nutrition*, *112*(11), 2020–2026. https://doi.org/10.1093/jn/112.11.2020

Bates, C. J. (2006). Thiamine. In B. A. Bowman & R. M. Russell (Eds.), *Present knowledge in nutrition* (9th ed) (pp. 242–249). International Life Sciences Institute. https://doi.org/10.1093/ajcn/85.5.1439a

Battisti, E. K., Marasca, S., Durigon, E. G., Villes, V. S., Schneider, T. L. S., Uczay, J., Peixoto, N. C., & Lazzari, R. (2017). Growth and oxidative parameters of *Rhamdia quelen* fed dietary levels of vitamin A. *Aquaculture*, *474*, 11–17. https://doi.org/10.1016/j.aquaculture.2017.03.025

Baugh, C. M., & Krumdieck, C. L. (1971). Naturally occurring folates. *Annals of the New York Academy of Sciences*, *186*, 7–28. https://doi.org/10.1111/j.1749-6632.1971.tb31123.x, PubMed: 4943577

Bayır, M., Arslan, G., Özdemir, E., & Bayır, A. (2022). Differential retention of duplicated retinoid-binding protein (crabp and rbp) genes in the rainbow trout genome after two whole genome duplications and their responses to dietary canola oil. *Aquaculture*, *549*, 737779. https://doi.org/10.1016/j.aquaculture.2021.737779

Bazyar Lakeh, A. A., Ahmadi, M. R., Safi, S., Ytrestøyl, T., & Bjerkeng, B. (2010). Growth performance, mortality and carotenoid pigmentation of fry offspring as affected by dietary supplementation of astaxanthin to female rainbow trout (*Oncorhynchus mykiss*) broodstock. *Journal of Applied Ichthyology*, *26*(1), 35–39. https://doi.org/10.1111/j.1439-0426.2009.01349.x

Becker, E. M., Christensen, J., Frederiksen, C. S., & Haugaard, V. K. (2003). Front-face fluorescence spectroscopy and chemometrics in analysis of yogurt: Rapid analysis of riboflavin. *Journal of Dairy Science*, *86*(8), 2508–2515. https://doi.org/10.3168/jds.S0022-0302(03)73845-4

Bell, J. G., Cowey, C. B., Adron, J. W., & Shanks, A. M. (1985). Some effects of vitamin E and selenium deprivation on tissue enzyme levels and indices of tissue peroxidation in rainbow trout (*Salmo gairdneri*). *British Journal of Nutrition*, *53*(1), 149–157. https://doi.org/10.1079/BJN19850019

Bell, J. G., McEvoy, J., Tocher, D. R., & Sargent, J. R. (2000). Depletion of α-tocopherol and astaxanthin in Atlantic salmon (*Salmo salar*) affects autoxidative defense and fatty acid metabolism. *Journal of Nutrition*, *130*(7), 1800–1808. https://doi.org/10.1093/jn/130.7.1800

Beltramo, E., Berrone, E., Tarallo, S., & Porta, M. (2008). Effects of thiamine and Benfotiamine on intracellular glucose metabolism and relevance in the prevention of diabetic complications. *Acta Diabetologica*, *45*(3), 131–141. https://doi.org/10.1007/s00592-008-0042-y

Bender, D. A. (1992). *Nutritional biochemistry of the vitamins*. Cambridge University Press. https://doi.org/10.1017/CBO9780511615191

Benedito-Palos, L., Navarro, J. C., Sitjà-Bobadilla, A., Bell, J. G., Kaushik, S., & Pérez-Sánchez, J. (2008). High levels of vegetable oils in plant protein-rich diets fed to gilthead sea bream (*Sparus aurata* L.): Growth performance, muscle fatty acid profiles and histological alterations of target tissues. *British Journal of Nutrition*, *100*(5), 992–1003. https://doi.org/10.1017/S0007114508966071

Berntssen, M. H. G., Julshamn, K., & Lundebye, A. K. (2010). Chemical contaminants in aquafeeds and Atlantic salmon (*Salmo salar*) following the use of traditional- versus alternative feed ingredients. *Chemosphere*, *78*(6), 637–646. https://doi.org/10.1016/j.chemosphere.2009.12.021

Berntssen, M. H. G., Ørnsrud, R., Rasinger, J., Søfteland, L., Lock, E. J., Kolås, K., Moren, M., Hylland, K., Silva, J., Johansen, J., & Lie, K. (2016). Dietary vitamin A supplementation ameliorates the effects of poly-aromatic hydrocarbons in Atlantic salmon (*Salmo salar*). *Aquatic Toxicology*, *175*, 171–183. https://doi.org/10.1016/j.aquatox.2016.03.016

Betancor, M. B., Atalah, E., Caballero, M., Benítez-santana, T., Roo, J., Montero, D., & Izquierdo, M. (2011). α-Tocopherol in weaning diets for European sea bass (*Dicentrarchus labrax*) improves survival and reduces tissue damage caused by excess dietary DHA contents. *Aquaculture Nutrition*, *17*(2), e112–e122. https://doi.org/10.1111/j.1365-2095.2009.00741.x

Betancor, M. B., Caballero, M. J., Terova, G., Saleh, R., Atalah, E., Benítez-Santana, T., Bell, J. G., & Izquierdo, M. (2012a). Selenium inclusion decreases oxidative stress indicators and muscle injuries in sea bass larvae fed high-DHA microdiets. *British Journal of Nutrition*, *108*(12), 2115–2128. https://doi.org/10.1017/S0007114512000311

Betancor, M. B., Caballero, M. J., Terova, G., Corà, S., Saleh, R., Benítez-Santana, T., Bell, J. G., Hernández-Cruz, C. M., & Izquierdo, M. (2012b). Vitamin C enhances vitamin E status and reduces oxidative stress indicators in sea bass larvae fed high DHA microdiets. *Lipids*, *47*(12), 1193–1207. https://doi.org/10.1007/s11745-012-3730-x

Bettendorff, L. (2013). Vitamin B1. In J. Zempleni, J. Suttie, J. Gregory & P. J. Stover (Eds.), *Handbook of vitamins* (5th ed) (pp. 267–324). CRC Press, Taylor & Francis Group, LLC ISBN 9781466515567. https://doi.org/10.1201/b15413

Bieber-Wlaschny, M. (1988). Vitamin E in swine nutrition. In update of vitamins and nutrition management in swine production technical conference, *7776*, 988.

Bilodeau, L., Dufresne, G., Deeks, J., Clément, G., Bertrand, J., Turcotte, S., Robichaud, A., Beraldin, F., & Fouquet, A. (2011). Determination of vitamin D3 and 25-hydroxyvitamin D3 in foodstuffs by HPLC UV-DAD and LC-MS/MS. *Journal of Food Composition and Analysis*, *24*(3), 441–448. https://doi.org/10.1016/j.jfca.2010.08.002

Birdsall, J. J. (1975). Technology of fortification of foods. *Proceedings of the National Academy of Sciences of the United States of America*. https://doi.org/10.17226/20201

Biswas, B. K., Biswas, A., Junichi, I., Kim, Y.-S., & Takii, K. (2013). The optimal dietary level of ascorbic acid for juvenile Pacific bluefin tuna, *Thunnus orientalis*. *Aquaculture International*, *21*(2), 327–336. https://doi.org/10.1007/s10499-012-9555-z

Blazer, V. S., & Wolke, R. E. (1984). The effects of α-tocopherol on the immune response and non-specific resistance factors of rainbow trout (*Salmo gairdneri* Richardson). *Aquaculture*, *37*(1), 1–9. https://doi.org/10.1016/0044-8486(84)90039-5

Blomhoff, R., Green, M. H., Green, J. B., Berg, T., & Norum, K. R. (1991). Vitamin A metabolism: New perspectives on absorption, transport, and storage. *Physiological Reviews*, *71*(4), 951–990. https://doi.org/10.1152/physrev.1991.71.4.951

Blondin, G. A., Kulkarni, B. D., & Nes, W. R. (1967). A study of the origin of vitamin-D from 7-dehydrocholesterol in fish. *Comparative Biochemistry and Physiology, 20*(2), 379–390. https://doi.org/10.1016/0010-406X(67)90254-X

Bonjour, J. P. (1991). Biotin. In. In L. J. Machlin (Ed.), *Handbook of vitamins* (2nd ed) (p. 393). ISBN 978-0824770518. Marcel Dekker, Inc.

Boonyapakdee, A., Pootangon, Y., Laudadio, V., & Tufarelli, V. (2015). Astaxanthin extraction from golden apple snail (*Pomacea canaliculata*) eggs to enhance colours in fancy carp (Cyprinus carpio). *Journal of Applied Animal Research, 43*(3), 291–294. https://doi.org/10.1080/09712119.2014.963102

Boonyaratpalin, M. (1998). Nutrition of *Penaeus merguiensis* and *Penaeus idicus. Reviews in Fisheries Science, 6*(1–2), 69–78. https://doi.org/10.1080/10641269891314203

Boyd, C. E., D'Abramo, L. R., Glencross, B. D., Huyben, D. C., Juarez, L. M., Lockwood, G. S., McNevin, A. A., Tacon, A. G. J., Teletchea, F., Tomasso, Jr., J. R., Tucker, C. S., & Valenti, W. C. (2020). Achieving sustainable aquaculture: Historical and current perspectives and future needs and challenges. *Journal of the World Aquaculture Society, 51*(3), 578–633. https://doi.org/10.1111/jwas.12714

Brass, E. P. (1993). Hydroxycobalamin[c-lactam] increases total coenzyme A content in primary culture hepatocytes by accelerating coenzyme A biosynthesis secondary to Acyl-CoA accumulation. *Journal of Nutrition, 123*(11), 1801–1807. https://doi.org/10.1093/jn/123.11.1801

Brody, T. (1999). *Nutritional biochemistry* (2nd ed). Academic Press.

Brønstad, I., Bjerkås, I., & Waagbø, R. (2002). The need for riboflavin supplementation in high and low energy diets for Atlantic salmon *Salmo salar* L. parr. *Aquaculture Nutrition, 8*(3), 209–220. https://doi.org/10.1046/j.1365-2095.2002.00210.x

Brown, M. R., & Farmer, C. L. (1994). Riboflavin content of six species of microalgae used in mariculture. *Journal of Applied Phycology, 6*(1), 61–65. https://doi.org/10.1007/BF02185905

Brown, M. R., Mular, M., Miller, I., Farmer, C., & Trenerry, C. (1999). The vitamin content of microalgae used in aquaculture. *Journal of Applied Phycology, 11*(3), 247–255. https://doi.org/10.1023/A:1008075903578

Brown, P. B., & Robinson, E. H. (1992). Vitamin D studies with channel catfish (*Ictalurus punctatus*) reared in calcium-free water. *Comparative Biochemistry and Physiology Part A, 103*(1), 213–219. https://doi.org/10.1016/0300-9629(92)90265-R

Brown, S. B., Honeyfield, D. C., & Vandenbyllaardt, L. (1998). Thiamine analysis in fish tissues. In G. McDonald, J. Fitzsimons & D. C. Honeyfield (Eds.), *Early life stage mortality syndrome in fishes of the Great Lakes and Baltic Sea* (pp. 73–81). ISBN 1888569085. American Fisheries Society.

Brownlee, N. R., Huttner, J. J., Panganamala, R. V., & Cornwell, D. G. (1977). Role of vitamin E in glutathione-induced oxidant stress: Methemoglobin, lipid peroxidation, and hemolysis. *Journal of Lipid Research, 18*(5), 635–644. https://doi.org/10.1016/S0022-2275(20)41605-0

Bureau, D. P., & Cho, C. Y. (1999). *Nutrition and feeding of fish.* OMNR Fish Culture Course. University of Guelph, 21-25 June 1999.

Burri, L., & Johnsen, L. (2015). Krill products: An overview of animal studies. *Nutrients, 7*(5), 3300–3321. https://doi.org/10.3390/nu7053300

Burtle, G. J., & Lovell, R. T. (1989). Lack of response of channel catfish (*Ictalurus punctatus*) to dietary myo-inositol. *Canadian Journal of Fisheries and Aquatic Sciences, 46*(2), 218–222. https://doi.org/10.1139/f89-030

Byrdwell, W. C., Horst, R. L., Phillips, K. M., Holden, J. M., Patterson, K. Y., Harnly, J. M., & Exler, J. (2013). Vitamin D levels in fish and shellfish determined by liquid chromatography with ultraviolet detection and mass spectrometry. *Journal of Food Composition and Analysis, 30*(2), 109–119. https://doi.org/10.1016/j.jfca.2013.01.005

Caballero, M. J., Izquierdo, M. S., Kjørsvik, E., Fernández, A. J., & Rosenlund, G. (2004). Histological alterations in the liver of sea bream, *Sparus aurata* L., caused by short- or long-term feeding with vegetable oils. Recovery of normal morphology after feeding fish oil as the sole lipid source. *Journal of Fish Diseases, 27*(9), 531–541. https://doi.org/10.1111/j.1365-2761.2004.00572.x

Cabello, F. C. (2006). Heavy use of prophylactic antibiotics in aquaculture: A growing problem for human and animal health and for the environment. *Environmental Microbiology, 8*(7), 1137–1144. https://doi.org/10.1111/j.1462-2920.2006.01054.x

Cahu, C. (1999). Nutrition et alimentation des larves de crevettes pénéides. In J. Guillaume, S. Kaushik, P. Bergot & R. Métailler (Eds.), *Nutrition et alimentation des poissons et crustacés*. INRA Éditions (pp. 313–324). Ifremer.

Calderón-Ospina, C. A., & Nava-Mesa, M. O. (2020). B vitamins in the nervous system: Current knowledge of the biochemical modes of action and synergies of thiamine, pyridoxine, and cobalamin. *CNS Neuroscience and Therapeutics, 26*(1), 5–13. https://doi.org/10.1111/cns.13207

Camporeale, G., & Zempleni, J. (2006). Biotin. In B. A. Bowman & R. M. Russell (Eds.), *Present knowledge in nutrition* (9th ed) (pp. 250–259). International Life Sciences Institute. https://doi.org/10.1093/ajcn/85.5.1439a

Canada Minister of Justice, 2021. (1983). *Feeds Regulations*, Schedule IV, SOR/83-593, Canada. https://laws-lois.justice.gc.ca/eng/regulations/SOR-83-593/page-9.html#h-879448

Cantatore, F. P., Loperfido, M. C., Magli, D. M., Mancini, L., & Carrozzo, M. (1991). The importance of vitamin C for hydroxylation of vitamin D3 to 1,25(OH)2D3 in man. *Clinical Rheumatology, 10*(2), 162–167. https://doi.org/10.1007/BF02207657

Canyurt, M. A., & Akhan, S. (2008). Effect of dietary vitamin E on the sperm quality of rainbow trout (*Oncorhynchus mykiss*). *Aquaculture Research, 39*(9), 1014–1018. https://doi.org/10.1111/j.1365-2109.2008.01952.x

Carmel, R. (1994). In vitro studies of gastric juice in patients with food-cobalamin malabsorption. *Digestive Diseases and Sciences, 39*(12), 2516–2522. https://doi.org/10.1007/BF02087684

Carpenter, K.J. (1981) "*Pellagra*". Hutchinson Ross, Stroudsburg, PA.

Carpenter, K. J., Schelstraete, M., Vilicich, V. C., & Wall, J. S. (1988). Immature corn as a source of niacin for rats. *Journal of Nutrition, 118*(2), 165–169. https://doi.org/10.1093/jn/118.2.165

Carson, D. A., Seto, S., & Wasson, D. B. (1987). Pyridine nucleotide cycling and poly(ADP-ribose) synthesis in resting human lymphocytes. *Journal of Immunology, 138*(6), 1904–1907. https://doi.org/10.4049/jimmunol.138.6.1904

Castledine, A. J., Cho, C. Y., Slinger, S. J., Hicks, B., & Bayley, H. S. (1978). Influence of dietary biotin level on growth, metabolism and pathology of rainbow trout. *Journal of Nutrition, 108*(4), 698–711. https://doi.org/10.1093/jn/108.4.698

Catacutan, M. R., & De la Cruz, M. (1989). Growth and mid-gut cells profile of *Penaeus monodon* juveniles fed water-soluble-vitamin deficient diets. *Aquaculture, 81*(2), 137–144. https://doi.org/10.1016/0044-8486(89)90239-1

Catacutan, M. R., Pagador, G. E., Doyola-Solis, E. F., Ishikawa, M., & Teshima, S. (2012). Level of L-ascorbyl-2-monophosphate-Mg as a vitamin C source in practical diets for the Asian sea bass, *Lates calcarifer*. *Israeli Journal of Aquaculture – Bamidgeh, 64*, 782–789. https://doi.org/10.46989/001c.20616

Caudill, M. A. (2010). Pre- and postnatal health: Evidence of increased choline needs. *Journal of the American Dietetic Association, 110*(8), 1198–1206. https://doi.org/10.1016/j.jada.2010.05.009

Cerezuela, R., Cuesta, A., Meseguer, J., & Angeles Esteban, M. Á. (2009). Effects of dietary vitamin D3 administration on innate immune parameters of seabream (*Sparus aurata* L.). *Fish and Shellfish Immunology, 26*(2), 243–248. https://doi.org/10.1016/j.fsi.2008.11.004

Cerolini, S., Maldjian, A., Surai, P., & Noble, R. (2000). Viability, susceptibility to peroxidation and fatty acid composition of boar semen during liquid storage. *Animal Reproduction Science, 58*(1–2), 99–111. https://doi.org/10.1016/S0378-4320(99)00035-4

Chaiyapechara, S., Casten, M. T., Hardy, R. W., & Dong, F. M. (2003). Fish performance, fillet characteristics, and health assessment index of rainbow trout (*Oncorhynchus mykiss*) fed diets containing adequate and high concentrations of lipid and vitamin E. *Aquaculture, 219*(1–4), 715–738. https://doi.org/10.1016/S0044-8486(03)00025-5

Chen, H. Y., Wu, F. C., & Tang, S. Y. (1991). Thiamin requirement of juvenile shrimp (*Penaeus monodon*). *Journal of Nutrition, 121*(12), 1984–1989. https://doi.org/10.1093/jn/121.12.1984

Chen, H. Y., & Hwang, G. (1992). Estimation of the dietary riboflavin required to maximize tissue riboflavin concentration in juvenile shrimp (*Penaeus monodon*). *Journal of Nutrition, 122*(12), 2474–2478. https://doi.org/10.1093/jn/122.12.2474

Chen, H. Y., Wu, F. C., & Tang, S. Y. (1994). Sensitivity of transketolase to the thiamin status of juvenile marine shrimp (*Penaeus monodon*). *Comparative Biochemistry and Physiology Part A, 109*(3), 655–659. https://doi.org/10.1016/0300-9629(94)90206-2

Chen, J. (1990). *Technical service internal reports*. BASF, Corp.

Chen, L., Feng, L., Jiang, W. D., Jiang, J., Wu, P., Zhao, J., Kuang, S. Y., Tang, L., Tang, W. N., Zhang, Y. A., Zhou, X. Q., & Liu, Y. (2015a). Dietary riboflavin deficiency decreases immunity and antioxidant capacity and changes tight junction proteins and related signaling molecules mRNA expression in the gills of young grass carp (*Ctenopharyngodon idella*). *Fish and Shellfish Immunology, 45*(2), 307–320. https://doi.org/10.1016/j.fsi.2015.04.004

Chen, S. J., Xie, S., Chen, M., Mi, Z., He, Q., Yang, F., Niu, J., Liu, Y., & Tian, L. (2019). Hypoxia-induced changes in survival, immune response and antioxidant status of the Pacific white shrimp (*Litopenaeus vannamei*) fed with graded levels of dietary myo -inositol. *Aquaculture Nutrition, 25*(2), 518–528. https://doi.org/10.1111/anu.12877

Chen, S. J., Guo, Y. C., Espe, M., Yang, F., Fang, W. P., Wan, M. G., Niu, J., Liu, Y. J., & Tian, L. X. (2018b). Growth performance, haematological parameters, antioxidant status and salinity stress tolerance of juvenile Pacific white shrimp (*Litopenaeus vannamei*) fed different levels of dietary myo-inositol. *Aquaculture Nutrition, 24*(5), 1527–1539. https://doi.org/10.1111/anu.12690

Chen, X. M., Gao, C. S., Du, X. Y., Yao, J. M., He, F. F., Niu, X. T., Wang, G. Q., & Zhang, D. M. (2020). Effects of dietary astaxanthin on the growth, innate immunity and antioxidant defence system of *Paramisgurnus dabryanus*. *Aquaculture Nutrition, 26*(5), 1453–1462. https://doi.org/10.1111/anu.13093

Chen, Y. F., Huang, C. F., Liu, L., Lai, C. H., & Wang, F. L. (2019). Concentration of vitamins in the 13 feed ingredients commonly used in pig diets. *Animal Feed Science and Technology, 247*, 1–8. https://doi.org/10.1016/j.anifeedsci.2018.10.011

Chen, Y. J., Liu, Y. J., Tian, L. X., Niu, J., Liang, G. Y., Yang, H. J., Yuan, Y., & Zhang, Y. Q. (2013). Effect of dietary vitamin E and selenium supplementation on growth, body composition, and antioxidant defense mechanism in juvenile largemouth bass (*Micropterus salmoides*) fed oxidized fish oil. *Fish Physiology and Biochemistry, 39*(3), 593–604. https://doi.org/10.1007/s10695-012-9722-1

Chen, Y. J., Yuan, R. M., Liu, Y. J., Yang, H. J., Liang, G. Y., & Tian, L. X. (2015). Dietary vitamin C requirement and its effects on tissue antioxidant capacity of juvenile largemouth bass, *Micropterus salmoides*. *Aquaculture, 435*, 431–436. https://doi.org/10.1016/j.aquaculture.2014.10.013

Chen, Y. L., Liu, W. S., Wang, X. D., Li, E. C., Qiao, F., Qin, J. G., & Chen, L. Q. (2018). Effect of dietary lipid source and vitamin E on growth, non-specific immune response and resistance to *Aeromonas hydrophila* challenge of Chinese mitten crab *Eriocheir sinensis*. *Aquaculture Research, 49*(5), 2023–2032. https://doi.org/10.1111/are.13659

Cheng, C. H., Liang, H. Y., Guo, Z. X., Wang, A. L., & Ye, C. X. (2017). Effect of dietary vitamin C on growth performance, antioxidant status and innate immunity of juvenile pufferfish (*Takifugu obscurus*). *Israeli Journal of Aquaculture – Bamidgeh, 69*, 1434. https://doi.org/10.46989/001c.20848

Cheng, K., Ma, C., Guo, X., Huang, Y., Tang, R., Karrow, N. A., & Wang, C. (2020a). Vitamin D3 modulates yellow catfish (*Pelteobagrus fulvidraco*) immune function *in vivo* and *in vitro* and this involves the vitamin D3/VDR-type I interferon axis. *Developmental and Comparative Immunology, 107*, 103644. https://doi.org/10.1016/j.dci.2020.103644

Cheng, K., Tang, Q., Huang, Y., Liu, X., Karrow, N. A., & Wang, C. (2020b). Effect of vitamin D3 on the immunomodulation of head kidney after *Edwardsiella ictaluri* challenge in yellow catfish (*Pelteobagrus fulvidraco*). *Fish and Shellfish Immunology, 99*, 353–361. https://doi.org/10.1016/j.fsi.2020.02.023

Cheng, K., Huang, Y., Wang, C., Ali, W., & Karrow, N. A. (2023). Physiological function of vitamin D3 in fish. *Reviews in Aquaculture, 15*(4), 1732–1748. https://doi.org/10.1111/raq.12814

Chew, B. P. (1995). Antioxidant vitamins affect food animal immunity and health. *Journal of Nutrition, 125*(6), Suppl., 1804S–1808S. https://doi.org/10.1093/jn/125.suppl_6.1804S

Chien, Y. H., & Shiau, W. C. (2005). The effects of dietary supplementation of algae and synthetic astaxanthin on body astaxanthin, survival, growth, and low dissolved oxygen stress resistance of kuruma prawn, *Marsupenaeus japonicus* Bate. *Journal of Experimental Marine Biology and Ecology, 318*(2), 201–211. https://doi.org/10.1016/j.jembe.2004.12.016

Chien, C. T., Chang, W. T., Chen, H. W., Wang, T. D., Liou, S. Y., Chen, T. J., Chang, Y. L., Lee, Y. T., & Hsu, S. M. (2004). Ascorbate supplement reduces oxidative stress in dyslipidemic patients undergoing apheresis. *Arteriosclerosis, Thrombosis, and Vascular Biology, 24*(6), 1111–1117. https://doi.org/10.1161/01. ATV.0000127620.12310.89

Chimsung, N., Tantikitti, C., Milley, J. E., Verlhac-Trichet, V., & Lall, S. P. (2014). Effects of various dietary factors on astaxanthin absorption in Atlantic salmon (*Salmo salar*). *Aquaculture Research, 45*(10), 1611–1620. https://doi.org/10.1111/are.12108

Cho, Y. S., Douglas, S. E., Gallant, J. W., Kim, K. Y., Kim, D. S., & Nam, Y. K. (2007). Isolation and characterization of cDNA sequences of L-gulono-gamma-lactone oxidase, a key enzyme for biosynthesis of ascorbic acid, from extant primitive fish groups. *Comparative Biochemistry and Physiology. Part B, Biochemistry and Molecular Biology, 147*(2), 178–190. https://doi.org/10.1016/j.cbpb.2007.01.001

Chow, C. K. (1979). Nutritional influence on cellular antioxidant defense systems. *American Journal of Clinical Nutrition, 32*(5), 1066–1081. https://doi.org/10.1093/ajcn/32.5.1066

Christensen, K. (1983). Pools of cellular nutrients. In P.M. Riis (Ed.), *Dynamic biochemistry of animal production.* Elsevier Science. ISBN 9780444420527.

Christiansen, R., & Torrissen, O. J. (1996). Growth and survival of Atlantic salmon, *Salmo salar* L. fed different dietary levels of astaxanthin. Juveniles. *Aquaculture Nutrition, 2*(1), 55–62. https://doi.org/10.1111/j.1365-2095.1996.tb00008.x

Christiansen, R., Lie, Ø., & Torrissen, O. J. (1994). Effect of astaxanthin and vitamin A on growth and survival during first feeding of Atlantic salmon, *Salmo salar* L. *Aquaculture Research, 25*(9), 903–914. https://doi.org/10.1111/j.1365-2109.1994.tb01352.x

Ciji, A., Sahu, N. P., Pal, A. K., & Akhtar, M. S. (2013). Nitrite-induced alterations in sex steroids and thyroid hormones of *Labeo rohita* juveniles: Effects of dietary vitamin E and l-tryptophan. *Fish Physiology and Biochemistry, 39*(5), 1297–1307. https://doi.org/10.1007/s10695-013-9784-8

Clare, C. E., Brassington, A. H., Kwong, W. Y., & Sinclair, K. D. (2019). One-carbon metabolism: Linking nutritional biochemistry to epigenetic programming of long-term development. *Annual Review of Animal Biosciences, 7*, 263–287. https://doi.org/10.1146/annurev-animal-020518-115206

Clifford, A. J., Jones, A. D., & Bills, N. D. (1990). Bioavailability of folates in selected foods incorporated into amino acid-based diets fed to rats. *Journal of Nutrition, 120*(12), 1640–1647. https://doi.org/10.1093/jn/120.12.1640

Coates, J., & Halver, J. (1958). Water-soluble vitamin requirements of silver salmon. Special scientific report-fisheries. https://spo.nmfs.noaa.gov/content/water-soluble-vitamin-requirements-silver-salmon. *U. S. Dept Interior.*

Cohen, N., Scott, C. G., Neukom, C., Lopresti, R. J., Weber, G., & Saucy, G. (1981). Total Synthesis of All Eight Stereoisomers of α-Tocopheryl Acetate. Determination of their diastereoisomeric and enantiomeric purity by gas chromatography. Helvetica Chimica Acta, 64(4), 1158–1173. https://doi.org/10.1002/hlca.19810640422

Combs, G. F., & McClung, J. P. (2017). *The vitamins: Fundamental aspects in nutrition and health* (5th ed). Elsevier Science. ISBN 978-0-12-802965-7.

Combs, Jr., G. F., & McClung, J. P. (2022). *The vitamins. Fundamental aspects in nutrition and health* (6th ed). Elsevier. ISBN: 9780323904735.

Conklin, D. E. (1997). Vitamin. In L. R. D'Abramo, D. E. Conklin & D. M. Akiyama (Eds.), *Crustacean nutrition* (pp. 130–131). World Aquaculture Society.

Conklin, D. E. (1989). Vitamin requirements of juvenile penaeid shrimp, advances in tropical aquaculture. https://archimer.ifremer.fr/doc/00000/1469/. Workshop at Tahiti, French Polynesia, 20 Feb – 4 Mar 1989.

Cooper, J. R., Roth, R. H., & Kini, M. M. (1963). Biochemical and physiological function of thiamine in nervous tissue. *Nature, 199*, 609–610. https://doi.org/10.1038/199609a0

Corbin, K. D., & Zeisel, S. H. (2012). The nutrigenetics and nutrigenomics of the dietary requirement for choline. *Progress in Molecular Biology and Translational Science, 108*, 159–177. https://doi.org/10.1016/B978-0-12-398397-8.00007-1

Courtot, E., Musson, D., Stratford, C., Blyth, D., Bourne, N. A., Rombenso, A. N., Simon, C. J., Wu, X., & Wade, N. M. (2022). Dietary fatty acid composition affects the apparent digestibility of algal carotenoids in diets for Atlantic salmon, *Salmo salar. Aquaculture Research, 53*(6), 2343–2353. https://doi.org/10.1111/are.15753

Cowey, C. B., Adron, J. W., Knox, D., & Ball, G. T. (1975) Studies on the nutrition of marine flatfish. The thiamin requirement of turbot (*Scophthalmus maximus*). *British Journal of Nutrition, 34*(3), 383–390. https://doi.org/10.1017/s000711457500044x

Cowey, C. B., Adron, J. W., Walton, M. J., Murray, J., Youngson, A., & Knox, D. (1981). Tissue distribution uptake and requirement for α-tocopherol of rainbow trout (*Salmo gairdneri*) fed diets with a minimal content of unsaturated fatty acids. *Journal of Nutrition, 111*(9), 1556–1567. https://doi.org/10.1093/jn/111.9.1556

Cowey, C. B., Adron, J. W., & Youngson, A. (1983). The vitamin E requirement of rainbow trout (*Salmo gairdneri*) given diets containing polyunsaturated fatty acids derived from fish oil. *Aquaculture, 30*(1–4), 85–93. https://doi.org/10.1016/0044-8486(83)90154-0

Craig, S. R., & Gatlin III, D. M. (1997). Growth and body composition of juvenile red drum (*Sciaenops ocellatus*) fed diets containing lecithin and supplemental choline. *Aquaculture, 151*(1–4), 259–267. https://doi.org/10.1016/S0044-8486(96)01479-2

Cui, P., Zhou, Q.-C., Huang, X.-L., & Xia, M.-H. (2016). Effect of dietary vitamin B6 on growth, feed utilization, health and non-specific immune of juvenile Pacific white shrimp, *Litopenaeus vannamei. Aquaculture Nutrition, 22*(5), 1143–1151. https://doi.org/10.1111/anu.12365

Cui, W., Ma, A., Farhadi, A., Saqib, H. S. A., Liu, S., Chen, H., & Ma, H. (2022). How myo-inositol improves the physiological functions of aquatic animals: A review. *Aquaculture, 553*, 738118. https://doi.org/10.1016/j.aquaculture.2022.738118

Czeczuga, B., & Czepak, R. (1976). Carotenoids in fish. VII. The kind of food and the content of carotenoids and vitamin A in *Carassius carassius* (L.) and *Leucaspius delineatus* (Heck). *Acta Hydrobiologica, 18*(2), 1–21.

D'Abramo, L. R., Moncreiff, C. A., Holcomb, F. P., Labrenty Montanez, J. L., & Buddington, R. K. (1994). Vitamin C requirement of the juvenile freshwater prawn, *Macrobrachium rosenbergii. Aquaculture, 128*(3–4), 269–275. https://doi.org/10.1016/0044-8486(94)90316-6

Dabrowski, K., & Köck, G. (1989). Absorption of ascorbic acid and ascorbic sulfate and their interactions with minerals in the digestive tract of rainbow trout (*Oncorhynchus mykiss*). *Canadian Journal of Fisheries and Aquatic Sciences, 46*(11), 1952–1957. https://doi.org/10.1139/f89-245

Dabrowski, K., El-Fiky, N., Köck, G., Frigg, M., & Wieser, W. (1990). Requirement and utilization of ascorbic acid and ascorbic sulfate in juvenile rainbow trout. *Aquaculture, 91*(3–4), 317–337. https://doi.org/10.1016/0044-8486(90)90197-U

da Costa, K. A., Niculescu, M. D., Craciunescu, C. N., Fischer, L. M., & Zeisel, S. H. (2006). Choline deficiency increases lymphocyte apoptosis and DNA damage in humans. *American Journal of Clinical Nutrition, 84*(1), 88–94. https://doi.org/10.1093/ajcn/84.1.88

Daghir, N. J., & Shah, M. A. (1973). Effect of dietary protein level on vitamin B6 requirement of chicks. *Poultry Science, 52*(4), 1247–1252. https://doi.org/10.3382/ps.0521247

Dagnelie, P. C., Van Staveren, W. A., & Van den Berg, H. (1991). Vitamin B12 from algae appears not to be bioavailable. *American Journal of Clinical Nutrition, 53*(3), 695–697. https://doi.org/10.1093/ajcn/53.3.695

Dai, T., Jiao, L., Tao, X., Lu, J., Jin, M., Sun, P., & Zhou, Q. (2022). Effects of dietary vitamin D3 supplementation on the growth performance, tissue Ca and P concentrations, antioxidant capacity, immune response and lipid metabolism in *Litopenaeus vannamei* larvae. *British Journal of Nutrition, 128*(5), 793–801. https://doi.org/10.1017/S0007114521004931

Dai, T., Lu, J., Tao, X., Zhang, X., Li, M., Jin, M., Sun, P., Liu, W., Jiao, L., & Zhou, Q. (2022a). Vitamin D3 alleviates high-fat induced hepatopancreas lipid accumulation and inflammation by activating AMPKα/PINK1/Parkin-mediated mitophagy in *Litopenaeus vannamei*. Rep. 24(August):101272. *Aquaculture Reports, 25*. https://doi.org/10.1016/j.aqrep.2022.101272

Dai, T., Zhang, X., Li, M., Tao, X., Jin, M., Sun, P., Zhou, Q., & Jiao, L. (2022b). Dietary vitamin K3 activates mitophagy, improves antioxidant capacity, immunity and affects glucose metabolism in *Litopenaeus vannamei. Food and Function, 13*(11), 6362–6372. https://doi.org/10.1039/d2fo00865c

Dakshinamurti, S., & Dakshinamurti, K. (2013). Vitamin B6. In J. Zempleni, J. Suttie, J. Gregory & P. J. Stover (Eds.), *Handbook of vitamins* (5th ed) (pp. 351–396). CRC Press, Taylor & Francis Group, LLC ISBN 9781466515567. https://doi.org/10.1201/b15413

Dalto, D. B., & Matte, J. J. (2017). Pyridoxine (vitamin B_6) and the glutathione peroxidase system; a link between one-carbon metabolism and antioxidation. *Nutrients, 9*(3), 189. https://doi.org/10.3390/nu90 30189

Dandapat, J., Chainy, G. B., & Rao, K. J. (2000). Dietary vitamin-E modulates antioxidant defence system in giant freshwater prawn, Macrobrachium rosenbergii. Comparative Biochemistry and Physiology. Toxicology and Pharmacology, 127(1), 101–115. https://doi.org/10.1016/s0742-8413(00)00132-8

Daniel, N., Muralidhar, A. P., Srivastava, P. P., Jain, K. K., Pani Prasad, K., Manish, J., & Sivaramakrishnan, T. (2018). Dietary ascorbic acid requirement for growth of striped catfish, *Pangasianodon hypophthalmus* (Sauvage, 1878) juveniles. *Aquaculture Nutrition, 24*(1), 616–624. https://doi.org/10.1111/anu.12596

Dantagnan, P., Gonzalez, K., Hevia, M., Betancor, M. B., Hernández, A. J., Borquez, A., & Montero, D. (2017). Effect of the arachidonic acid/vitamin E interaction on the immune response of juvenile Atlantic salmon (*Salmo salar*) challenged against Piscirickettsia salmonis. *Aquaculture Nutrition, 23*(4), 710–720. https://doi.org/10.1111/anu.12438

Darby, W. J., McNutt, K. W., & Todhunter, E. N. (1975). Niacin. *Nutrition Reviews, 33*(10), 289–297. https://doi.org/10.1111/j.1753-4887.1975.tb05075.x

Darias, M. J., Mazurais, D., Koumoundouros, G., Glynatsi, N., Christodoulopoulou, S., Huelvan, C., Desbruyeres, E., Le Gall, M. M., Quazuguel, P., Cahu, C. L., & Zambonino-Infante, J. L. (2010). Dietary vitamin D3 affects digestive system ontogenesis and ossification in European sea bass (*Dicentrachus labrax*, Linnaeus, 1758). *Aquaculture, 298*(3–4), 300–307. https://doi.org/10.1016/j.aquaculture.2009.11.002

Darias, M. J., Mazurais, D., Koumoundouros, G., Le Gall, M. M., Huelvan, C., Desbruyeres, E., Quazuguel, P., Cahu, C. L., & Zambonino-Infante, J. L. (2011). Imbalanced dietary ascorbic acid alters molecular pathways involved in skeletogenesis of developing European sea bass (*Dicentrarchus labrax*). *Comparative Biochemistry and Physiology. Part A, Molecular and Integrative Physiology, 159*(1), 46–55. https://doi.org/10.1016/j.cbpa.2011.01.013

Dawood, M. A. O., & Koshio, S. (2018). Vitamin C supplementation to optimize growth, health and stress resistance in aquatic animals. *Reviews in Aquaculture, 10*(2), 334–350. https://doi.org/10.1111/raq.12163

Dawood, M. A. O., Koshio, S., Ishikawa, M., & Yokoyama, S. (2016). Immune responses and stress resistance in red sea bream, *Pagrus major*, after oral administration of heat-killed Lactobacillus plantarum and vitamin C. *Fish and Shellfish Immunology, 54*, 266–275. https://doi.org/10.1016/j.fsi.2016.04.017

Dawood, M. A. O., Koshio, S., El-Sabagh, M., Billah, M. M., Zaineldin, A. I., Zayed, M. M., & Omar, A. A. E. D. (2017). Changes in the growth, humoral and mucosal immune responses following β-glucan and vitamin C administration in red sea bream, *Pagrus major. Aquaculture, 470*, 214–222. https://doi.org/10.1016/j.aquaculture.2016.12.036

Dawood, M. A. O., Koshio, S., Abdel-Daim, M. M., & Van Doan, H. (2019). Probiotic application for sustainable aquaculture. *Reviews in Aquaculture, 11*(3), 907–924. https://doi.org/10.1111/raq.12272

Dawood, M. A. O., Zommara, M., Eweedah, N. M., & Helal, A. I. (2020a). Synergistic effects of selenium nanoparticles and vitamin E on growth, immune-related gene expression, and regulation of antioxidant status of Nile tilapia (*Oreochromis niloticus*). *Biological Trace Element Research, 195*(2), 624–635. https://doi.org/10.1007/s12011-019-01857-6

Dawood, M. A. O., Zommara, M., Eweedah, N. M., Helal, A. I., & Aboel-Darag, M. A. (2020b). The potential role of nano-selenium and vitamin C on the performances of Nile tilapia (*Oreochromis niloticus*). *Environmental Science and Pollution Research International, 27*(9), 9843–9852. https://doi.org/10.1007/s11356-020-07651-5

de Lima, M. B., da Silva, E. P., Pereira, R., Romano, G. G., de Freitas, L. W., Dias, C. T. S., & Menten, J. F. M. (2018). Estimate of choline nutritional requirements for chicks from 1 to 21 days of age. *Journal of Animal Physiology and Animal Nutrition, 102*(3), 780–788. https://doi.org/10.1111/jpn.12881

DeLuca, H.F. (2014) History of the discovery of vitamin D and its active metabolites. *BoneKEy Rep. 3,479*. https://doi.org/10.1038/bonekey.2013.213

Del Tito, Jr., B. J. (1983). Role of beta-carotene and lutein in the synthesis of vitamin A in goldfish. *Progressive Fish-Culturist, 45*(2), 94–97. https://doi.org/10.1577/1548-8659(1983)45[94:ROBALI]2.0.CO;2

Deng, D. F., & Wilson, R. P. (2003). Dietary riboflavin requirement of juvenile sunshine bass (*Morone chrysops♀ X Morone saxatilis♂*). *Aquaculture, 218*(1–4), 695–701. https://doi.org/10.1016/S0044-8486(02)00513-6

Deng, J. M., Zhang, X. D., Zhang, J. W., Bi, B. L., Wang, H. Z., Zhang, L., & Mi, H. F. (2019). Effects of dietary ascorbic acid levels on cholesterol metabolism in rainbow trout (*Oncorhynchus mykiss*). *Aquaculture Nutrition, 25*(6), 1345–1353. https://doi.org/10.1111/anu.12955

Desai, I. D. (1984). Vitamin E analysis methods for animal tissues. *Methods in Enzymology, 105*, 138–147. https://doi.org/10.1016/S0076-6879(84)05019-9

Deshimaru, O., & Kuroki, K. (1976). Studies on a purified diet for prawn. VII. Adequate dietary levels of ascorbic acid and inositol. *NIPPON SUISAN GAKKAISHI, 42*(5), 571–576. https://doi.org/10.2331/suisan.42.571

Deshimaru, O., & Kuroki, K. (1979). Requirement of prawn for dietary thiamine, pyridoxine, and choline chloride. *Bulletin of the Japanese Society of Scientific Fisheries, 45*(3), 370–373.

Deshimaru, O., & Yone, Y. (1978). Requirement of prawn for dietary minerals. *NIPPON SUISAN GAKKAISHI, 44*(8), 907–910. https://doi.org/10.2331/suisan.44.907

Dhur, A., Galan, P., & Hercberg, S. (1991). Folate status and the immune system. *Progress in Food and Nutrition Science, 15*(1–2), 43–60. PubMed: 1887065

Dioguardi, M., Guardiola, F. A., Vazzana, M., Cuesta, A., Esteban, M. A., & Cammarata, M. (2017). Vitamin D3 affects innate immune status of European sea bass (*Dicentrarchus labrax L.*). *Fish Physiology and Biochemistry, 43*(4), 1161–1174. https://doi.org/10.1007/s10695-017-0362-3

Dove, C. R., & Ewan, R. C. (1991). Effect of trace minerals on the stability of vitamin E in swine grower diets. *Journal of Animal Science, 69*(5), 1994–2000. https://doi.org/10.2527/1991.6951994x

Draut, H., Liebenstein, T., & Begemann, G. (2019). New insights into the control of cell fate choices and differentiation by retinoic acid in cranial, axial and caudal structures. *Biomolecules, 9*(12), 860. https://doi.org/10.3390/biom9120860

Drouin, G., Godin, J. R., & Pagé, B. (2011). The genetics of vitamin C loss in vertebrates. *Current Genomics, 12*(5), 371–378. https://doi.org/10.2174/138920211796429736

dsm-firmenich Animal Nutrition and Health. (2022a). *dsm-firmenich Product Forms: Quality feed additives for more sustainable farming.* https://www.dsm.com/anh/news/downloads/infographics-checklists-and-guides/quality-feed-additives-for-more-sustainable-farming.html

dsm-firmenich Animal Nutrition and Health. (2022b). *OVN Optimum Vitamin Nutrition® guidelines.* https://www.dsm.com/anh/products-and-services/tools/ovn.html

Duan, Y., Zhu, X., Han, D., Yang, Y., & Xie, S. (2012). Dietary choline requirement in slight methionine-deficient diet for juvenile gibel carp (*Carassius auratus gibelio*). *Aquaculture Nutrition, 18*(6), 620–627. https://doi.org/10.1111/j.1365-2095.2011.00930.x

Dumas, A. (2022). Optimal vitamin Nutrition to foster sustainable salmonid aquaculture. International Symposium of Fish Nutrition and Feeding, 2022; June 5–9. Sorrento, Italy.

Duncan, P. L., Lovell, R. T., Butterworth, C. E., Jr., Freeberg, L. E., & Tamura, T. (1993). Dietary folate requirement determined for channel catfish, *Ictalurus punctatus. Journal of Nutrition, 123*(11), 1888–1897. https://doi.org/10.1093/jn/123.11.1888

Dunning, M. W. (2018). *Quantification and profiling of hepatic retinoids in freshwater fishes by liquid chromatography – Tandem mass spectrometry.* http://hdl.handle.net/10012/13059 [MSc Thesis]. University of Waterloo.

Dupree, H. K. (1966). Vitamins essential for growth of channel catfish. https://pubs.usgs.gov/publication/tp7. United States Department of the Interior, Fish and Wildlife Service, Bureau of Sport Fisheries and Wildlife.

Dupree, H. (1970). *Dietary requirement of vitamin A acetate and beta carotene. Progress in sport fishery research. Resource Publ., 88* (pp. 148–150). Bureau of Sport, Fisheries and Wildlife.

Ebadi, H., Zakeri, M., Mousavi, S. M., Yavari, V., & Souri, M. (2021). The interaction effects of dietary lipid, vitamin E and vitamin C on growth performance, feed utilization, muscle proximate composition and antioxidant enzyme activity of white leg shrimp (*Litopenaeus vannamei*). *Aquaculture Research, 52*(5), 2048–2060. https://doi.org/10.1111/are.15056

EFSA FEEDAP Panel. (2005). Opinion of the Scientific Panel on additives and products or substances used in animal feed (FEEDAP) on the safety of use of colouring agents in animal nutrition – Part I. General principles and astaxanthin. *EFSA Journal, 3*(12), 291. https://doi.org/10.2903/j.efsa.2005.291

EFSA FEEDAP Panel. (2007). Safety and efficacy of Panaferd-AX (red carotenoid rich bacterium *Paracoccus carotinifaciens*) as feed additive for salmon and trout. *EFSA Journal, 546*, 1–30. https://doi.org/10.2903/j.efsa.2007.546

EFSA FEEDAP Panel. (2012). Scientific Opinion on the safety and efficacy of folic acid as a feed additive for all animal species. *EFSA Journal, 10*(5), 2674. https://doi.org/10.2903/j.efsa.2012.2674

EFSA FEEDAP Panel. (2016). Safety and efficacy of lecithins for all animal species. *EFSA Journal, 14*(8), 13. https://doi.org/10.2903/j.efsa.2016.4561

EFSA FEEDAP Panel. (2017). Scientific opinion on the safety of vitamin D3 addition to feeding stuffs for fish. *EFSA Journal*, 2017(15):4713. https://doi.org/10.2903/j.efsa.2017.4713

EFSA FEEDAP Panel. (2023). Scientific Opinion on the safety and efficacy of a feed additive consisting of vitamin B12 (cyanocobalamin) produced by fermentation with *Ensifer adhaerens* CGMCC 19596 for all animal species (Hebei Huarong Pharmaceutical Co. ltd). *EFSA Journal, 21*(4), 7972, 14 pp. https://doi.org/10.2903/j.efsa.2023.7972

EFSA NDA Panel. (2014). Scientific Opinion on the safety of astaxanthin-rich ingredients (AstaReal A1010 and AstaReal L10) as novel food ingredients. *EFSA Journal, 12*(7), 3757. https://doi.org/10.2903/j.efsa.2014.3757

Eggersdorfer, M., Laudert, D., Létinois, U., McClymont, T., Medlock, J., Netscher, T. and Bonrath, W. (2012) One hundred years of vitamins a success story of the natural sciences. *Angew. Chem. Int. Ed. Engl.* 51(52):12960–12990. https://doi.org/10.1002/anie.201205886

El Basuini, M. F., Shahin, S. A., Teiba, I. I., Zaki, M. A. A., El-Hais, A. M., Sewilam, H., Almeer, R., Abdelkhalek, N., & Dawood, M. A. O. (2021). The influence of dietary coenzyme Q10 and vitamin C on the growth rate, immunity, oxidative-related genes, and the resistance against *Streptococcus agalactiae* of Nile tilapia (*Oreochromis niloticus*). *Aquaculture, 531*, 735862. https://doi.org/10.1016/j.aquaculture.2020.735862

Elbahnaswy, S., & Elshopakey, G. E. (2023). Recent progress in practical applications of a potential carotenoid astaxanthin in aquaculture industry: A review. *Fish Physiology and Biochemistry*. https://doi.org/10.1007/s10695-022-01167-0

Elkaradawy, A., Abdel-Rahim, M. M., Albalawi, A. E., Althobaiti, N. A., Abozeid, A. M., & Mohamed, R. A. (2021). Synergistic effects of the soapbark tree, *Quillaja saponaria* and vitamin E on water quality, growth performance, blood health, gills and intestine histomorphology of Nile tilapia, *Oreochromis niloticus* fingerlings. *Aquaculture Reports, 20*, 100733. https://doi.org/10.1016/j.aqrep.2021.100733

Ellis, N. R., & Madsen, L. L. (1944). The thiamine requirements of pigs as related to the fat content of the diet. *Journal of Nutrition, 27*(3), 253–262. https://doi.org/10.1093/jn/27.3.253

El-Sayed, A. M., & Izquierdo, M. (2022). The importance of vitamin E for farmed fish–A review. *Reviews in Aquaculture, 14*(2), 688–703. https://doi.org/10.1111/raq.12619

El-Sayed, Y. S., El-Gazzar, A. M., El-Nahas, A. F., & Ashry, K. M. (2016). Vitamin C modulates cadmium-induced hepatic antioxidants' gene transcripts and toxicopathic changes in Nile tilapia, *Oreochromis niloticus*. *Environmental Science and Pollution Research International, 23*(2), 1664–1670. https://doi.org/10.1007/s11356-015-5412-8

Elvehjem, C. A., Madden, R.J., Strong, F.M. and Wolley, D.W. (1974) The isolation and identification of the anti-black tongue factor. *Nutr. Rev.* 32(2):48–50. https://doi.org/10.1111/j.1753-4887.1974.tb06263.x

Ensminger, A. H., Ensminger, M. E., Konlande, J. E., & Robson, J. R. K. (1983) In Ensminger, A. H. (Ed.). Foods and nutrition encyclopedia. https://www.academia.edu/58476183/Foods_and_nutrition_encyclopedia_2nd_edn_Vols_1_and_2_A_H_Ensminger_M_E_Ensminger_J_E_Konlande_and_J_R_K_Robson_CRC_Press_Boca_Raton_Florida_1994_2415_pp_ISBN_0_8493_8980_1_set_. Ensminger Publishing, USA.

Eskelinen, P. (1989). Effects of different diets on egg production and egg quality of Atlantic salmon (*Salmo salar* L.). *Aquaculture, 79*(1–4), 275–281. https://doi.org/10.1016/0044-8486(89)90468-7

Eskola, M., Kos, G., Elliott, C.T., Hajšlová, J., Mayar, S.& Krska, R. (2020) Worldwide contamination of food-crops with mycotoxins: Validity of the widely cited 'FAO estimate' of 25. Crit. Rev. Food Sci. Nutr.60(16): 2773–2789. https://doi.org/10.1080/10408398.2019.1658570

Espe, M., Zerrahn, J.-E., Holen, E., Rønnestad, I., Veiseth-Kent, E., & Aksnes, A. (2016). Choline supplementation to low methionine diets increase phospholipids in Atlantic salmon, while taurine supplementation had no effects on phospholipid status, but improved taurine status. *Aquaculture Nutrition, 22*(4), 776–785. https://doi.org/10.1111/anu.12298

Espe, M., Vikeså, V., Helgøy Thomsen, T., Adam, A.-C., Saito, T., & Skjærven, K. H. (2020). Atlantic salmon fed a nutrient package of surplus methionine, vitamin B12, folic acid and vitamin B6 improved growth and reduced the relative liver size, but when in excess growth reduced. *Aquaculture Nutrition, 26*(2), 477–489. https://doi.org/10.1111/anu.13010

Esteban-Pretel, G., Marín, M. P., Renau-Piqueras, J., Barber, T., & Timoneda, J. (2010). Vitamin A deficiency alters rat lung alveolar basement membrane: Reversibility by retinoic acid. *Journal of Nutritional Biochemistry, 21*(3), 227–236. https://doi.org/10.1016/j.jnutbio.2008.12.007

European Commission. (2002). *Report of the Scientific committee for animal nutrition on the use of astaxanthin-rich Phaffia rhodozyma in feedingstuffs for salmon and trout. Health and consumer protection directorate-general, directorate C – Scientific opinions* https://food.ec.europa.eu/system/files/2020-12/sci-com_scan-old_report_out76.pdf

European Commission. (2020). Commission implementing regulation (EU) 2020/998 of 9 July 2020 concerning the renewal of the authorisation of astaxanthin–dimethyldisuccinate as a feed additive for fish and crustaceans and repealing Regulation (EC) no 393/2008. *Official Journal of the European Union, L221*(96–98). https://eur-lex.europa.eu/legal-content/EN/TXT/PDF/?uri=CELEX:32020R0998&from=EN

Evans, H.M. & Bishop, K.S. (1922) On the existence of a hitherto unrecognized dietary factor essential for reproduction. Science 56(1458), 650-651. https://doi.org/10.1126/science.56.1458.650

Fagone, P., & Jackowski, S. (2013). Phosphatidylcholine and the CDP-choline cycle. *Biochimica et Biophysica Acta, 1831*(3), 523–532. https://doi.org/10.1016/j.bbalip.2012.09.009

Faizan, M., Stubhaug, I., Menoyo, D., Esatbeyoglu, T., Wagner, A. E., Struksnæs, G., Koppe, W., & Rimbach, G. (2013). Dietary alpha-tocopherol affects tissue vitamin E and malondialdehyde levels but does not change antioxidant enzymes and fatty acid composition in farmed Atlantic salmon (*Salmo salar L.*). *International Journal for Vitamin and Nutrition Research. Internationale Zeitschrift Fur Vitamin- und Ernahrungsforschung. Journal International de Vitaminologie et de Nutrition, 83*(4), 238–245. https://doi.org/10.1024/0300-9831/a000166

Falah, F., Rajabi Islami, H., & Shamsaie Mehrgan, M. (2020). Dietary folic acid improved growth performance, immuno-physiological response and antioxidant status of fingerling Siberian sturgeon, *Acipenser baerii* (Brandt 1896). *Aquaculture Reports, 17*, 100391. https://doi.org/10.1016/j.aqrep.2020.100391

Falahatkar, B., Dabrowski, K., & Arslan, M. (2011). Ascorbic acid turnover in rainbow trout, *Oncorhynchus mykiss*: Is there a vitamin enrichment effect during embryonic period on the juvenile fish "sensitivity" to deficiency? *Aquaculture, 320*(1–2), 99–105. https://doi.org/10.1016/j.aquaculture.2011.08.012

Falcon, D. R., Barros, M. M., Pezzato, L. E., Sampaio, F. G., & Hisano, H. (2007). Physiological responses of Nile tilapia, *Oreochromis niloticus*, fed vitamin C- and lipid-supplemented diets and submitted to low-temperature stress. *Journal of the World Aquaculture Society, 38*(2), 287–295. https://doi.org/10.1111/j.1749-7345.2007.00098.x

Food and Agriculture Organization (FAO). (2022). Sustainability in action. *State of World Fisheries and Aquaculture* 2020. In: FAO Rome, Italy. https://doi.org/10.4060/ca9229en

Farquharson, C., Berry, E. B., Barbara-Mawer, E., Seawright, E., & Whitehead, C. C. (1998). Ascorbic acid-induced chondrocyte differentiation: The role of the extracellular matrix and 1,25 dihydroxycholecalciferols. *European Journal of Cell Biology, 76*(2), 110–118. https://doi.org/10.1016/S0171-9335(98)80023-X

Fatima, M., Afzal, M., & Shah, S. Z. H. (2019). Effect of dietary oxidized oil and vitamin E on growth performance, lipid peroxidation and fatty acid profile of *Labeo rohita* fingerlings. *Aquaculture Nutrition, 25*(2), 281–291. https://doi.org/10.1111/anu.12851

Fawzy, S., Wang, W., Zhou, Y., Xue, Y., Yi, G., Wu, M., & Huang, X. (2022). Can dietary β-carotene supplementation provide an alternative to astaxanthin on the performance of growth, pigmentation, biochemical, and immuno-physiological parameters of Litopenaeus vannamei ?. Rep. 23. *Aquaculture Reports, 23*. https://doi.org/10.1016/j.aqrep.2022.101054

Feng, L., He, W., Jiang, J., Liu, Y., & Zhou, X.-Q. (2010). Effects of dietary pyridoxine on disease resistance, immune responses and intestinal microflora in juvenile Jian carp (*Cyprinus carpio* var. Jian). *Aquaculture Nutrition, 16*(3), 254–261. https://doi.org/10.1111/j.1365-2095.2009.00660.x

Feng, L., Huang, H.-H., Liu, Y., Jiang, J., Jiang, W.-D., Hu, K., Li, S.-H., & Zhou, X.-Q. (2011). Effect of dietary thiamin supplement on immune responses and intestinal microflora in juvenile Jian carp (*Cyprinus carpio* var. Jian). *Aquaculture Nutrition, 17*(5), 557–569. https://doi.org/10.1111/j.1365-2095.2011.00851.x

Feng, L., Li, S. Q., Jiang, W. D., Liu, Y., Jiang, J., Wu, P., Zhao, J., Kuang, S. Y., Tang, L., Tang, W. N., Zhang, Y. A., & Zhou, X. Q. (2016). Deficiency of dietary niacin impaired intestinal mucosal immune function via regulating intestinal NF-κB, Nrf2 and MLCK signaling pathways in young grass carp (*Ctenopharyngodon idella*). *Fish and Shellfish Immunology, 49*, 177–193. https://doi.org/10.1016/j.fsi.2015.12.015

Feng, T., Bai, J., Xu, X., Guo, Y., Huang, Z., & Liu, Y. (2018). Supplementation with N-carbamylglutamate and vitamin C: Improving gestation and lactation outcomes in sows under heat stress. *Animal Production Science, 58*(10), 1854–1859. https://doi.org/10.1071/AN15562

Fenstermacher, D. K., & Rose, R. C. (1986). Absorption of pantothenic acid in rat and chick intestine. *American Journal of Physiology, 250*(2 Pt 1), G155–G160. https://doi.org/10.1152/ajpgi.1986.250.2.G155

Ferland, G. (2006). Vitamin K. In B. A. Bowman & R. M. Russell (Eds.), *Present knowledge in nutrition* (9th ed) (pp. 220–230). International Life Sciences Institute. https://doi.org/10.1093/ajcn/85.5.1439a

Fernández, I., & Gisbert, E. (2011). The effect of vitamin A on flatfish development and skeletogenesis: A review. *Aquaculture, 315*(1–2), 34–48. https://doi.org/10.1016/j.aquaculture.2010.11.025

Ferreira, M., Larsen, B. K., Granby, K., Cunha, S. C., Monteiro, C., Fernandes, J. O., Nunes, M. L., Marques, A., Dias, J., Cunha, I., Castro, L. F. C., & Valente, L. M. P. (2020). Diets supplemented with *Saccharina latissima* influence the expression of genes related to lipid metabolism and oxidative stress modulating rainbow trout (*Oncorhynchus mykiss*) fillet composition. *Food and Chemical Toxicology, 140*, 111332. https://doi.org/10.1016/j.fct.2020.111332

Fichter, S. A., & Mitchell, G. E. (1997). Sheep blood response to orally supplemented vitamin K dissolved in coconut oil. *Journal of Animal Science, 72*(1), 266 [Abstr.].

Fisher, L. R., & Kon, S. K. (1959). Vitamin A in the invertebrates. *Biological Reviews, 34*(1), 1–36. https://doi.org/10.1111/j.1469-185X.1959.tb01300.x

Fitzsimons, J. D., Honeyfield, D. C., & Rush, S. (2021). Ontogenetic dietary changes and thiamine status of Lake Ontario chinook salmon, 2005–2006. *North American Journal of Fisheries Management, 41*(5), 1499–1513. https://doi.org/10.1002/nafm.10659

Flores, M., Díaz, F., Medina, R., Re, A. D., & Licea, A. (2007). Physiological, metabolic and haematological responses in white shrimp *Litopenaeus vannamei* (Boone) juveniles fed diets supplemented with astaxanthin acclimated to low-salinity water. *Aquaculture Research, 38*(7), 740–747. https://doi.org/10.1111/j.1365-2109.2007.01720.x

Fontagné-Dicharry, S., Lataillade, E., Surget, A., Brèque, J., Zambonino-Infante, J. L., & Kaushik, S. J. (2010). Effects of dietary vitamin A on broodstock performance, egg quality, early growth and retinoid nuclear receptor expression in rainbow trout (*Oncorhynchus mykiss*). *Aquaculture, 303*(1–4), 40–49. https://doi.org/10.1016/j.aquaculture.2010.03.009

Fox, H. M. (1991). Pantothenic acid. In L. J. Machlin (Ed.), *Handbook of vitamins* (2nd ed) (pp. 429–451). Marcel Dekker, Inc.

Franceschi, R. T. (1992). The role of ascorbic acid in mesenchymal differentiation. *Nutrition Reviews, 50*(3), 65–70. https://doi.org/10.1111/j.1753-4887.1992.tb01271.x

Fraser, M. R., & De Nys, R. (2011). A quantitative determination of deformities in barramundi (*Lates calcarifer*; Bloch) fed a vitamin deficient diet. *Aquaculture Nutrition, 17*(3), 235–243. https://doi.org/10.1111/j.1365-2095.2009.00734.x

Freiser, H., & Jiang, Q. (2009). Optimization of the enzymatic hydrolysis and analysis of plasma conjugated gamma-CEHC and sulfated long-chain carboxychromanols, metabolites of vitamin E. *Analytical Biochemistry, 388*(2), 260–265. https://doi.org/10.1016/j.ab.2009.02.027

Friesecke, H. (1980). *Vitamin B12, 1728.* F. Hoffmann-La Roche, and Co., Ltd.

Frigg, M. (1987). Biotin in poultry and swine rations and its significance for optimum performance. *Proceedings of the Maryland Nutrition Conference* (pp. 101–108).

Frigg, M., Prabucki, A. L., & Ruhdel, E. U. (1990). Effect of dietary vitamin E levels on oxidative stability of trout fillets. *Aquaculture, 84*(2), 145–158. https://doi.org/10.1016/0044-8486(90)90344-M

Frigg, M., & Volker, L. (1994). Biotin inclusion helps optimize animal performance. *Feedstuffs, 66*(1), 12–13.

Frye, T. M. (1994). The performance of vitamins in multicomponent premixes. In *Proceedings of the Roche Technical Symposium.*

Funada, U., Wada, M., Kawata, T., Mori, K., Tamai, H., Isshiki, T., Onoda, J., Tanaka, N., Tadokoro, T., & Maekawa, A. (2001). Vitamin B12 deficiency affects immunoglobulin production and cytokine levels in mice. *International Journal for Vitamin and Nutrition Research: Internationale Zeitschrift Fur Vitamin- und Ernahrungsforschung. Journal International de Vitaminologie et de Nutrition, 71*(1), 60–65. https://doi.org/10.1024/0300-9831.71.1.60

Funk, C. and Dubin, H.E. (1922) The vitamins. Williams & Wilkins, Co., Baltimore, MD.

Furuita, H., Tanaka, H., Yamamoto, T., Suzuki, N., & Takeuchi, T. (2003). Supplemental effect of vitamin A in diet on the reproductive performance and egg quality of the Japanese flounder *Paralichthys olivaceus* (T and S). *Aquaculture Research, 34*(6), 461–468. https://doi.org/10.1046/j.1365-2109.2003.00831.x

Gabaudan, J., & Hardy, R. (2000). *Vitamins sources for fish feeds. Encyclopedia of aquaculture* (pp. 961–965). Wiley & Sons, Inc.

Gadient, M. (1986). Effect of pelleting on nutritional quality of feed. In *Proceedings of the Maryland Nutr. Conf.* Feed Manuf. College Park, MD.

Gadient, M., & Schai, E. (1994). Leaching of various vitamins from shrimp feed. *Aquaculture, 124*(1–4), 201–205. https://doi.org/10.1016/0044-8486(94)90378-6

Gao, J., Koshio, S., Ishikawa, M., Yokoyama, S., Mamauag, R. E. P., & Han, Y. (2012a). Effects of dietary oxidized fish oil with vitamin E supplementation on growth performance and reduction of lipid peroxidation in tissues and blood of red sea bream *Pagrus major. Aquaculture, 356–357*, 73–79. https://doi.org/10.1016/j.aquaculture.2012.05.034

Gao, J., Koshio, S., Ishikawa, M., Yokoyama, S., Abdul, Md., K., & Asda, L. (2012b). Interaction between vitamin C and E on oxidative stress in tissues and blood of red sea bream (*Pagrus major*). *Aquaculture Science, 60*(1), 119–126. https://doi.org/10.11233/aquaculturesci.60.119

Gao, J., Koshio, S., Ishikawa, M., Yokoyama, S., & Mamauag, R. E. P. (2014). Interactive effects of vitamin C and E supplementation on growth performance, fatty acid composition and reduction of oxidative stress in juvenile Japanese flounder *Paralichthys olivaceus* fed dietary oxidized fish oil. *Aquaculture, 422–423*, 84–90. https://doi.org/10.1016/j.aquaculture.2013.11.031

Gardner, H. W. (1989). Oxygen radical chemistry of polyunsaturated fatty acids. *Free Radical Biology and Medicine, 7*(1), 65–86. https://doi.org/10.1016/0891-5849(89)90102-0

Gasperi, V., Sibilano, M., Savini, I., & Catani, M. V. (2019). Niacin in the central nervous system: An update of biological aspects and clinical applications. *International Journal of Molecular Sciences, 20*(4), 974. https://doi.org/10.3390/ijms20040974

Georga, I., Glynatsi, N., Baltzois, A., Karamanos, D., Mazurais, D., Darias, M. J., Cahu, C. L., Zambonino-Infante, J. L., & Koumoundouros, G. (2011). Effect of vitamin A on the skeletal morphogenesis of European sea bass, *Dicentrarchus labrax* (Linnaeus, 1758). *Aquaculture Research, 42*(5), 684–692. https://doi.org/10.1111/j.1365-2109.2010.02676.x

George, J. C., Barnett, B. J., Cho, C. Y., & Slinger, S. J. (1981). Vitamin D3 and muscle function in the rainbow trout. *Cytobios, 31*(121), 7–18. ISSN 0011-4529.

Gesto, M., Castro, L. F. C., Reis-Henriques, M. A., & Santos, M. M. (2012). Tissue-specific distribution patterns of retinoids and didehydroretinoids in rainbow trout *Oncorhynchus mykiss. Comparative Biochemistry and Physiology. Part B, Biochemistry and Molecular Biology, 161*(1), 69–78. https://doi.org/10.1016/j.cbpb.2011.09.006

Geurden, I., Kaushik, S., & Corraze, G. (2008). Dietary phosphatidylcholine affects postprandial plasma levels and digestibility of lipid in common carp (*Cyprinus carpio*). *British Journal of Nutrition, 100*(3), 512–517. https://doi.org/10.1017/S0007114508904396

Ghosh, H. P., Sarkar, P. K., & Guha, B. C. (1963). Distribution of the bound form of nicotinic acid in natural materials. *Journal of Nutrition, 79*(4), 451–453. https://doi.org/10.1093/jn/79.4.451

Ghotbi, M., Ghotbi, M., & Azari Takami, G. (2011). Contribution of vitaton (β-carotene) to the rearing factors, survival rate and visual flesh color of rainbow trout fish in comparison with astaxanthin. *World Academy of Science, Engineering and Technology, 59*, 1649–1654. https://doi.org/10.5281/zenodo.1063214

Gibson, G. E., & Zhang, H. (2002). Interactions of oxidative stress with thiamine homeostasis promote neurodegeneration. *Neurochemistry International, 40*(6), 493–504. https://doi.org/10.1016/S0197-0186(01)00120-6

Gjedrem, T., Robinson, N., & Rye, M. (2012). The importance of selective breeding in aquaculture to meet future demands for animal protein: A review. *Aquaculture, 350–353*, 117–129. https://doi.org/10.1016/j.aquaculture.2012.04.008

Glencross, B., Fracalossi, D. M., Hua, K., Izquierdo, M., Mai, K., Øverland, M., Robb, D., Roubach, R., Schrama, J., Small, B., Tacon, A., Valente, L. M. P., Viana, M., Xie, S., & Yakupityage, A. (2023). Harvesting the benefits of nutritional research to address global challenges in the 21st century Journal of the World Aquaculture Society, 54(2), 343–363. https://doi.org/10.1111/jwas.12948

Godoy-Parejo, C., Deng, C., Zhang, Y., Liu, W., & Chen, G. (2020). Roles of vitamins in stem cells. *Cellular and Molecular Life Sciences, 77*(9), 1771–1791. https://doi.org/10.1007/s00018-019-03352-6

Goldsmith, T. H., & Cronin, T. W. (1993). The retinoids of seven species of mantis shrimp. *Visual Neuroscience, 10*(5), 915–920. https://doi.org/10.1017/s095252380000612x

Golub, M. S., & Gershwin, M. E. (1985). Stress-induced immunomodulation: What is it if it is? In G. P. Moberg (Ed.), *Animal stress*. Springer. https://doi.org/10.1007/978-1-4614-7544-6_11

Gonçalves, A., Roi, S., Nowicki, M., Dhaussy, A., Huertas, A., Amiot, M. J., & Reboul, E. (2015). Fat-soluble vitamin intestinal absorption: Absorption sites in the intestine and interactions for absorption. *Food Chemistry, 172*, 155–160. https://doi.org/10.1016/j.foodchem.2014.09.021

Gonçalves, R. A., Naehrer, K., & Santos, G. A. (2018). Occurrence of mycotoxins in commercial aquafeeds in Asia and Europe: A real risk to aquaculture? *Reviews in Aquaculture, 10*(2), 263–280. https://doi.org/10.1111/raq.12159

Gong, H., Lawrence, A. L., Jiang, D.-H., & Gatlin III, D. M. (2003). Effect of dietary phospholipids on the choline requirement of *Litopenaeus vannamei* juveniles. *Journal of the World Aquaculture Society, 34*(3), 289–299. https://doi.org/10.1111/j.1749-7345.2003.tb00067.x

Graff, I. E., Lie, Ø., & Aksnes, L. (1999). In vitro hydroxylation of vitamin D3 and 25-hydroxyvitamin D3 in tissues of Atlantic salmon *Salmo salar*, Atlantic mackerel *Scomber scombrus*, Atlantic halibut *Hippoglossus hippoglossus* and Atlantic cod *Gadus morhua*. *Aquaculture Nutrition, 5*(1), 23–32. https://doi.org/10.1046/j.1365-2095.1999.00084.x

Graff, I. E., Høie, S., Totland, G. K., & Lie, Ø. (2002b) Three different levels of dietary vitamin D 3 fed to first-feeding fry of Atlantic salmon (*Salmo salar L.*): Effect on growth, mortality, calcium content and bone formation. *Aquaculture Nutrition, 8*(2), 103–111. https://doi.org/10.1046/j.1365-2095.2002.00197.x

Graff, I. E., Stefansson, S. O., Aksnes, L., & Lie, Ø. (2004). Plasma levels of vitamin D3 metabolites during parr-smolt transformation of Atlantic salmon *Salmo salar L*. *Aquaculture, 240*(1–4), 617–622. https://doi.org/10.1016/j.aquaculture.2004.06.025

Green, A. S., & Fascetti, A. J. (2016). Meeting the vitamin A requirement: The efficacy and importance of β-carotene in animal species. *The Scientific World Journal, 2016*, 7393620. http://doi.org/10.1155/2016/7393620

Green, R., & Miller, J. W. (2013). Vitamin B12. In J. Zempleni, J. Suttie, J. Gregory & P. J. Stover (Eds.), *Handbook of vitamins* (5th ed) (pp. 447–490). CRC Press, Taylor & Francis Group, LLC ISBN 9781466515567. https://doi.org/10.1201/b15413

Gregory, J. F. (1989). Chemical and nutritional aspects of folate research: Analytical procedures, methods of folate synthesis, stability and bioavailability of dietary folates. *Advances in Food and Nutrition Research, 33*, 1–101. https://doi.org/10.1016/S1043-4526(08)60126-6

Gregory, J. F. (2001). Case study: Folate bioavailability. *Journal of Nutrition, 131*(4), Suppl., 1376S–1382S. https://doi.org/10.1093/jn/131.4.1376S

Gregory, J. F., Bhandari, S. D., Bailey, L. B., Toth, J. P., Baumgartner, T. G., & Cerda, J. J. (1991). Relative bioavailability of deuterium-labeled monoglutamyl and hexaglutamyl folates in human subjects. *American Journal of Clinical Nutrition, 53*(3), 736–740. https://doi.org/10.1093/ajcn/53.3.736

Griboff, J., Morales, D., Bertrand, L., Bonansea, R. I., Monferrán, M. V., Asis, R., Wunderlin, D. A., & Amé, M. V. (2014). Oxidative stress response induced by atrazine in *Palaemonetes argentinus*: The protective effect of vitamin E. *Ecotoxicology and Environmental Safety*, 108, 1–8. https://doi.org/10.1016/j.ecoenv.2014.06.025

Griffin, M. E., Wilson, K. A., White, M. R., & Brown, P. B. (1994). Dietary choline requirement of juvenile hybrid striped bass. *Journal of Nutrition*, 124(9), 1685–1689. https://doi.org/10.1093/jn/124.9.1685

Griminger, P. (1984). Vitamin K in animal nutrition: Deficiency can be fatal. Part 1. *Feedstuffs*, 56(38), 24–25.

Griminger, P., & Donis, O. (1960). Potency of vitamin K1 and analogues in counteracting the effects of dicumarol and sulfaquinoxaline in the chick. *Journal of Nutrition*, 70(3), 361–368. https://doi.org/10.1093/jn.70.3.361

Grisdale-Helland, B., Helland, S. J., & Åsgård, T. (1991). Problems associated with the present use of menadione sodium bisulfite and vitamin A in diets for Atlantic salmon. *Aquaculture*, 92, 351–358. https://doi.org/10.1016/0044-8486(91)90040-E

Gross, J., & Budowski, P. (1966). Conversion of carotenoids into vitamins A1 and A2 in two species of freshwater fish. *Biochemical Journal*, 101(3), 747–754. https://doi.org/10.1042/bj1010747

Guary, M., Kanazawa, A., Tanaka, N., & Ceccaldi, H. J. (1976). Nutritional requirement of prawn: VI. Requirement for ascorbic acid. *Memoirs of Faculty of Fisheries Kagoshima University*, 25, 53–57. https://ir.kagoshima-u.ac.jp/?action=repository_action_common_download&item_id=4655&item_no=1&attribute_id=16&file_no=1

Guggenheim, K.Y. (1995) Basic issues of the history of nutrition. Magnes Press, Hebrew University.

Guillou, A., Choubert, G., Storebakken, T., de la Noüet, J., & Kaushik, S. (1989). Bioconversion pathway of astaxanthin into retinol2 in mature rainbow trout (*Salmo gairdneri* Rich.). *Comparative Biochemistry and Physiology Part B*, 94(3), 481–485. https://doi.org/10.1016/0305-0491(89)90185-5

Guimarães, I. G., Lim, C., Yildirim-Aksoy, M., Li, M. H., & Klesius, P. H. (2014). Effects of dietary levels of vitamin A on growth, hematology, immune response and resistance of Nile tilapia (*Oreochromis niloticus*) to *Streptococcus iniae*. *Animal Feed Science and Technology*, 188, 126–136. https://doi.org/10.1016/j.anifeedsci.2013.12.003

Guimarães, I. G., Pezzato, L. E., Santos, V. G., Orsi, R. O., & Barros, M. M. (2016). Vitamin A affects haematology, growth and immune response of Nile tilapia (*Oreochromis niloticus*, L.), but has no protective effect against bacterial challenge or cold-induced stress. *Aquaculture Research*, 47(6), 2004–2018. https://doi.org/10.1111/are.12656

Hadas, E., Koven, W., Sklan, D., & Tandler, A. (2003). The effect of dietary phosphatidylcholine on the assimilation and distribution of ingested free oleic acid (18:1n-9) in gilthead seabream (*Sparus aurata*) larvae. *Aquaculture*, 217(1–4), 577–588. https://doi.org/10.1016/S0044-8486(02)00431-3

Halver, J. E. (1957). Nutrition of salmonoid fishes: III. Water-soluble vitamin requirements of chinook salmon. *Journal of Nutrition*, 62(2), 225–243. https://doi.org/10.1093/jn/62.2.225

Halver, J. E., Ashley, L. M., & Smith, R. R. (1969). Ascorbic acid requirements of coho salmon and rainbow trout. *Transactions of the American Fisheries Society*, 98(4), 762–771. https://doi.org/10.1577/1548-8659(1969)98[762:AAROCS]2.0.CO;2

Halver, J. E. (2002). The vitamins. In J. E. Halver & R. W. Hardy (Eds.). https://faculty.ksu.edu.sa/sites/default/files/Fish%20Nutrition.pdf, *Fish nutrition* (3rd ed) (pp. 61–141). Academic Press.

Hamidoghli, A., Bae, J., Won, S., Lee, S., Kim, D.-J., & Bai, S. C. (2019). A review on Japanese eel (*Anguilla japonica*) aquaculture, with special emphasis on nutrition. *Reviews in Fisheries Science and Aquaculture*, 27(2), 226–241. https://doi.org/10.1080/23308249.2019.1583165

Hamre, K. (2011). Metabolism, interactions, requirements and functions of vitamin E in fish. *Aquaculture Nutrition*, 17(1), 98–115. https://doi.org/10.1111/j.1365-2095.2010.00806.x

Hamre, K., & Lie, Ø. (1995). Minimum requirement of vitamin E for Atlantic salmon, *Salmo salar* L., at first feeding. *Aquaculture Research*, 26(3), 175–184. https://doi.org/10.1111/j.1365-2109.1995.tb00900.x

Hamre, K., & Lie, Ø. (1997). Retained levels of dietary α-, γ- and δ-tocopherol in tissues and body fluids of Atlantic salmon, *Salmo salar*, L. *Aquaculture Nutrition*, 3(2), 99–107. https://doi.org/10.1046/j.1365-2095.1997.00077.x

Hamre, K., Waagbø, R., Berge, R. K., & Lie, Ø. (1997). Vitamins C and E interact in juvenile Atlantic salmon (*Salmo salar, L.*). *Free Radical Biology and Medicine, 22*(1–2), 137–149. https://doi.org/10.1016/S0891-5849(96)00281-X

Hamre, K., Sissener, N. H., Lock, E. J., Olsvik, P. A., Espe, M., Torstensen, B. E., Silva, J., Johansen, J., Waagbø, R., & Hemre, G. I. (2016). Antioxidant nutrition in Atlantic salmon (*Salmo salar*) parr and post-smolt, fed diets with high inclusion of plant ingredients and graded levels of micronutrients and selected amino acids. *PeerJ, 4*, e2688. https://doi.org/10.7717/peerj.2688

Hamre, K., Bjørnevik, M., Espe, M., Conceição, L. E. C., Johansen, J., Silva, J., Hillestad, M., Prabhu, A. J., Taylor, J. F., Tocher, D. R., Lock, E. J., & Hemre, G. I. (2020). Dietary micronutrient composition affects fillet texture and muscle cell size in Atlantic salmon (*Salmo salar*). *Aquaculture Nutrition, 26*(3), 936–945. https://doi.org/10.1111/anu.13051

Hamre, K., Micallef, G., Hillestad, M., Johansen, J., Remø, S., Zhang, W., Ødegård, E., Araujo, P., Prabhu Philip, A. J., & Waagbø, R. (2022). Changes in daylength and temperature from April until August for Atlantic salmon (*Salmo salar*) reared in sea cages, increase growth, and may cause consumption of antioxidants, onset of cataracts and increased oxidation of fillet astaxanthin. *Aquaculture, 551*. https://doi.org/10.1016/j.aquaculture.2022.737950

Hankes, L. V. (1984). Nicotinic acid and nicotinamide. In L. J. Machlin (Ed.), *Handbook of vitamins*. Marcel Dekker, Inc.

Hansen, A.-C., Olsvik, P. A., & Hemre, G.-I. (2013). Effect of different dietary vitamin B12 levels on their retention in the body of zebrafish *Danio rerio* and on the gene expression of vitamin B12 binding proteins. *Aquaculture Nutrition, 19*(3), 413–420. https://doi.org/10.1111/j.1365-2095.2012.00975.x

Hansen, A.-C., Waagbø, R., & Hemre, G.-I. (2015). New B vitamin recommendations in fish when fed plant-based diets. *Aquaculture Nutrition, 21*(5), 507–527. https://doi.org/10.1111/anu.12342

Hansen, A. K. G., Kortner, T. M., Krasnov, A., Björkhem, I., Penn, M., & Krogdahl, Å. (2020a). Choline supplementation prevents diet induced gut mucosa lipid accumulation in post-smolt Atlantic salmon (*Salmo salar L.*). *BMC Veterinary Research, 16*(1), 32. https://doi.org/10.1186/s12917-020-2252-7

Hansen, A. K. G., Kortner, T. M., Denstadli, V., Måsøval, K., Björkhem, I., Grav, H. J., & Krogdahl, Å. (2020b). Dose–response relationship between dietary choline and lipid accumulation in pyloric enterocytes of Atlantic salmon (*Salmo salar L.*) in seawater. *British Journal of Nutrition, 123*(10), 1081–1093. https://doi.org/10.1017/S0007114520000434

Harabawy, A. S. A., & Mosleh, Y. Y. I. (2014). The role of vitamins A, C, E and selenium as antioxidants against genotoxicity and cytotoxicity of cadmium, copper, lead and zinc on erythrocytes of Nile tilapia, *Oreachromis niloticus*. *Ecotoxicology and Environmental Safety, 104*, 28–35. http://doi.org/10.1016/j.ecoenv.2014.02.015

Harare, K., Berge, R. K., & Lie, Ø. (1998). Oxidative stability of Atlantic salmon (*Salmo salar L.*) fillet enriched in α-, γ-, and δ-tocopherol through dietary supplementation. *Food Chemistry, 62*(2), 173–178. https://doi.org/10.1016/S0308-8146(97)00209-4

Harder, A. M., Ardren, W. R., Evans, A. N., Futia, M. H., Kraft, C. E., Marsden, J. E., Richter, C. A., Rinchard, J., Tillitt, D. E., & Christie, M. R. (2018). Thiamine deficiency in fishes: Causes, consequences, and potential solutions. *Reviews in Fish Biology and Fisheries, 28*(4), 865–886. https://doi.org/10.1007/s11160-018-9538-x

Hardy, R. W., Halver, J. E., & Brannon, E. L. (1979). Effect of dietary protein level on the pyridoxine requirement and disease resistance of chinook salmon (*Oncorhynchus tshawytscha*). In *Proceedings of the World Symposium on Finfish Nutrition and Feed Technology, 1* (pp. 253–260). Hamburg, Germany, 20–23 June 1978.

Hardy, R. W., & Brezas, A. (2022). Diet formulation and manufacture. In R. W. Hardy & S. J. Kaushik (Eds.), *Fish nutrition* (4th ed) (pp. 643–708). Academic Press.

Hari, B., & Kurup, B. M. (2002). Vitamin C (ascorbyl 2 polyphosphate) requirement of freshwater prawn *Macrobrachium rosenbergii* (de Man). *Asian Fisheries Science, 15*(2), 145–154. https://doi.org/10.33997/j.afs.2002.15.2.006

Harlıoğlu, M. M., & Barım, Ö. (2004). The effect of dietary vitamin E on the pleopod egg and stage-1 juvenile numbers of freshwater crayfish *Astacus leptodactylus* (Eschscholtz, 1823). *Aquaculture, 236*(1–4), 267–276. https://doi.org/10.1016/j.aquaculture.2004.01.022

Harris, H.F. (1919). Pellagra. Macmillan Co. New York, New York.

Harrison, E. H. (2012). Mechanisms involved in the intestinal absorption of dietary vitamin A and provitamin A carotenoids. *Biochimica et Biophysica Acta, 1821*(1), 70–77. https://doi.org/10.1016/j.bbalip.2011.06.002

Harsij, M., Gholipour Kanani, H., & Adineh, H. (2020). Effects of antioxidant supplementation (nano-selenium, vitamin C and E) on growth performance, blood biochemistry, immune status and body composition of rainbow trout (*Oncorhynchus mykiss*) under sub-lethal ammonia exposure. *Aquaculture, 521*, 734942. https://doi.org/10.1016/j.aquaculture.2020.734942

Hasanthi, M., & Lee, K. J. (2023). Dietary niacin requirement of Pacific white shrimp (*Litopenaeus vannamei*). *Aquaculture, 566*, 739169. https://doi.org/10.1016/j.aquaculture.2022.739169

Hassaan, M. S., Goda, A. M. A.-S., Mahmoud, S. A., & Tayel, S. I. (2014). Protective effect of dietary vitamin E against fungicide copper oxychloride stress on Nile tilapia, *Oreochromis niloticus* (L.) fingerlings. *International Aquatic Research, 6*(1), 58. https://doi.org/10.1007/s40071-014-0058-6

Hatlen, B., Larsson, T., Østbye, T. K., Romarheim, O. H., Rubio, L. M., & Ruyter, B. (2022). Improved fillet quality in harvest-size Atlantic salmon fed high n-3 canola oil as a DHA-source. *Aquaculture, 560*, 738555. https://doi.org/10.1016/j.aquaculture.2022.738555

He, C. S., Gleeson, M., & Fraser, W. D. (2013). Measurement of circulating 25-hydroxy vitamin D using three commercial enzyme-linked immunosorbent assay kits with comparison to liquid chromatography: Tandem mass spectrometry method. *ISRN Nutrition, 2013*, 723139. https://doi.org/10.5402/2013/723139

He, H., & Lawrence, A. L. (1993a). Vitamin C requirements of the shrimp *Penaeus vannamei*. *Aquaculture, 114*(3–4), 305–316. https://doi.org/10.1016/0044-8486(93)90305-I

He, H., & Lawrence, A. L. (1993b). Vitamin E requirement of the shrimp *Penaeus vannamei*. *Aquaculture, 118*(3–4), 245–255. https://doi.org/10.1016/0044-8486(93)90460-G

He, H., Lawrence, A. L., & Liu, R. (1992). Evaluation of dietary essentiality of fat-soluble vitamins, A, D, E and K for penaeid shrimp (*Penaeus vannamei*). *Aquaculture, 103*(2), 177–185. https://doi.org/10.1016/0044-8486(92)90411-D

He, S., Ding, M., Watson Ray, G. W., Yang, Q., Tan, B., Dong, X., Chi, S., Liu, H., & Zhang, S. (2021). Effect of dietary vitamin D levels on growth, serum biochemical parameters, lipid metabolism enzyme activities, fatty acid synthase and hepatic lipase mRNA expression for orange-spotted grouper (*Epinephelus coioides*) in growth mid-stage. *Aquaculture Nutrition, 27*(3), 655–665. https://doi.org/10.1111/anu.13212

Head, B., Ramsey, S. A., Kioussi, C., Tanguay, R. L., & Traber, M. G. (2021). Vitamin E deficiency disrupts gene expression networks during zebrafish development. *Nutrients, 13*(2), 468. https://doi.org/10.3390/nu13020468

Hemre, G. I., Deng, D. F., Wilson, R. P., & Berntssen, M. H. G. (2004). Vitamin A metabolism and early biological responses in juvenile sunshine bass (*Morone chrysops × M. saxatilis*) fed graded levels of vitamin A. *Aquaculture, 235*(1–4), 645–658. https://doi.org/10.1016/j.aquaculture.2004.01.031

Hemre, G. I., Lock, E. J., Olsvik, P. A., Hamre, K., Espe, M., Torstensen, B. E., Silva, J., Hansen, A. C., Waagbø, R., Johansen, J. S., Sanden, M., & Sissener, N. H. (2016). Atlantic salmon (*Salmo salar*) require increased dietary levels of B-vitamins when fed diets with high inclusion of plant based ingredients. *PeerJ, 4*, e2493. https://doi.org/10.7717/peerj.2493

Henderson, L. M. (1984). Vitamin B6. In R. E. Olson et al. (Eds.), *Nutrition reviews, present knowledge in nutrition* (5th ed) The Nutrition Foundation, Inc.

Henderson, L. M., & Gross, C. J. (1979). Metabolism of niacin and niacinamide in perfused rat intestine. *Journal of Nutrition, 109*(4), 654–662. https://doi.org/10.1093/jn/109.4.654

Hernandez, L. H., & Hardy, R. W. (2020). Vitamin A functions and requirements in fish. *Aquaculture Research, 51*(8), 3061–3071. https://doi.org/10.1111/are.14667

Hernandez, L. H. H., Teshima, S. I., Ishikawa, M., Koshio, S., Gallardo-Cigarroa, F. J., Uyan, O., & Alam, S. (2009). Vitamin A effects and requirements on the juvenile Kuruma prawn *Marsupenaeus japonicus*. *Hidrobiológica, 19*, 217–223.

Hien, D., & Doolgindachbaporn, S. (2011). Effect of dietary pantothenic acid supplementary on culturing of green catfish (*Mystus nemurus*, Valenciennes, 1840). *Journal of Public Health Research, 4*(1), 1–6.

Hilton, J. W. (1983). Hypervitaminosis A in rainbow trout (*Salmo gairdneri*): Toxicity signs and maximum tolerable level. *Journal of Nutrition, 113*(9), 1737–1745. https://doi.org/10.1093/jn/113.9.1737

Hilton, J. W., & Ferguson, H. W. (1982). Effect of excess vitamin D3 on calcium metabolism in rainbow trout *Salmo gairdneri* Richardson. *Journal of Fish Biology, 21*(4), 373–379. https://doi.org/10.1111/j.1095-8649.1982.tb02842.x

Hilton, J. W., Cho, C. Y., & Slinger, S. J. (1978). Effect of graded levels of supplemental ascorbic acid in practical diets fed to rainbow trout (*Salmo gairdneri*). *Journal of the Fisheries Research Board of Canada, 35*(4), 431–436. https://doi.org/10.1139/f78-075

Hodges, R. E., Ohlson, M. A., & Bean, W. B. (1958). Pantothenic acid deficiency in man. *Journal of Clinical Investigation, 37*(11), 1642–1657. https://doi.org/10.1172/JCI103756

Hoffmann–La Roche. (1984). Roche technical Bulletin-Vitamin B12. *RCD 6723.* Hoffmann–La Roche, Inc. Nutley, NJ.

Hollander, D. (1973). Vitamin K1 absorption by everted intestinal sacs of the rat. *American Journal of Physiology, 225*(2), 360–364. https://doi.org/10.1152/ajplegacy.1973.225.2.360

Höller, U., Bakker, S. J. L., Düsterloh, A., Frei, B., Köhrle, J., Konz, T., Lietz, G., McCann, A., Michels, A. J., Molloy, A. M., Murakami, H., Rein, D., Saris, W. H. M., Schmidt, K., Shimbo, K., Schumacher, S., Vermeer, C., Kaput, J., . . . Rezzi, S. (2018). Micronutrient status assessment in humans: Current methods of analysis and future trends. *TrAC Trends in Analytical Chemistry, 102,* 110–122. https://doi.org/10.1016/j.trac.2018.02.001

Holst, A. & Frolich, T. (1907) Experimental studies relating to ship-beri-beri and scurvy. *The Journal of Hygiene* 7: 634–671. Reprinted (1974) *Nutrition Reviews, 32*(9), 273–275. https://doi.org/10.1111/j.1753-4887.1974.tb00973.x

Horvli, O., Lie, Ø., & Aksnes, L. (1998). Tissue distribution of vitamin D3 in Atlantic salmon *Salmo salar*: Effect of dietary level. *Aquaculture Nutrition, 4*(2), 127–131. https://doi.org/10.1046/j.1365-2095.1998.00062.x

Hosokawa, H., Shimeno, S., & Takeda, M. (2001). Dietary choline requirement of fingerling yellowtail. *Aquaculture Science, 49,* 231–235.

Hossain, M. A., Almatar, S. M., & Yaseen, S. B. (2016). Effects of varying levels of dietary vitamin E (α-tocopherol) on growth performance, proximate and fatty acid composition of juvenile silver pomfret (*Pampus argenteus* Euphrasen, 1788). *Journal of Applied Ichthyology, 32*(1), 55–60. https://doi.org/10.1111/jai.12956

Hsu, T. S., & Shiau, S. Y. (1999a). Tissue storage of vitamin E in grass shrimp *Penaeus monodon* fed dietary DL-α-tocopheryl acetate. *Fisheries Science, 65*(1), 169–170. https://doi.org/10.2331/fishsci.65.169

Hsu, T. S., & Shiau, S. Y. (1999b). Influence of dietary ascorbate derivatives on tissue copper, iron and zinc concentrations in grass shrimp, *Penaeus monodon. Aquaculture, 179*(1–4), 457–464. https://doi.org/10.1016/S0044-8486(99)00179-9

Hu, C. J., Chen, S. M., Pan, C. H., & Huang, C. H. (2006). Effects of dietary vitamin A or β-carotene concentrations on growth of juvenile hybrid tilapia, *Oreochromis niloticus × O. aureus. Aquaculture, 253*(1–4), 602–607. https://doi.org/10.1016/j.aquaculture.2005.09.003

Huang, F., Jiang, M., Wen, H., Wu, F., Liu, W., Tian, J., & Shao, H. (2016). Dietary vitamin C requirement of genetically improved farmed tilapia, *Oreochromis niloticus. Aquaculture Research, 47*(3), 689–697. https://doi.org/10.1111/are.12527

Huang, F., Wu, F., Zhang, S., Jiang, M., Liu, W., Tian, J., Yang, C., & Wen, H. (2017). Dietary vitamin C requirement of juvenile Chinese sucker (*Myxocyprinus asiaticus*). *Aquaculture Research, 48*(1), 37–46. https://doi.org/10.1111/are.12858

Huang, H.-H., Feng, L., Liu, Y., Jiang, J., Jiang, W.-D., Hu, K., Li, S.-H., & Zhou, X.-Q. (2011). Effects of dietary thiamin supplement on growth, body composition and intestinal enzyme activities of juvenile Jian carp (*Cyprinus carpio* var. Jian). *Aquaculture Nutrition, 17*(2), e233–e240. https://doi.org/10.1111/j.1365-2095.2010.00756.x

Huang, M., Lin, H., Xu, C., Yu, Q., Wang, X., Qin, J. G., Chen, L., Han, F., & Li, E. (2020). Growth, metabolite, antioxidative capacity, transcriptome, and the metabolome response to dietary choline chloride in Pacific white shrimp *Litopenaeus vannamei. Animals: An Open Access Journal from MDPI, 10*(12), 2246. https://doi.org/10.3390/ani10122246

Huang, X.-L., Xia, M.-H., Wang, H.-L., Jin, M., Wang, T., & Zhou, Q.-C. (2015). Dietary thiamine could improve growth performance, feed utilisation and non-specific immune response for juvenile Pacific white shrimp (*Litopenaeus vannamei*). *Aquaculture Nutrition, 21*(3), 364–372. https://doi.org/10.1111/anu.12167

Hughes, S. G. (1984). Effect of excess dietary riboflavin on growth of rainbow trout. *Journal of Nutrition, 114*(9), 1660–1663. https://doi.org/10.1093/jn/114.9.1660

Hung, S. S. O., Berge, G. M., & Storebakken, T. (1997). Growth and digestibility effects of soya lecithin and choline chloride on juvenile Atlantic salmon. *Aquaculture Nutrition, 3*(2), 141–144. https://doi.org/10.1046/j.1365-2095.1997.00080.x

Hung, S. S. O. (1989). Choline requirement of hatchery-produced juvenile white sturgeon (*Acipenser transmontanus*). *Aquaculture, 78*(2), 183–194. https://doi.org/10.1016/0044-8486(89)90031-8

Hungerford, D. M., & Linder, M. C. (1983). Interactions of pH and ascorbate in intestinal iron absorption. *Journal of Nutrition, 113*(12), 2615–2622. https://doi.org/10.1093/jn/113.12.2615

Huyghebaert, A. (1991). Stability of vitamin K in a mineral premix. *World Poultry, 7*, 71.

IAFFD - *International Aquaculture Feed Formulation Database* https://app.iaffd.com/ficd.

Ibrahem, M. D., Fathi, M., Mesalhy, S., & Abd El-Aty, A. M. (2010). Effect of dietary supplementation of inulin and vitamin C on the growth, hematology, innate immunity, and resistance of Nile tilapia (*Oreochromis niloticus*). *Fish and Shellfish Immunology, 29*(2), 241–246. https://doi.org/10.1016/j.fsi.2010.03.004

Ibrahim, R. E., Ahmed, S. A. A., Amer, S. A., Al-Gabri, N. A., Ahmed, A. I., Abdel-Warith, A.-W. A., Younis, E.-S. M. I., & Metwally, A. E. (2020). Influence of vitamin C feed supplementation on the growth, antioxidant activity, immune status, tissue histomorphology, and disease resistance in Nile tilapia, Oreochromis niloticus. *Aquaculture Reports, 18*, 100545. https://doi.org/10.1016/j.aqrep.2020.100545

Ikeda, S., Takasu, M., Matsuda, T., Kakinuma, A., & Horio, F. (1997). Ascorbic acid deficiency decreases the renal level of kidney fatty acid-binding protein by lowering the alpha2u-globulin gene expression in liver in scurvy-prone ods rats. *Journal of Nutrition, 127*(11), 2173–2178. https://doi.org/10.1093/jn/127.11.2173

Irías-Mata, A., Sus, N., Hug, M. L., Müller, M., Vetter, W., & Frank, J. (2020). α-tocomonoenol is bioavailable in mice and may partly be regulated by the function of the hepatic α-tocopherol transfer protein. *Molecules, 25*(20), 4803. https://doi.org/10.3390/molecules25204803

Irie, T., & Seki, T. (2002). Retinoid composition and retinal localization in the eggs of teleost fishes. *Comparative Biochemistry and Physiology. Part B, Biochemistry and Molecular Biology, 131*(2), 209–219. https://doi.org/10.1016/S1096-4959(01)00496-1

Iskakova, M., Karbyshev, M., Piskunov, A., & Rochette-Egly, C. R. (2015, December). Nuclear and extranuclear effects of vitamin A. *Canadian Journal of Physiology and Pharmacology, 93*(12), 1065–1075. https://doi.org/10.1139/cjpp-2014-0522

IUPAC-IUB Joint Commission on Biochemical Nomenclature (JCBN). (1982). *European Journal of Biochemistry, 123*, 473–475; Pure Appl. Chem. 54, 1507–1510. https://pubmed.ncbi.nlm.nih.gov/3743566/

Izquierdo, M., Domínguez, D., Jiménez, J. I., Saleh, R., Hernández-Cruz, C. M., Zamorano, M. J., & Hamre, K. (2019). Interaction between taurine, vitamin E and vitamin C in microdiets for gilthead seabream (*Sparus aurata*) larvae. *Aquaculture, 498*, 246–253. https://doi.org/10.1016/j.aquaculture.2018.07.010

Jacob, R. A. (1995). The integrated antioxidant system. *Nutrition Research, 15*(5), 755–766. https://doi.org/10.1016/0271-5317(95)00041-G

Jacob, R. A. (2006) Niacin. In B. A. Bowman & R. M. Russell (Eds.). *Present knowledge in nutrition* (9th ed) (pp. 261–268). International Life Sciences Institute. https://doi.org/10.1093/ajcn/85.5.1439a

Jacob, R. A., Wu, M. M., Henning, S. M., & Swendseid, M. E. (1994). Homocysteine increases as folate decreases in plasma of healthy men during short-term dietary folate and methyl group restriction. *Journal of Nutrition, 124*(7), 1072–1080. https://doi.org/10.1093/jn/124.7.1072

Jacobs, D. R., & Tapsell, L. C. (2013). Food synergy: The key to a healthy diet. *Proceedings of the Nutrition Society, 72*(2), 200–206. https://doi.org/10.1017/S0029665112003011

Jacobs, D. R., Jr., Gross, M. D., & Tapsell, L. C. (2009). Food synergy: An operational concept for understanding nutrition. *American Journal of Clinical Nutrition, 89*(5), 1543S–1548S. https://doi.org/10.3945/ajcn.2009.26736B

Jagruthi, C., Yogeshwari, G., Anbazahan, S. M., Mari, L. S., Arockiaraj, J., Mariappan, P., Sudhakar, G. R., Balasundaram, C., & Harikrishnan, R. (2014). Effect of dietary astaxanthin against *Aeromonas hydrophila* infection in common carp, *Cyprinus carpio*. *Fish and Shellfish Immunology, 41*(2), 674–680. https://doi.org/10.1016/j.fsi.2014.10.010

Janoušek, J., Pilařová, V., Macáková, K., Nomura, A., Veiga-Matos, J., da Silva, D. D. D., Remião, F., Saso, L., Malá-Ládová, K., Malý, J., Nováková, L., & Mladěnka, P. (2022). Vitamin D: Sources, physiological role, biokinetics, deficiency, therapeutic use, toxicity, and overview of analytical methods for detection of vitamin D and its metabolites. *Critical Reviews in Clinical Laboratory Sciences, 59*(8), 517–554. https://doi.org/10.1080/10408363.2022.2070595

Jensen, S. K., Nørgaard, J. V., & Lauridsen, C. (2006). Bioavailability of α-tocopherol stereoisomers in rats depends on dietary doses of all-rac- or RRR-α-tocopheryl acetate. *British Journal of Nutrition, 95*(3), 477–487. https://doi.org/10.1079/BJN20051667

Jia, G., Feng, J., & Qin, Z. (2006). Studies on the fatty liver diseases of *Sciaenops ocellatus* caused by different ether extract levels in diets. *Frontiers of Biology in China, 1*(1), 9–12. https://doi.org/10.1007/s11515-005-0002-7

Jiang, G.-Z., Wang, M., Liu, W.-B., Li, G.-F., & Qian, Y. (2013). Dietary choline requirement for juvenile blunt snout bream, *Megalobrama amblycephala*. *Aquaculture Nutrition, 19*(4), 499–505. https://doi.org/10.1111/anu.12001

Jiang, J., Shi, D., Zhou, X. Q., Yin, L., Feng, L., Jiang, W. D., Liu, Y., Tang, L., Wu, P., & Zhao, Y. (2015). Vitamin D inhibits lipopolysaccharide-induced inflammatory response possibly through the toll-like receptor 4 signaling pathway in the intestine and enterocytes of juvenile Jian carp (*Cyprinus carpio* var. Jian). *British Journal of Nutrition, 114*(10), 1560–1568. https://doi.org/10.1017/S0007114515003256

Jiang, M., Huang, F., Wen, H., Yang, C., Wu, F., Liu, W., & Tian, J. (2014a). Dietary niacin requirement of GIFT tilapia, *Oreochromis niloticus*, reared in freshwater. *Journal of the World Aquaculture Society, 45*(3), 333–341. https://doi.org/10.1111/jwas.12119

Jiang, M., Huang, F., Zhao, Z., Wen, H., Wu, F., Liu, W., Yang, C., & Wang, W. (2014b). Dietary thiamin requirement of juvenile grass carp, *Ctenopharyngodon idella*. *Journal of the World Aquaculture Society, 45*(4), 461–468. https://doi.org/10.1111/jwas.12132

Jiang, M., Ma, L., Shao, H., Wu, F., Liu, W., Tian, J., Yu, L., Lu, X., & Wen, H. (2020). Dietary vitamin E requirement of sub-adult genetically improved farmed tilapia strain of Nile tilapia (*Oreochromis niloticus*) reared in freshwater. *Aquaculture Nutrition, 26*(2), 233–241. https://doi.org/10.1111/anu.12983

Jiang, M., Wang, W. M., Wen, H., Wu, F., Zhao, Z. Y., & Liu, A. l. and Liu, W. (2007). Effects of dietary vitamin K3 on growth, carcass composition and blood coagulation time for grass carp fingerling (*Ctenopharyngodon idellus*). *Freshwater Fisheries, 37*(2), 61–64.

Jiang, S., Liu, D., Zhou, F., Mo, X., Yang, Q., Huang, J., Yang, L., & Jiang, S. (2020). Effect of vitamin E on spermatophore regeneration and quality of pond-reared, black tiger shrimp (*Penaeus monodon*). *Aquaculture Research, 51*(6), 2197–2204. https://doi.org/10.1111/are.14524

Jiang, W. D., Feng, L., Liu, Y., Jiang, J., & Zhou, X.-Q. (2009). Growth, digestive capacity and intestinal microflora of juvenile Jian carp (*Cyprinus carpio* var. Jian) fed graded levels of dietary inositol. *Aquaculture Research, 40*(8), 955–962. https://doi.org/10.1111/j.1365-2109.2009.02191.x

Jiang, W. D., Hu, K., Liu, Y., Jiang, J., Wu, P., Zhao, J., Zhang, Y. A., Zhou, X. Q., & Feng, L. (2016). Dietary myo-inositol modulates immunity through antioxidant activity and the Nrf2 and E2F4/cyclin signalling factors in the head kidney and spleen following infection of juvenile fish with *Aeromonas hydrophila*. *Fish and Shellfish Immunology, 49*, 374–386. https://doi.org/10.1016/j.fsi.2015.12.017

Jiang, W. D., Zhou, X. Q., Zhang, L., Liu, Y., Wu, P., Jiang, J., Kuang, S. Y., Tang, L., Tang, W. N., Zhang, Y. A., Shi, H. Q., & Feng, L. (2019). Vitamin A deficiency impairs intestinal physical barrier function of fish. *Fish and Shellfish Immunology, 87*, 546–558. https://doi.org/10.1016/j.fsi.2019.01.056

Jiang, W. D., Zhang, L., Feng, L., Wu, P., Liu, Y., Jiang, J., Kuang, S. Y., Tang, L., & Zhou, X. Q. (2020). Inconsistently impairment of immune function and structural integrity of head kidney and spleen by vitamin A deficiency in grass carp (*Ctenopharyngodon idella*). *Fish and Shellfish Immunology, 99*, 243–256. https://doi.org/10.1016/j.fsi.2020.02.019

Jiang, X., Yan, J., West, A. A., Perry, C. A., Malysheva, O. V., Devapatla, S., Pressman, E., Vermeylen, F., & Caudill, M. A. (2012). Maternal choline intake alters the epigenetic state of fetal cortisol-regulating genes in humans. *FASEB Journal, 26*(8), 3563–3574. https://doi.org/10.1096/fj.12-207894

Jin, M., Pan, T., Tocher, D. R., Betancor, M. B., Monroig, Ó., Shen, Y., Zhu, T., Sun, P., Jiao, L., & Zhou, Q. (2019). Dietary choline supplementation attenuated high-fat diet-induced inflammation through regulation of lipid metabolism and suppression of NFκB activation in juvenile black seabream (*Acanthopagrus schlegelii*). *Journal of Nutritional Science, 8*, e38. https://doi.org/10.1017/jns.2019.34

John, T. M., George, J. C., Hilton, J. W., & Slinger, S. J. (1979). Influence of dietary ascorbic acid on plasma lipid levels in the rainbow trout. *International Journal for Vitamin and Nutrition Research. Internationale Zeitschrift Fur Vitamin- und Ernahrungsforschung. Journal International de Vitaminologie et de Nutrition, 49*(4), 400–405. PubMed: 549878

Johnston, C. S. (2006). Vitamin C. In B. A. Bowman & R. M. Russell (Eds.), *Present knowledge in nutrition* (9th ed) (pp. 233–241). International Life Sciences Institute. https://doi.org/10.1093/AJCN/85.5.1439A

Johnston, C. S., Steinberg, F. M., & Rucker, R. B. (2013). Ascorbic acid. In J. Zempleni, J. Suttie, J. Gregory & P. J. Stover (Eds.), *Handbook of vitamins* (5th ed) (pp. 515–550). CRC Press, Taylor & Francis Group. https://doi.org/10.1201/b15413

Jolly, D. W., Craig, C., & Nelson, T. E. (1977). Estrogen and prothrombin synthesis: Effect of estrogen on absorption of vitamin K1. *American Journal of Physiology, 232*(1), H12–H17. https://doi.org/10.1152/ajpheart.1977.232.1.H12

Ju, Z. Y., Deng, D. F., Dominy, W. G., & Forster, I. P. (2011). Pigmentation of Pacific white shrimp, *Litopenaeus vannamei*, by dietary astaxanthin extracted from Haematococcus pluvialis. *Journal of the World Aquaculture Society, 42*(5), 633–644. https://doi.org/10.1111/j.1749-7345.2011.00511.x

Kaempf-Rotzoll, D. E., Traber, M. G., & Arai, H. (2003). Vitamin E and transfer proteins. *Current Opinion in Lipidology, 14*(3), 249–254. https://doi.org/10.1097/00041433-200306000-00004

Kalinowski, C. T., Robaina, L. E., & Izquierdo, M. S. (2011). Effect of dietary astaxanthin on the growth performance, lipid composition and post-mortem skin colouration of red porgy *Pagrus pagrus*. *Aquaculture International, 19*(5), 811–823. https://doi.org/10.1007/s10499-010-9401-0

Kamireddy, N., Jittinandana, S., Kenney, P. B., Slider, S. D., Kiser, R. A., Mazik, P. M., & Hankins, J. A. (2011). Effect of dietary vitamin E supplementation and refrigerated storage on quality of rainbow trout fillets. *Journal of Food Science, 76*(4), S233–S241. https://doi.org/10.1111/j.1750-3841.2011.02121.x

Kanazawa, A., Teshima, S., & Tanaka, N. (1976). Nutritional requirements of prawn - V: Requirements for choline and inositol. *Memoirs of Faculty of Fisheries Kagoshima University, 25*, 47–51.

Kao, C., & Robinson, R. J. (1972). *Aspergillus flavus* deterioration of grain: Its effect on amino acids and vitamins in whole wheat. *Journal of Food Science, 37*(2), 261–263. https://doi.org/10.1111/j.1365-2621.1972.tb05831.x

Kao, T. T., Chu, C. Y., Lee, G. H., Hsiao, T. H., Cheng, N. W., Chang, N. S., Chen, B. H., & Fu, T. F. (2014). Folate deficiency-induced oxidative stress contributes to neuropathy in young and aged zebrafish – Implication in neural tube defects and Alzheimer's diseases. *Neurobiology of Disease, 71*, 234–244. http://doi.org/10.1016/j.nbd.2014.08.004

Karidis, N. P., Kouraklis, G., & Theocharis, S. E. (2006). Platelet-activating factor in liver injury: A relational scope. *World Journal of Gastroenterology, 12*(23), 3695–3706. https://doi.org/10.3748/wjg.v12.i23.3695

Kashiwada, K. (1971). Studies on the production of B vitamins by intestinal bacteria of carp. VI. Production of folic acid by intestinal bacteria of carp. *Mem. Fac. Fish. Kagoshima, 20*, 185–189.

Kasner, P., Chambon, P., & Leid, M. (1994). Role of nuclear retinoic acid receptors in the regulation of gene expression. In *Vitamin A in health and disease Blomhoff, R.* (ed.) (p. 189). CRC Press.

Kaisuyama, M., & Matsuno, T. (1988). Carotenoid and vitamin A, and metabolism of carotenoids, β-carotene, canthaxanthin, astaxanthin, zeaxanthin, lutein and tunaxanthin in tilapia *Tilapia nilotica*. *Comparative Biochemistry and Physiology Part B, 90*(1), 131–139. https://doi.org/10.1016/0305-0491(88)90049-1

Kaushik, S. J., Gouillou-Coustans, M. F., & Cho, C. Y. (1998). Application of the recommendations on vitamin requirements of finfish by NRC (1993) to salmonids and sea bass using practical and purified diets. *Aquaculture, 161*(1–4), 463–474. https://doi.org/10.1016/S0044-8486(97)00293-7

Ke, Z. J., & Gibson, G. E. (2004). Selective response of various brain cell types during neurodegeneration induced by mild impairment of oxidative metabolism. *Neurochemistry International, 45*(2–3), 361–369. https://doi.org/10.1016/j.neuint.2003.09.008

Kesavan, V., & Noronha, J. M. (1983). Folate malabsorption in aged rats related to low levels of pancreatic folyl conjugase. *American Journal of Clinical Nutrition, 37*(2), 262–267. https://doi.org/10.1093/ajcn/37.2.262

Ketola, H. G. (1976). Choline metabolism and nutritional requirement of lake trout (*Salvelinus namaycush*) 1. *Journal of Animal Science, 43*(2), 474–477. https://doi.org/10.2527/jas1976.432474x

Ketola, H. G., Isaacs, G. R., Robins, J. S., & Lloyd, R. C. (2008). Effectiveness and retention of thiamine and its analogs administered to steelhead and landlocked Atlantic salmon. *Journal of Aquatic Animal Health, 20*(1), 29–38. https://doi.org/10.1577/H07-012.1

Khan, Y. M., & Khan, M. A. (2019). Dietary biotin requirement of fingerling *Catla catla* (Hamilton) based on growth, feed conversion efficiency, and liver biotin concentration. *Journal of the World Aquaculture Society, 50*(3), 674–683. https://doi.org/10.1111/jwas.12529

Khan, Y. M., & Khan, M. A. (2021a). Optimization of dietary pyridoxine improved growth performance, hematological indices, antioxidant capacity, intestinal enzyme activity, non-specific immune response, and liver pyridoxine concentration of fingerling major *carp Catla catla* (Hamilton). *Aquaculture, 541*, 736815. https://doi.org/10.1016/j.aquaculture.2021.736815

Khan, M. Y., & Khan, M. A. (2021b). Dietary thiamin requirement of fingerling major carp *Catla catla* (Hamilton). *Journal of Animal Physiology and Animal Nutrition, 106*(4), 939–946. https://doi.org/10.1111/jpn.13644

Kheirabadi, E. P., Shekarabi, P. H., Yadollahi, F., Soltani, M., Najafi, E., von Hellens, J., Flores, C. L., Salehi, K., & Faggio, C. (2022). Red yeast (*Phaffia rhodozyma*) and its effect on growth, antioxidant activity and color pigmentation of rainbow trout (*Oncorhynchus mykiss*). *Aquaculture Reports, 23*, 101082. https://doi.org/10.1016/j.aqrep.2022.101082

King, K. (1978). Distribution of γ-carboxyglutamic acid in calcified tissues. *Biochimica et Biophysica Acta, 542*(3), 542–546. https://doi.org/10.1016/0304-4165(78)90384-7

Kirkland, J. B. (2013). Niacin. In J. Zempleni, J. Suttie, J. Gregory & P. J. Stover (Eds.), *Handbook of vitamins* (5th ed) (pp. 149–190). Taylor & Francis Group. https://doi.org/10.1201/b15413

Kitamura, S., Suwa, T., Ohara, S., & Nakagawa, K. (1967a). Studies on vitamin requirements of rainbow trout-III. *Nippon Suisan Gakkaishi, 33*(12), 1126–1131. https://doi.org/10.2331/suisan.33.1126

Kitamura, S., Suwa, T., Ohara, S., & Nakagawa, K. (1967b). Studies on Vitamin Requirements of Rainbow Trout-II. Nippon Suisan Gakkaishi, 33(12), 1120–1125. https://doi.org/10.2331/suisan.33.1120

Kliewer, S. A., Umesono, K., Mangelsdorf, D. J., & Evans, R. M. (1992). Retinoid X receptor interacts with nuclear receptors in retinoic acid, thyroid hormone and vitamin D3 signalling. *Nature, 355*(6359), 446–449. https://doi.org/10.1038/355446a0

Kocabas, G. III, & Gatlin. (1999). Dietary vitamin E requirement of hybrid striped bass (*Morone chrysops* female × *M. saxatilis* male). *Aquaculture Nutrition, 5*(1), 3–7. https://doi.org/10.1046/j.1365-2095.1999.00074.x

Koch, J. F. A., Sabioni, R. E., Aguilar Aguilar, F. A., Lorenz, E. K., & Cyrino, J. E. P. (2018). Vitamin A requirements of dourado (*Salminus brasiliensis*): Growth performance and immunological parameters. *Aquaculture, 491*, 86–93. https://doi.org/10.1016/j.aquaculture.2018.03.017

Kodentsova, V. M., Iakushina, L. M., Vrzhesinskaia, O. A., Beketova, N. A., & Spirichev, V. B. (1993). Effect of riboflavin administration on vitamin B6 metabolism. *Voprosy Pitaniia, 5*(5), 32–36. PubMed: 8042309

Koehn, J. Z., Allison, E. H., Golden, C. D., & Hilborn, R. (2022). The role of seafood in sustainable diets. *Environmental Research Letters, 17*(3), 035003. https://doi.org/10.1088/1748-9326/ac3954

Kominato, T. (1971). Speed of vitamin B12 turnover and its relation to the intestine in the rat. *Vitamins, 44*, 76–83.

Kopinski, J. S., & Leibholz, J. (1989). Biotin studies in pigs. 2. The biotin requirement of the growing pig. *British Journal of Nutrition, 62*(3), 761–766. https://doi.org/10.1079/bjn19890076

Koskela, J., Leskinen, H., Mattila, P., Airaksinen, S., Rinne, M., Pihlava, J. M., & Pihlanto, A. (2021). The effect of gradual addition of Camelina seeds in the diet of rainbow trout (*Oncorhynchus mykiss*) on growth, feed efficiency and meat quality. *Aquaculture Research, 52*(10), 4681–4692. https://doi.org/10.1111/are.15302

Krasnov, A., Reinisalo, M., Pitkänen, T. I., Nishikimi, M., & Mölsä, H. (1998). Expression of rat gene for L-gulono-γ-lactone oxidase, the key enzyme of L-ascorbic acid biosynthesis, in guinea pig cells and in teleost fish rainbow trout (*Oncorhynchus mykiss*). *Biochimica et Biophysica Acta, 1381*(2), 241–248. https://doi.org/10.1016/S0304-4165(98)00037-3

Krinke, G. J., & Fitzgerald, R. E. (1988). The pattern of pyridoxine-induced lesion: Difference between the high and the low toxic level. *Toxicology, 49*(1), 171–178. https://doi.org/10.1016/0300-483X(88)90190-4

Krogdahl, Å., Hansen, A. K. G., Kortner, T. M., Björkhem, I., Krasnov, A., Berge, G. M., & Denstadli, V. (2020). Choline and phosphatidylcholine, but not methionine, cysteine, taurine and taurocholate, eliminate excessive gut mucosal lipid accumulation in Atlantic salmon (*Salmo salar L*). *Aquaculture*, 528, 735552. https://doi.org/10.1016/j.aquaculture.2020.735552

Krogdahl, Å., Chikwati, E. M., Krasnov, A., Dhanasiri, A., Berge, G. M., Aru, V., Khakimov, B., Engelsen, S. B., Vinje, H., & Kortner, T. M. (2023). Dietary fishmeal level and a package of choline, β-glucan, and nucleotides modulate gut function, microbiota, and health in Atlantic salmon (*Salmo salar L.*). *Aquaculture Nutrition*, 2023, 5422035. https://doi.org/10.1155/2023/5422035

Krossøy, C., Waagbø, R., Fjelldal, P.-G., Wargelius, A., Lock, E.-J., Graff, I. E., & Ørnsrud, R. (2009). Dietary menadione nicotinamide bisulphite (vitamin K3) does not affect growth or bone health in first-feeding fry of Atlantic salmon (*Salmo salar L.*). *Aquaculture Nutrition*, 15(6), 638–649. https://doi.org/10.1111/j.1365-2095.2008.00633.x

Krossøy, C., Lock, E. J., & Ørnsrud, R. (2010). Vitamin K-dependent γ-glutamyl carboxylase in Atlantic salmon (*Salmo salar L.*). *Fish Physiology and Biochemistry*, 36(3), 627–635. https://doi.org/10.1007/s10695-009-9335-5

Krossøy, C., Waagbø, R., & Ørnsrud, R. (2011). Vitamin K in fish nutrition. *Aquaculture Nutrition*, 17(6), 585–594. https://doi.org/10.1111/j.1365-2095.2011.00904.x

Krumdieck, C. L. (1990). Folic acid. In M. L. Brown (Ed.) Int., *Present knowledge in nutrition* (6th ed) (pp. 179–188) Life Sci. Institute/Nutrition Foundation.

Kumar, N., Ambasankar, K., Krishnani, K. K., Kumar, P., Akhtar, M. S., Bhushan, S., & Minhas, P. S. (2016). Dietary pyridoxine potentiates thermal tolerance, heat shock protein and protect against cellular stress of milkfish (*Chanos chanos*) under endosulfan-induced stress. *Fish and Shellfish Immunology*, 55(7), 407–414. https://doi.org/10.1016/j.fsi.2016.06.011

Kumar, V., Pillai, B. R., Sahoo, P. K., Mohanty, J., & Mohanty, S. (2009). Effect of dietary astaxanthin on growth and immune response of Giant freshwater prawn *Macrobrachium rosenbergii* (De Man). *Asian Fisheries Science*, 22(1), 61–69. https://doi.org/10.33997/j.afs.2009.22.1.007

Kumlu, M., Fletcher, D. J., & Fisher, C. M. (1998). Larval pigmentation, survival and growth of Penaeus indicus fed the nematode *Panagrellus redivivus* enriched with astaxanthin and various lipids. *Aquaculture Nutrition*, 4(3), 193–200. https://doi.org/10.1046/j.1365-2095.1998.00071.x

La Frano, M. R., Cai, Y., Burri, B. J., & Thilsted, S. H. (2018). Discovery and biological relevance of 3,4-didehydroretinol (vitamin A2) in small indigenous fish species and its potential as a dietary source for addressing vitamin A deficiency. *International Journal of Food Sciences and Nutrition*, 69(3), 253–261. https://doi.org/10.1080/09637486.2017.1358358

Lall, S. P., & Dumas, A. (2022). Chapter 3. Nutritional requirements of cultured fish: Formulating nutritionally adequate feeds. In D. A. Davis (Ed.), *Feed and feeding practices in aquaculture* (2nd ed) (pp. 53–109). Woodhead Publishing.

Lall, S. P., & Lewis-McCrea, L. M. (2007). Role of nutrients in skeletal metabolism and pathology in fish – An overview. *Aquaculture*, 267(1–4), 3–19. https://doi.org/10.1016/j.aquaculture.2007.02.053

Lall, S. P., & Weerakoon, D. E. M. (1990). Vitamin B6 requirement of Atlantic salmon (*Salmo salar*) [Abstract]. *FASEB Journal*, 4, 3749.

Lambertsen, G., & Braekkan, O. R. (1969). In vivo conversion of vitamin A1 to vitamin A2. *Acta Chemica Scandinavica*, 23(3), 1063–1064. https://oi.org/10.3891/acta.chem.scand.23-1063

Lapatra, S. E. (1998). Factors affecting pathogenicity of infectious hematopoietic necrosis virus (IHNV) for salmonid fish. *Journal of Aquatic Animal Health*, 10(2), 121–131. https://doi.org/10.1577/1548-8667(1998)010<0121:FAPOIH>2.0.CO;2

Larsson, D., Björnsson, B. T., & Sundell, K. (1995). Physiological concentrations of 24,25-dihydroxyvitamin D3 rapidly decrease the in vitro intestinal calcium uptake in the Atlantic cod, *Gadus morhua*. *General and Comparative Endocrinology*, 100(2), 211–217. https://doi.org/10.1006/gcen.1995.1150

Lawrance, P. (2015). Niacin (vitamin B3) – A review of analytical methods for use in food. https://assets.publishing.service.gov.uk/government/uploads/system/uploads/attachment_data/file/409990/Niacin_review-FINAL_06032015.pdf. National Measurement Office.

Le, K. T., Dao, T. T. T., Fotedar, R., & Partrigde, G. J. (2014a). Effects of variation in dietary contents of selenium and vitamin E on growth and physiological and haematological responses of yellowtail kingfish, *Seriola lalandi. Aquaculture International, 22*(2), 435–446. https://doi.org/10.1007/s10499-013-9651-8

Le, K. T., Fotedar, R., & Partridge, G. (2014b). Selenium and vitamin E interaction in the nutrition of yellowtail kingfish (*Seriola lalandi*): Physiological and immune responses. *Aquaculture Nutrition, 20*(3), 303–313. https://doi.org/10.1111/anu.12079

Leatherland, J. F., Barnett, B. J., Cho, C. Y., & Slinger, S. J. (1980). Effect of dietary cholecalciferol deficiency on plasma thyroid hormone concentrations in rainbow trout, *Salmo gairdneri* (Pisces, Salmonidae). *Environmental Biology of Fishes, 5*(2), 167–173. https://doi.org/10.1007/BF02391624

Lee, B. J., Jaroszewska, M., Dabrowski, K., Czesny, S., & Rinchard, J. (2012). Effects of dietary vitamin B1 (thiamine) and magnesium on the survival, growth and histological indicators in lake trout (*Salvelinus namaycush*) juveniles. *Comparative Biochemistry and Physiology. Part A, Molecular and Integrative Physiology, 162*(3), 219–226. https://doi.org/10.1016/j.cbpa.2012.03.008

Lee, J. (2003). *Vitamin K requirements of juvenile hybrid tilapia (Oreochromis niloticus× O. aureus) and grouper (Epinephelus malabaricus).* https://nutricionacuicola.uanl.mx/index.php/acu/article/view/165 [Masters Thesis]. National Taiwan Ocean University.

Lee, K. J., & Dabrowski, K. (2004). Long-term effects and interactions of dietary vitamins C and E on growth and reproduction of yellow perch, *Perca flavescens. Aquaculture, 230*(1–4), 377–389. https://doi.org/10.1016/S0044-8486(03)00421-6

Lee, L. C., Carlson, R. W., Judge, D. L., & Ogawa, M. (1973) The absorption cross sections of N2, O2, CO, NO, CO2, N2O, CH4, C2H4, C2H6 and C4H10 from 180 to 700Å. *Journal of Quantitative Spectroscopy and Radiative Transfer, 13*(10), 1023–1031. https://doi.org/10.1016/0022-4073(73)90075-7

Lee, M. H., & Shiau, S. Y. (2004). Vitamin E requirements of juvenile grass shrimp, *Penaeus monodon*, and effects on non-specific immune responses. *Fish and Shellfish Immunology, 16*(4), 475–485. https://doi.org/10.1016/j.fsi.2003.08.005

Lee, R. F., & Puppione, D. L. (1978). Serum lipoproteins in the spiny lobster, *Panulirus interruptus. Comparative Biochemistry and Physiology. B, Comparative Biochemistry, 59*(3), 239–243. https://doi.org/10.1016/0305-0491(78)90253-5

Leeson, S., & Summers, J. D. (2008). Commercial poult. Nutr. *3rd Nottingham University Press (digital edition).* https://www.agropustaka.id/wp-content/uploads/2020/04/agropustaka.id_buku_Commercial-Poultry-Nutrition-3rd-Edition-by-S.-Leeson-J.-D.-Summers.pdf

Lemos, D., & Weissman, D. (2021). Moulting in the grow-out of farmed shrimp: A review. *Reviews in Aquaculture, 13*(1), 5–17. https://doi.org/10.1111/raq.12461

Lewis, L. L., Stark, C. R., Fahrenholz, A. C., Bergstrom, J. R., & Jones, C. K. (2015). Evaluation of conditioning time and temperature on gelatinized starch and vitamin retention in a pelleted swine diet. *Journal of Animal Science, 93*(2), 615–619. https://doi.org/10.2527/jas.2014-8074

Li, E., Yu, N., Chen, L., Zeng, C., Liu, L., & Qin, J. G. (2010). Dietary vitamin B6 requirement of the Pacific white shrimp, *Litopenaeus vannamei*, at low salinity. *Journal of the World Aquaculture Society, 41*(5), 756–763. https://doi.org/10.1111/j.1749-7345.2010.00417.x

Li, J., Liang, X. F., Tan, Q., Yuan, X., Liu, L., Zhou, Y., & Li, B. (2014). Effects of vitamin E on growth performance and antioxidant status in juvenile grass carp *Ctenopharyngodon idellus. Aquaculture, 430*, 21–27. https://doi.org/10.1016/j.aquaculture.2014.03.019

Li, J. Y., Zhang, D. D., Xu, W. N., Jiang, G. Z., Zhang, C. N., Li, X. F., & Liu, W. B. (2014). Effects of dietary choline supplementation on growth performance and hepatic lipid transport in blunt snout bream (*Megalobrama amblycephala*) fed high-fat diets. *Aquaculture, 434*, 340–347. https://doi.org/10.1016/j.aquaculture.2014.08.006

Li, J.-Y., Li, X.-F., Xu, W.-N., Zhang, C.-N., & Liu, W.-B. (2016). Effects of dietary choline supplementation on growth performance, lipid deposition and intestinal enzyme activities of blunt snout bream *Megalobrama amblycephala* fed high-lipid diet. *Aquaculture Nutrition, 22*(1), 181–190. https://doi.org/10.1111/anu.12231

Li, L., Feng, L., Jiang, W. D., Jiang, J., Wu, P., Zhao, J., Kuang, S. Y., Tang, L., Tang, W. N., Zhang, Y. A., Zhou, X. Q., & Liu, Y. (2015a). Dietary pantothenic acid depressed the gill immune and physical barrier function via NF-κB, TOR, Nrf2, p38MAPK and MLCK signalling pathways in grass carp (*Ctenopharyngodon idella*). *Fish and Shellfish Immunology, 47*(1), 500–510. https://doi.org/10.1016/j.fsi.2015.09.038

Li, L., Feng, L., Jiang, W. D., Jiang, J., Wu, P., Kuang, S. Y., Tang, L., Tang, W. N., Zhang, Y. A., Zhou, X. Q., & Liu, Y. (2015b). Dietary pantothenic acid deficiency and excess depress the growth, intestinal mucosal immune and physical functions by regulating NF-κB, TOR, Nrf2, and MLCK signaling pathways in grass carp (*Ctenopharyngodon idella*). *Fish and Shellfish Immunology, 45*(2), 399–413. https://doi.org/10.1016/j. fsi.2015.04.030

Li, M., Chen, L., Qin, J. G., Li, E., Yu, N., & Du, Z. (2013). Growth performance, antioxidant status and immune response in darkbarbel catfish *Pelteobagrus vachelli* fed different PUFA/vitamin E dietary levels and exposed to high or low ammonia. *Aquaculture, 406–407*, 18–27. https://doi.org/10.1016/j.aquaculture. 2013.04.028

Li, M. Y., Gao, C. S., Du, X. Y., Zhao, L., Niu, X. T., Wang, G. Q., & Zhang, D. M. (2020). Effect of sub-chronic exposure to selenium and astaxanthin on *Channa argus*: Bioaccumulation, oxidative stress and inflammatory response. *Chemosphere, 244*, 125546. https://doi.org/10.1016/j.chemosphere.2019. 125546

Li, S., Lian, X., Chen, N., Wang, M., & Sang, C. (2018a). Effects of dietary vitamin E level on growth performance, feed utilization, antioxidant capacity and nonspecific immunity of largemouth bass, *Micropterus salmoides*. *Aquaculture Nutrition, 24*(6), 1679–1688. https://doi.org/10.1111/anu.12802

Li, S. A., Jiang, W. D., Feng, L., Liu, Y., Wu, P., Jiang, J., Kuang, S. Y., Tang, L., Tang, W. N., Zhang, Y. A., Yang, J., Tang, X., Shi, H. Q., & Zhou, X. Q. (2018b). Dietary myo-inositol deficiency decreased intestinal immune function related to NF-κB and TOR signaling in the intestine of young grass carp (*Ctenopharyngodon idella*). *Fish and Shellfish Immunology, 76*, 333–346. https://doi.org/10.1016/j.fsi.2018.03.017

Li, S.-Q., Feng, L., Jiang, W.-D., Liu, Y., Wu, P., Zhao, J., Kuang, S.-Y., Jiang, J., Tang, L., Tang, W.-N., Zhang, Y.-A., & Zhou, X.-Q. (2016). Deficiency of dietary niacin decreases digestion and absorption capacities via declining the digestive and brush border enzyme activities and downregulating those enzyme gene transcription related to TOR pathway of the hepatopancreas and intestine in young. *Aquaculture Nutrition, 22*(6), 1267–1282. https://doi.org/10.1111/anu.12333

Li, W., Zhou, X.-Q., Feng, L., Liu, Y., & Jiang, J. (2010). Effect of dietary riboflavin on growth, feed utilization, body composition and intestinal enzyme activities of juvenile Jian carp (*Cyprinus carpio* var. Jian). *Aquaculture Nutrition, 16*(2), 137–143. https://doi.org/10.1111/j.1365-2095.2008.00645.x

Li, X., Hua, X., Wei, X., Guo, X., Li, N., & Yao, J. (2021). Effects of dietary vitamin D3 on the growth and antioxidant capacity of largemouth bass (*Micropterus salmoides*). *Journal of Shanghai Ocean University, 3091*, 94–102.

Li, X. F., Wang, F., Qian, Y., Jiang, G. Z., Zhang, D. D., & Liu, W. B. (2016). Dietary vitamin B12 requirement of fingerling blunt snout bream *Megalobrama amblycephala* determined by growth performance, digestive and absorptive capability and status of the GH-IGF-I axis. *Aquaculture, 464*, 647–653. https://doi. org/10.1016/j.aquaculture.2016.08.019

Li, X.-F., Wang, T.-J., Qian, Y., Jiang, G.-Z., Zhang, D.-D., & Liu, W.-B. (2017). Dietary niacin requirement of juvenile blunt snout bream *Megalobrama amblycephala* based on a dose–response study. *Aquaculture Nutrition, 23*(6), 1410–1417. https://doi.org/10.1111/anu.12516

Li, X. Y., Huang, H. H., Hu, K., Liu, Y., Jiang, W. D., Jiang, J., Li, S. H., Feng, L., & Zhou, X. Q. (2014). The effects of dietary thiamin on oxidative damage and antioxidant defence of juvenile fish. *Fish Physiology and Biochemistry, 40*(3), 673–687. https://doi.org/10.1007/s10695-013-9875-6

Li, Y., Fan, B., Huang, Y., Wu, D., Zhang, M., & Zhao, Y. (2018). Effects of dietary vitamin E on reproductive performance and antioxidant capacity of *Macrobrachium nipponense* female shrimp. *Aquaculture Nutrition, 24*(6), 1698–1708. https://doi.org/10.1111/anu.12804

Li, Y., Huang, Y., Zhang, M., Chen, Q., Fan, W., & Zhao, Y. (2019). Effect of dietary vitamin E on growth, immunity and regulation of hepatopancreas nutrition in male oriental river prawn, *Macrobrachium nipponense*. *Aquaculture Research, 50*(7), 1741–1751. https://doi.org/10.1111/are.14070

Li, Z., & Vance, D. E. (2008). Phosphatidylcholine and choline homeostasis. *Journal of Lipid Research, 49*(6), 1187–1194. https://doi.org/10.1194/jlr.R700019-JLR200

Liang, X. P., Li, Y., Hou, Y. M., Qiu, H., & Zhou, Q. C. (2017). Effect of dietary vitamin C on the growth performance, antioxidant ability and innate immunity of juvenile yellow catfish (*Pelteobagrus fulvidraco* Richardson). *Aquaculture Research, 48*(1), 149–160. https://doi.org/10.1111/are.12869

Lie, O., Sandvin, A., & Waagbø, R. (1994). Transport of alpha-tocopherol in Atlantic salmon (*Salmo salar*) during vitellogenesis. *Fish Physiology and Biochemistry, 13*(3), 241–247. https://doi.org/10.1007/BF00 004362

Lightner, D. V., Colvin, L. B., Brand, C., & Danald, D. A. (1977). Black death, a disease syndrome of Penaeid shrimp related to a dietary deficiency of ascorbic acid. *Proceedings of the World Maricult. Proceedings of the Annual Meeting – World Mariculture Society, 8*(1–4), 611–624. https://doi.org/10.1111/j.1749-7345.1977. tb00149.x

Lightner, D. V., Hunter, B., Magarelli, Jr., P. C., & Colvin, L. B. (1979). Ascorbic acid: Nutritional requirement and role in wound repair in Penaeid shrimp. *Proceedings of the World Maricult. Proceedings of the World Mariculture Society, 10*(1–4), 513–528. https://doi.org/10.1111/j.1749-7345.1979.tb00047.x

Lim, C., Yildirim-Aksoy, M., Shelby, R., Li, M. H., & Klesius, P. H. (2010a). Growth performance, vitamin E status, and proximate and fatty acid composition of channel catfish, *Ictalurus punctatus*, fed diets containing various levels of fish oil and vitamin E. *Fish Physiology and Biochemistry, 36*(4), 855–866. https://doi.org/10.1007/s10695-009-9360-4

Lim, C., Yildirim-Aksoy, M., Welker, T., Klesius, P. H., & Li, M. H. (2010b). Growth performance, immune Response, and resistance to *Streptococcus iniae* of Nile tilapia, *Oreochromis niloticus*, fed diets containing various levels of vitamins C and E. *Journal of the World Aquaculture Society, 41*(1), 35–48. https://doi.org/10.1111/j.1749-7345.2009.00311.x

Lim, C., Yildirim-Aksoy, M., Barros, M. M., & Klesius, P. (2011). Thiamin requirement of Nile tilapia, *Oreochromis niloticus. Journal of the World Aquaculture Society, 42*(6), 824–833. https://doi.org/10.1111/j.1749-7345.2011.00531.x

Lim, K. C., Yusoff, F. M., Shariff, M., & Kamarudin, M. S. (2018). Astaxanthin as feed supplement in aquatic animals. *Reviews in Aquaculture, 10*(3), 738–773. https://doi.org/10.1111/raq.12200

Lim, K. C., Yusoff, F. M., Shariff, M., & Kamarudin, M. S. (2019a). Dietary administration of astaxanthin improves feed utilization, growth performance and survival of Asian seabass, *Lates calcarifer* (Bloch, 1790). *Aquaculture Nutrition, 25*(6), 1410–1421. https://doi.org/10.1111/anu.12961

Lim, K. C., Yusoff, F. M., Shariff, M., Kamarudin, M. S., & Nagao, N. (2019b). Dietary supplementation of astaxanthin enhances hemato-biochemistry and innate immunity of Asian seabass, *Lates calcarifer* (Bloch, 1790). *Aquaculture, 512*, 734339. https://doi.org/10.1016/j.aquaculture.2019.734339

Lim, K. C., Yusoff, F. M., Shariff, M., & Kamarudin, M. S. (2021). Dietary astaxanthin augments disease resistance of Asian seabass, *Lates calcarifer* (Bloch, 1790), against *Vibrio alginolyticus* infection. *Fish and Shellfish Immunology, 114*, 90–101. https://doi.org/10.1016/j.fsi.2021.03.025

Lin, M.-F., & Shiau, S.-Y. (2004). Requirements of vitamin C (L-ascorbyl-2-monophosphate-Mg and L-ascorbyl-2-monophosphate-Na) and its effects on immune responses of grouper, *Epinephelus malabaricus. Aquaculture Nutrition, 10*(5), 327–333. https://doi.org/10.1111/j.1365-2095.2004.00307.x

Lin, Y. H., Lin, H. Y., & Shiau, S. Y. (2011). Dietary folic acid requirement of grouper, *Epinephelus malabaricus*, and its effects on non-specific immune responses. *Aquaculture, 317*(1–4), 133–137. https://doi.org/10.1016/j.aquaculture.2011.04.010

Lin, Y. H., Lin, H. Y., & Shiau, S. Y. (2012). Estimation of dietary pantothenic acid requirement of grouper, *Epinephelus malabaricus* according to physiological and biochemical parameters. *Aquaculture, 324–325*, 92–96. https://doi.org/10.1016/j.aquaculture.2011.10.020

Liñán-Cabello, M. A., Paniagua-Michel, J., & Hopkins, P. M. (2002). Bioactive roles of carotenoids and retinoids in crustaceans. *Aquaculture Nutrition, 8*(4), 299–309. https://doi.org/10.1046/j.1365-2095.2002.00221.x

Linn, S. M., Ishikawa, M., Koshio, S., & Yokoyama, S. (2014). Effect of dietary vitamin E supplementation on growth performance and oxidative condition of red sea bream Pagrus major. *Aquaculture Sci., 62*(4), 329–339.

Liu, A. (2020) *The choline requirement of juvenile yellowtail kingfish (Seriola lalandi) and the interactive effects of choline and water temperature on nutrient assimilation* (Doctoral dissertation, UNSW Sydney). https://doi.org/10.26190/unsworks/3976

Liu, A., Pirozzi, I., Codabaccus, B., Hines, B., Simon, C., Sammut, J., & Booth, M. (2019). Digestible choline requirement of juvenile yellowtail kingfish (*Seriola lalandi*). *Aquaculture, 509*, 209–220. https://doi.org/10.1016/j.aquaculture.2019.05.020

Liu, A., Pirozzi, I., Codabaccus, B. M., Stephens, F., Francis, D. S., Sammut, J., & Booth, M. A. (2020). Effects of dietary choline on liver lipid composition, liver histology and plasma biochemistry of juvenile yellowtail kingfish (*Seriola lalandi*). *British Journal of Nutrition*, 1–15. https://doi.org/10.1017/S0007114520003669

Liu, A., Mazumder, D., Pirozzi, I., Sammut, J., & Booth, M. (2021a). The effect of dietary choline and water temperature on the contribution of raw materials to the muscle tissue of juvenile yellowtail kingfish (*Seriola lalandi*): An investigation using a stable isotope mixing model. *Animal Feed Science and Technology*, *280*, 115087. https://doi.org/10.1016/j.anifeedsci.2021.115087

Liu, A., Pirozzi, I., Codabaccus, B. M., Sammut, J., & Booth, M. A. (2021b). The interactive effect of dietary choline and water temperature on the liver lipid composition, histology, and plasma biochemistry of juvenile yellowtail kingfish (*Seriola lalandi*). *Aquaculture*, *531*, 735893. https://doi.org/10.1016/j.aquaculture.2020.735893

Liu, A., To, V. P. T. H., Santigosa, E., Dumas, A., & Hernandez, J. M. (2022). Vitamin nutrition in salmonid aquaculture: From avoiding deficiencies to enhancing functionalities. *Aquaculture*, *561*, 738654. https://doi.org/10.1016/j.aquaculture.2022.738654

Liu, A., To, V. P. T. H., Dumas, A., Hernandez, J. M., & Santigosa, E. (2024). Vitamin nutrition in shrimp aquaculture: A review focusing on the last decade. *Aquaculture*, *578*, 740004. https://doi.org/10.1016/j.aquaculture.2023.740004

Liu, B., Zhao, Z., Brown, P. B., Cui, H., Xie, J., Habte-Tsion, H.-M., & Ge, X. (2016). Dietary vitamin A requirement of juvenile Wuchang bream (*Megalobrama amblycephala*) determined by growth and disease resistance. *Aquaculture*, *450*, 23–30. https://doi.org/10.1016/j.aquaculture.2015.06.042

Liu, F., Shi, H. Z., Guo, Q. S., Yu, Y. B., Wang, A. M., Lv, F., & Shen, W. B. (2016). Effects of astaxanthin and emodin on the growth, stress resistance and disease resistance of yellow catfish (*Pelteobagrus fulvidraco*). *Fish and Shellfish Immunology*, *51*, 125–135. https://doi.org/10.1016/j.fsi.2016.02.020

Liu, J., Shao, R., Lan, Y., Liao, X., Zhang, J., Mai, K., Ai, Q., & Wan, M. (2021). Vitamin D3 protects turbot (*Scophthalmus maximus* L.) from bacterial infection. *Fish and Shellfish Immunology*, *118*, 25–33. https://doi.org/10.1016/j.fsi.2021.08.024

Liu, M. Y., Sun, M., Zhang, L., Ge, Y. P., Liu, B., & Li, X. F. (2023). The essentiality of dietary myo-inositol to oriental river prawn (*Macrobrachium nipponense*): Evidence in growth performance, lipid metabolism and mitochondrial function. *Aquaculture*, *567*, 739304. https://doi.org/10.1016/j.aquaculture.2023.739304

Liu, S., Wang, X., Bu, X., Lin, Z., Li, E., Shi, Q., Zhang, M., Qin, J. G., & Chen, L. (2021). Impact of vitamin D3 supplementation on growth, molting, antioxidant capability, and immunity of juvenile Chinese mitten crabs (*Eriocheir sinensis*) by metabolites and vitamin D receptors. *Journal of Agricultural and Food Chemistry*, *69*(43), 12794–12806. https://doi.org/10.1021/acs.jafc.1c04204

Liu, T., Zhang, J., & Li, A. (1995). Studies on the optimal requirements of pantothenic acid, biotin, folic acid and vitamin B12 in the shrimp *Peneaus chinensis*. *Journal of Fishery Sciences of China*, *2*, 48–55.

Liu, T. B., Li, A. K., & Zhang, J. M. (1993). Studies on vitamin nutrition for the shrimp *Penaeus chinensis*: 10. Studies on the choline chloride and inositol requirements in the shrimp *Penaeus chinensis*. *Journal of Ocean University of Qingdao*, *23*, 67–74.

Liu, W., Lu, X., Jiang, M., Wu, F., Tian, J., Yu, L., & Wen, H. (2020). Effects of dietary niacin on liver health in genetically improved farmed tilapia (*Oreochromis niloticus*). *Aquaculture Reports*, *16*, 100243. https://doi.org/10.1016/j.aqrep.2019.100243

Liu, Y., Chi, L., Feng, L., Jiang, J., Jiang, W. D., Hu, K., Li, S. H., & Zhou, X. Q. (2011). Effects of graded levels of dietary vitamin C on the growth, digestive capacity and intestinal microflora of juvenile Jian carp (*Cyprinus carpio* var. Jian). *Aquaculture Research*, *42*(4), 534–548. https://doi.org/10.1111/j.1365-2109.2010.02649.x

Liu, Y., Wang, W. N., Wang, A. L., Wang, J. M., & Sun, R. Y. (2007). Effects of dietary vitamin E supplementation on antioxidant enzyme activities in *Litopenaeus vannamei* (Boone, 1931) exposed to acute salinity changes. *Aquaculture*, *265*(1–4), 351–358. https://doi.org/10.1016/j.aquaculture.2007.02.010

Lock, E.-J., Waagbø, R., Wendelaar Bonga, S., & Flik, G. (2010). The significance of vitamin D for fish: A review. *Aquaculture Nutrition*, *16*(1), 100–116. https://doi.org/10.1111/j.1365-2095.2009.00722.x

Lock, E. J., Ørnsrud, R., Aksnes, L., Spanings, F. A. T., Waagbø, R., & Flik, G. (2007). The vitamin D receptor and its ligand 1α,25-dihydroxyvitamin D3 in Atlantic salmon (*Salmo salar*). *Journal of Endocrinology*, *193*(3), 459–471. https://doi.org/10.1677/JOE-06-0198

Lovell, R. T., & Li, Y. P. (1978). Essentiality of vitamin D in diets of channel catfish (*Ictalurus punctatus*). *Transactions of the American Fisheries Society, 107*(6), 809–811. https://doi.org/10.1577/1548-8659 (1978)107<809:EOVDID>2.0.CO;2

Lovell, R. T., & Lim, C. (1978). Vitamin C in pond diets for channel catfish. *Transactions of the American Fisheries Society, 107*(2), 321–325. https://doi.org/10.1577/1548-8659(1978)107<321:VCIPDF>2.0.CO;2

Lozano, A. R., Borges, P., Robaina, L., Betancor, M., Hernández-Cruz, C. M., García, J. R., Caballero, M. J., Vergara, J. M., & Izquierdo, M. (2017). Effect of different dietary vitamin E levels on growth, fish composition, fillet quality and liver histology of meagre (*Argyrosomus regius*). *Aquaculture, 468*, 175–183. https://doi.org/10.1016/j.aquaculture.2016.10.006

Lu, J., Tao, X., Luo, J., Zhu, T., Jiao, L., Jin, M., & Zhou, Q. (2022). Dietary choline promotes growth, antioxidant capacity and immune response by modulating p38MAPK/p53 signaling pathways of juvenile Pacific white shrimp (*Litopenaeus vannamei*). *Fish and Shellfish Immunology, 131*, 827–837. https://doi.org/10.1016/j.fsi.2022.10.062

Lu, X., Zhang, Z. Q., Wen, H., Jiang, M., & Du, H. (2020). Dietary choline requirement of juvenile Chinese sucker (*Myxocyprinus asiaticus*). *Aquaculture Reports, 18*, 100484. https://doi.org/10.1016/j.aqrep.2020.100484

Lu, Y., Liang, X. P., Jin, M., Sun, P., Ma, H. N., Yuan, Y., & Zhou, Q. C. (2016). Effects of dietary vitamin E on the growth performance, antioxidant status and innate immune response in juvenile yellow catfish (*Pelteobagrus fulvidraco*). *Aquaculture, 464*, 609–617. https://doi.org/10.1016/j.aquaculture.2016.08.009

Lu, Z., Chen, T. C., Zhang, A., Persons, K. S., Kohn, N., Berkowitz, R., Martinello, S., & Holick, M. F. (2007). An evaluation of vitamin D3 content in fish: Is the vitamin D content adequate to satisfy the dietary requirement for vitamin D? *Journal of Steroid Biochemistry and Molecular Biology, 103*(3–5), 642–644. https://doi.org/10.1016/j.jsbmb.2006.12.010

Luo, Z., Wei, C. C., Ye, H. M., Zhao, H. P., Song, Y. F., & Wu, K. (2016). Effect of dietary choline levels on growth performance, lipid deposition and metabolism in juvenile yellow catfish *Pelteobagrus fulvidraco*. *Comparative Biochemistry and Physiology. Part B, Biochemistry and Molecular Biology, 202*, 1–7. https://doi.org/10.1016/j.cbpb.2016.07.005

Lushchak, V. I., & Bagnyukova, T. V. (2006). Temperature increases results in oxidative stress in goldfish tissues. 1. Indices of oxidative stress. *Comparative Biochemistry and Physiology. Toxicology and Pharmacology, 143*(1), 30–35. https://doi.org/10.1016/j.cbpc.2005.11.017

Lygren, B., Hjeltnes, B., & Waagbø, R. (2001). Immune response and disease resistance in Atlantic salmon (*Salmo salar* L.) fed three levels of dietary vitamin E and the effect of vaccination on the liver status of antioxidant vitamins. *Aquaculture International, 9*(5), 401–411. https://doi.org/10.1023/A:1020509308400

Lyon, P., Strippoli, V., Fang, B., & Cimmino, L. (2020). B vitamins and one-carbon metabolism: Implications in human health and disease. *Nutrients, 12*(9). https://doi.org/10.3390/nu12092867

Maeda, Y., Kawata, S., Inui, Y., Fukuda, K., Igura, T., & Matsuzawa, Y. (1996). Biotin deficiency decreases ornithine transcarbamylase activity and mRNA in rat liver. *Journal of Nutrition, 126*(1), 61–66. https://doi.org/10.1093/jn/126.1.61

Maeland, A., Waagbø, R., Sandnes, K., & Hjeltnes, B. (1998). Biotin in practical fish-meal based diet for Atlantic salmon *Salmo salar* L. fry. *Aquaculture Nutrition, 4*(4), 241–247. https://doi.org/10.1046/j.1365-2095.1998.00076.x

Mai, K., Waagbø, R., Zhou, X. Q., Ai, Q. H., & Feng, L. (2022) Chapter 3. Vitamins. In R. W. Hardy & S. J. Kaushik (Eds.), *Fish nutrition* (4th ed) (pp. 57–179). Academic Press. ISBN: 9780128195871.

Malhotra, P., Karande, A. A., Prasadan, T. K., & Adiga, P. R. (1991). Riboflavin carrier protein from carp (*C. carpio*) eggs: Comparison with avian riboflavin carrier protein. *Biochemistry International, 23*(1), 127–136. PubMed: 1713763

Maningas, M. B., Kondo, H., & Hirono, I. (2013). Molecular mechanisms of the shrimp clotting system. *Fish and Shellfish Immunology, 34*(4), 968–972. https://doi.org/10.1016/j.fsi.2012.09.018

Manthey, K. C., Griffin, J. B., & Zempleni, J. (2002). Biotin supply affects expression of biotin transporters, biotinylation of carboxylases and metabolism of interleukin-2 in Jurkat cells. *Journal of Nutrition, 132*(5), 887–892. https://doi.org/10.1093/jn/132.5.887

Maranesi, M., Marchetti, M., Bochicchio, D., & Cabrini, L. (2005). Vitamin B6 supplementation increases the docosahexaenoic acid concentration of muscle lipids of rainbow trout (*Oncorhynchus mykiss*). *Aquaculture Research, 36*(5), 431–438. https://doi.org/10.1111/j.1365-2109.2005.01215.x

Marchetti, M., Tossani, N., Marchetti, S., & Bauce, G. (1999). Leaching of crystalline and coated vitamins in pelleted and extruded feeds. *Aquaculture, 171*(1–2), 83–92. https://doi.org/10.1016/S0044-8486(98)00412-8

Marchetti, M., Tassinari, M., & Marchetti, S. (2000). Menadione nicotinamide bisulphite as a source of vitamin K and niacin activities for the growing pig. *Animal Science, 71*(1), 111–117. https://doi.org/10.1017/S135772980005493X

Marks, J. (1975). *A Guide to the vitamins. Their role in health and disease.* Springer.

Martin, P. R., Singleton, C. K., & Hiller-Sturmhöfel, S. (2003). The role of thiamine deficiency in alcoholic brain disease. *Alcohol Research and Health, 27*(2), 134–142. PubMed: 15303623

Masumoto, T. (2002). Yellowtail, *Seriola quinqueradiata*. Nutrient requirements and feeding of finfish for aquaculture C. D. Webster & C. Lim (Eds.) (pp. 131–146). https://scholar.google.it/scholar_url?url=https://kenanaonline.com/files/0062/62625/0851995195ch10.pdf&hl=it&sa=X&ei=8GwgZaaYH9K-y9YPqMy92A0&scisig=AFWwaealcmMJylfg5YGgvLjvzfhZ&oi=scholarr

Masumoto, T., Hardy, R. W., & Casillas, E. (1987). Comparison of transketolase activity and thiamin pyrophosphate levels in erythrocytes and liver of rainbow trout (*Salmo gairdneri*) as indicators of thiamin status. *Journal of Nutrition, 117*(8), 1422–1426. https://doi.org/10.1093/jn/117.8.1422

Masumoto, T., Hardy, R. W., & Stickney, R. R. (1994). Pantothenic Acid deficiency detection in rainbow trout (*Oncorhynchus mykiss*). *Journal of Nutrition, 124*(3), 430–435. https://doi.org/10.1093/jn/124.3.430

Masumoto, T., Miki, T., Itoh, Y., Hosokawa, H., & Shimeno, S. (1999). Effect of pantothenic acid deficiency on plasma free amino acid profile in rainbow trout *Oncorhynchus mykiss*. *Fisheries Science, 65*(5), 794–795. https://doi.org/10.2331/fishsci.65.794

Matsuo, M., Ogata, Y., Yamanashi, Y., & Takada, T. (2023). ABCG5 and ABCG8 are involved in vitamin K transport. *Nutrients, 15*(4), 998. https://doi.org/10.3390/nu15040998

Mattila, P., Piironen, V., Uusi-Rauva, E., & Koivistoinen, P. (1995). Cholecalciferol and 25-hydroxycholecalciferol contents in fish and fish products. *Journal of Food Composition and Analysis, 8*(3), 232–243. https://doi.org/10.1006/jfca.1995.1017

Maurya, V. K., & Aggarwal, M. (2017). Factors influencing the absorption of vitamin D in GIT: An overview. *Journal of Food Science and Technology, 54*(12), 3753–3765. https://doi.org/10.1007/s13197-017-2840-0

McCay, P. B. (1985). Vitamin E: Interactions with free radicals and ascorbate. *Annual Review of Nutrition, 5*(1), 323–340. https://doi.org/10.1146/annurev.nu.05.070185.001543

McCollum, E.V. (1957) "A history of nutrition". Houghton Mifflin Co., Boston, MA.

McCormick, D. B. (1990). Riboflavin. In M. L. Brown (Ed.), *Nutrition reviews, present knowledge in nutrition.* International Life Sciences Institute.

McCormick, D. B. (2006). Vitamin B6. In B. A. Bowman & R. M. Russell (Eds.), *Present knowledge in nutrition* (9th ed) (pp. 269–277). International Life Sciences Institute. https://doi.org/10.1093/ajcn/85.5.1439a

McDowell, L. R. (1996). *Benefits of supplemental vitamin E including relationship to gossypol.* University of Florida.

McDowell, L. R. (2000). *Vitamins in animal and human nutrition.* Iowa State University Press. https://doi.org/10.1002/9780470376911

McDowell, L. R. (2006). Vitamins and minerals functioning as antioxidants and vitamin and mineral supplementation considerations. In *ARPAS California Chapter Conference Proceedings* (pp. 1–22).

McDowell, L.R. (2013) "*Vitamin history, the early years*". Design Publishing Inc. Sarasora, FL, USA.

McDowell, L. R., & Ward, N. (2008). Optimum vitamin nutrition for poultry. *International Poultry Production, 16*(4), 27–34. http://www.positiveaction.info/pdfs/articles/pp16.4p27.pdf

McGinnis, C. H. (1986). Vitamin stability and activity of water-soluble vitamins as influenced by manufacturing processes and recommendations for the water-soluble vitamin. In *Bioavailability of nutrients in feed ingredients.* National Feed Ingredient Association.

McGinnis, C. H. (1988). New concepts in vitamin nutrition. In *Proceedings of the 1988 Georgia Nutrition Conference of the Feed Industry.* University of Wisconsin.

McLaren, B. A., Keller, E., O'Donnell, D. J., & Elvehjem, C. A. (1947). The nutrition of rainbow trout. 1. Studies of vitamin requirements. *Archives of Biochemistry, 15*(2), 169–178. PubMed: 20270758

Mehansho, H., & Henderson, L. M. (1980). Transport and accumulation of pyridoxine and pyridoxal by erythrocytes. *Journal of Biological Chemistry, 255*(24), 11901–11907. https://doi.org/10.1016/S0021-9258(19)70220-8

Menoyo, D., Sanz-Bayón, C., Nessa, A. H., Esatbeyoglu, T., Faizan, M., Pallauf, K., De Diego, N., Wagner, A. E., Ipharraguerre, I., Stubhaug, I., & Rimbach, G. (2014). Atlantic salmon (*Salmo salar L.*) as a marine functional source of gamma-tocopherol. *Marine Drugs, 12*(12), 5944–5959. https://doi.org/10.3390/md12125944

Mentch, S. J., & Locasale, J. W. (2016). One-carbon metabolism and epigenetics: Understanding the specificity. *Annals of the New York Academy of Sciences, 1363*(1), 91–98. https://doi.org/10.1111/nyas.12956

Miao, L. H., Xian-ping, G., Jun, X., Bo, L., Ke-bao, W., Jian, Z., Ming-chun, R., Qun-lan, Z., Liang-kun, P., & Ru-li, C. (2015a). Dietary vitamin D3 requirement of Wuchang bream (*Megalobrama amblycephala*). *Aquaculture, 436*, 104–109. https://doi.org/10.1016/j.aquaculture.2014.10.049

Miao, L. H., Xie, J., Ge, X. P., Wang, K. B., Zhu, J., Liu, B., Ren, M. C., Zhou, Q. L., & Pan, L. K. (2015b). Chronic stress effects of high doses of vitamin D3 on *Megalobrama amblycephala*. *Fish and Shellfish Immunology, 47*(1), 205–213. https://doi.org/10.1016/j.fsi.2015.09.012

Michael, F. R., Koshio, S., Teshima, S. I., Ishikawa, M., & Uyan, O. (2006). Effect of choline and methionine as methyl group donors on juvenile kuruma shrimp, *Marsupenaeus japonicus* bate. *Aquaculture, 258*(1–4), 521–528. https://doi.org/10.1016/j.aquaculture.2006.04.019

Michael, F. R., Teshima, S.-I., & Koshio, S. (2011). Interactive effect of two methyl-group sources on the growth and phospholipids' content of kuruma shrimp *Marsupenaeus japonicus* Bate post-larvae. *Aquaculture Nutrition, 17*(3), e701–e707. https://doi.org/10.1111/j.1365-2095.2010.00830.x

Michael, F. R., & Koshio, S. (2016). Effect of choline chloride as an osmoregulator as well as its role in growth and the biochemical content of post larval kuruma shrimp; *Marsupenaeus japonicus* (Bate). *Aquaculture Nutrition, 22*(3), 597–605. https://doi.org/10.1111/anu.12283

Miller, J. W., Roger, L. M., & Rucker, R. B. (2006). Pantothenic acid. In B. A. Bowman & R. M. Russell (Eds.), *Present knowledge in nutrition* (9th ed) (pp. 327–339). International Life Sciences Institute. https://doi.org/10.1093/ajcn/85.5.1439a

Ming, J., Xie, J., Xu, P., Ge, X., Liu, W., & Ye, J. (2012). Effects of emodin and vitamin C on growth performance, biochemical parameters and two HSP70s mRNA expression of Wuchang bream (*Megalobrama amblycephala* Yih) under high temperature stress. *Fish and Shellfish Immunology, 32*(5), 651–661. https://doi.org/10.1016/j.fsi.2012.01.008

Mirvaghefi, A., Ali, M., & Asadi, F. (2015). Effects of vitamin E, selenium and vitamin C on various biomarkers following oxidative stress caused by diazinon exposure in rainbow trout. *Journal of Aquactic Marine Biology 2*, 00035.

Misir, R., & Blair, R. (1988). Biotin bioavailability of protein supplements and cereal grains for starting turkey poults. Poultry Science, 67(9), 1274–1280. https://doi.org/10.3382/ps.0671274

Mladěnka P, Macáková K, Kujovská Krčmová L, Javorská L, Mrštná K, Carazo A, Protti M, Remião F, Nováková L and the OEMONOM researchers and collaborators (2022). Vitamin K - sources, physiological role, kinetics, deficiency, detection, therapeutic use, and toxicity. Nutr Rev.80(4):677-698 https://doi.org/10.1093/nutrit/nuab061

Mobarhan, S., Greenberg, B., Mehta, R., Friedman, H., & Barch, D. (1992). Zinc deficiency reduces hepatic cellular retinol-binding protein in rats. *International Journal for Vitamin and Nutrition Research. Internationale Zeitschrift Fur Vitamin- und Ernahrungsforschung. Journal International de Vitaminologie et de Nutrition, 62*(2), 148–154.

Moccia, R. D., Hung, S. S. O., Slinger, S. J., & Ferguson, H. W. (1984). Effect of oxidized fish oil, vitamin E and ethoxyquin on the histopathology and haematology of rainbow trout, *Salmo gairdneri* Richardson. *Journal of Fish Diseases, 7*(4), 269–282. https://doi.org/10.1111/j.1365-2761.1984.tb00932.x

Mochizuki, M., Kim, H. J., Kasai, H., Nishizawa, T., & Yoshimizu, M. (2009). Virulence changes of infectious hematopoietic necrosis virus against rainbow trout (*Oncorhynchus mykiss*) with Viral Molecular Evolution. *Fish Pathology, 44*(4), 159–165. https://doi.org/10.3147/jsfp.44.159

Mock, D. M. (1990). Biotin. In R. E. Olson (Ed.) *Nutrition reviews, present knowledge in nutrition*. Nutrition Foundation.

Mock, D. M. (2013). Biotin. In J. Zempleni, J. Suttie, J. Gregory & P. J. Stover (Eds.), *Handbook of vitamins* (5th ed) (pp. 397–420). CRC Press, Taylor & Francis Group. https://doi.org/10.1201/b15413

Mock, D. M., & Malik, M. I. (1992). Distribution of biotin in human plasma: Most of the biotin is not bound to protein. *American Journal of Clinical Nutrition, 56*(2), 427–432. https://doi.org/10.1093/ajcn/56.2.427

Mock, N. I., & Mock, D. M. (1992). Biotin deficiency in rats: Disturbances of leucine metabolism are detectable early. *Journal of Nutrition, 122*(7), 1493–1499. https://doi.org/10.1093/jn/122.7.1493

Moe, Y. Y., Koshio, S., Teshima, S., Ishikawa, M., Matsunaga, Y., & Panganiban, A. (2004). Effect of vitamin C derivatives on the performance of larval kuruma shrimp, *Marsupenaeus japonicus*. *Aquaculture, 242*(1–4), 501–512. https://doi.org/10.1016/j.aquaculture.2004.08.028

Mohamed, J. S., & Ibrahim, A. (2001). Quantifying the dietary niacin requirement of the Indian catfish, *Heteropneustes fossilis* (Bloch), fingerlings. *Aquaculture Research, 32*(3), 157–162. https://doi.org/10.1046/j.1365-2109.2001.00530.x

Mohamed, S. H., El-Leithy, E. M. M., Ghandour, R. A., & Galal, M. K. (2019). Molecular, biochemical and histopathological studies on the ameliorative effect of vitamin C on the renal and muscle tissues of Nile tilapia fish (*Oreochromis niloticus*) affected by the usage of engine oil. *Aquaculture Research, 50*(11), 3357–3368. https://doi.org/10.1111/are.14294

Mohiseni, M., Sepidnameh, M., Bagheri, D., Banaee, M., & Nematdust Haghi, B. N. (2017). Comparative effects of Shirazi thyme and vitamin E on some growth and plasma biochemical changes in common carp (*Cyprinus carpio*) during cadmium exposure. *Aquaculture Research, 48*(9), 4811–4821. https://doi.org/10.1111/are.13301

Mondal, H., & Thomas, J. (2022). A review on the recent advances and application of vaccines against fish pathogens in aquaculture. *Aquaculture International, 30*(4), 1971–2000. https://doi.org/10.1007/s10499-022-00884-w

Moniruzzaman, M., Park, G., Yun, H., Lee, S., Park, Y., & Bai, S. C. (2017). Synergistic effects of dietary vitamin E and selenomethionine on growth performance and tissue methylmercury accumulation on mercury-induced toxicity in juvenile olive flounder, *Paralichthys olivaceus* (Temminck et Schlegel). *Aquaculture Research, 48*(2), 570–580. https://doi.org/10.1111/are.12904

Montrucchio, G., Alloatti, G., & Camussi, G. (2000). Role of platelet-activating factor in cardiovascular pathophysiology. *Physiological Reviews, 80*(4), 1669–1699. https://doi.org/10.1152/physrev.2000.80.4.1669

Moradi, S., Javanmardi, S., Gholamzadeh, P., & Tavabe, K. R. (2022). The ameliorative role of ascorbic acid against blood disorder, immunosuppression, and oxidative damage of oxytetracycline in rainbow trout (*Oncorhynchus mykiss*). *Fish Physiology and Biochemistry, 48*(1), 201–213. https://doi.org/10.1007/s10695-022-01045-9

Moreau, R., & Dabrowski, K. (2000). Biosynthesis of ascorbic acid by extant actinopterygians. *Journal of Fish Biology, 57*(3), 733–745. https://doi.org/10.1111/j.1095-8649.2000.tb00271.x

Morito, C. L. H., Conrad, D. H., & Hilton, J. W. (1986). The thiamin deficiency signs and requirement of rainbow trout (*Salmo gairdneri*, Richardson). *Fish Physiology and Biochemistry, 1*(2), 93–104. https://doi.org/10.1007/BF02290209

Morris, M. S., Sakakeeny, L., Jacques, P. F., Picciano, M. F., & Selhub, J. (2010). Vitamin B6 intake is inversely related to, and the requirement is affected by, inflammation status. *Journal of Nutrition, 140*(1), 103–110. https://doi.org/10.3945/jn.109.114397

Morris, P. C., & Davies, S. J. (1995). Thiamin supplementation of diets containing varied lipid: Carbohydrate ratio given to gilthead seabream (*Sparus aurata* L.). *Animal Science, 61*(3), 597–603. https://doi.org/10.1017/S1357729800014193

Morshedian, A., Toomey, M. B., Pollock, G. E., Frederiksen, R., Enright, J. M., McCormick, S. D., Cornwall, M. C., Fain, G. L., & Corbo, J. C. (2017). Cambrian origin of the CYP27C1-mediated vitamin A1-to-A2 switch, a key mechanism of vertebrate sensory plasticity. *Royal Society Open Science, 4*(7), 170362. https://doi.org/10.1098/rsos.170362

Morton, R. A., & Creed, R. H. (1939). The conversion of carotene into vitamin A2 by some freshwater fishes. *Biochemical Journal, 33*(3), 318–324. https://doi.org/10.1042/bj0330318

Moser, U., & Bendich, A. (1991). Vitamin C. In L. J. Machlin (Ed.), *Handbook of vitamins* (2nd ed) (p. 195). Marcel Dekker.

Moss, S. M., Forster, I. P., & Tacon, A. G. J. (2006). Sparing effect of pond water on vitamins in shrimp diets. *Aquaculture, 258*(1–4), 388–395. https://doi.org/10.1016/j.aquaculture.2006.04.008

Mowi. (2023). Salmon farming industry handbook. https://mowi.com/wp-content/uploads/2023/06/2023-Salmon-Farming-Industry-Handbook-2023.pdf

Mozaffarian, D., Rosenberg, I., & Uauy, R. (2018). History of modern nutrition science–implications for current research, dietary guidelines, and food policy. *BMJ, 361*, k2392. https://doi.org/10.1136/bmj.k2392

Murai, T., & Andrews, J. W. (1977). Vitamin K and anticoagulant relationships in catfish diets. *Nippon Suisan Gakkaishi, 43*(7), 785–794. https://doi.org/10.2331/suisan.43.785

Murai, T., & Andrews, J. W. (1978). Thiamin requirement of channel catfish fingerlings. *Journal of Nutrition, 108*(1), 176–180. https://doi.org/10.1093/jn/108.1.176

Murai, T., & Andrews, J. W. (1979). Pantothenic Acid requirements of channel catfish fingerlings. *Journal of Nutrition, 109*(7), 1140–1142. https://doi.org/10.1093/jn/109.7.1140

Muralt, von, A. (1962). The role of thiamine in neurophysiology. *Annals of the New York Academy of Sciences, 98*(2), 499–507. https://doi.org/10.1111/j.1749-6632.1962.tb30571.x

Mustafa, A., Hayat, S. A., & Quarrar, P. (2013). Stress modulated physiological responses in Nile tilapia (*Oreochromis niloticus*) treated with non-ascorbic acid supplemented feed. *Advances in Zoology and Botany, 1*(2), 39–45. https://doi.org/10.13189/azb.2013.010204

Nabokina, S. M., Kashyap, M. L., & Said, H. M. (2005). Mechanism and regulation of human intestinal niacin uptake. *American Journal of Physiology. Cell Physiology, 289*(1), C97–C103. https://doi.org/10.1152/ajpcell.00009.2005

Naderi, N., & House, J. D. (2018). Recent developments in folate nutrition. In *Advances in Food and Nutrition Research* N. A. M. Eskin (Ed.). Academic Press, 195–213.

Naderi, M., Keyvanshokooh, S., Salati, A. P., & Ghaedi, A. (2017a). Effects of dietary vitamin E and selenium nanoparticles supplementation on acute stress responses in rainbow trout (*Oncorhynchus mykiss*) previously subjected to chronic stress. *Aquaculture, 473*, 215–222. https://doi.org/10.1016/j.aquaculture.2017.02.020

Naderi, M., Keyvanshokooh, S., Salati, A. P., & Ghaedi, A. (2017b). Proteomic analysis of liver tissue from rainbow trout (*Oncorhynchus mykiss*) under high rearing density after administration of dietary vitamin E and selenium nanoparticles. *Comparative Biochemistry and Physiology. Part D, Genomics and Proteomics, 22*, 10–19. https://doi.org/10.1016/j.cbd.2017.02.001

Naderi, M., Keyvanshokooh, S., Salati, A. P., & Ghaedi, A. (2017c). Combined or individual effects of dietary vitamin E and selenium nanoparticles on humoral immune status and serum parameters of rainbow trout (*Oncorhynchus mykiss*) under high stocking density. *Aquaculture, 474*, 40–47. https://doi.org/10.1016/j.aquaculture.2017.03.036

Naderi, M., Keyvanshokooh, S., Ghaedi, A., & Salati, A. P. (2019). Interactive effects of dietary Nano selenium and vitamin E on growth, haematology, innate immune responses, antioxidant status and muscle composition of rainbow trout under high rearing density. *Aquaculture Nutrition, 25*(5), 1156–1168. https://doi.org/10.1111/anu.12931

Nagaraj, R. Y., Wu, W. D., & Vesonder, R. F. (1994). Toxicity of corn culture material of *Fusarium proliferatum* M-7176 and nutritional intervention in chicks. *Poultry Science, 73*(5), 617–626. https://doi.org/10.3382/ps.0730617

Naiel, M. A. E., Ismael, N. E. M., Abd El-Hameed, S. A. A., & Amer, M. S. (2020). The antioxidative and immunity roles of chitosan nanoparticle and vitamin C-supplemented diets against Imidacloprid toxicity on *Oreochromis niloticus*. *Aquaculture, 523*, 735219. https://doi.org/10.1016/j.aquaculture.2020.735219

Nakagawa, K., Shibata, A., Yamashita, S., Tsuzuki, T., Kariya, J., Oikawa, S., & Miyazawa, T. (2007). In vivo angiogenesis is suppressed by unsaturated vitamin E, tocotrienol. *Journal of Nutrition, 137*(8), 1938–1943. https://doi.org/10.1093/jn/137.8.1938

Nasar, M. F., Shah, S. Z. H., Aftab, K., Fatima, M., Bilal, M., & Hussain, M. (2021). Dietary vitamin C requirement of juvenile grass carp (*Ctenopharyngodon idella*) and its effects on growth attributes, organ indices, whole-body composition and biochemical parameters. *Aquaculture Nutrition, 27*(6), 1903–1911. https://doi.org/10.1111/anu.13327

Navarre, O., & Halver, J. E. (1989). Disease resistance and humoral antibody production in rainbow trout fed high levels of vitamin C. *Aquaculture, 79*(1–4), 207–221. https://doi.org/10.1016/0044-8486(89)90462-6

Naylor, R. L., Hardy, R. W., Buschmann, A. H., Bush, S. R., Cao, L., Klinger, D. H., Little, D. C., Lubchenco, J., Shumway, S. E., & Troell, M. (2021). A 20-year retrospective review of global aquaculture. *Nature, 591*(7851), 551–563. https://doi.org/10.1038/s41586-021-03308-6

Neidhart, M. (2016). Methyl donors. In *DNA methylation and complex human disease* (1st ed) (pp. 429–439). Academic Press, Elsevier, Inc.

Ng, W. K., Keembiyehetty, C. N., & Wilson, R. P. (1998). Bioavailability of niacin from feed ingredients commonly used in feeds for channel catfish, *Ictalurus punctatus*. *Aquaculture, 161*(1–4), 393–404. https://doi.org/10.1016/S0044-8486(97)00287-1

Ng, W. K., Wang, Y., Ketchimenin, P., & Yuen, K. H. (2004). Replacement of dietary fish oil with palm fatty acid distillate elevates tocopherol and tocotrienol concentrations and increases oxidative stability in the muscle of African catfish, *Clarias gariepinus*. *Aquaculture, 233*(1–4), 423–437. https://doi.org/10.1016/j.aquaculture.2003.10.013

Nguyen, B. T., Koshio, S., Sakiyama, K., Harakawa, S., Gao, J., Mamauag, R. E., Ishikawa, M., & Yokoyama, S. (2012). Effects of dietary vitamins C and E and their interactions on reproductive performance, larval quality and tissue vitamin contents in kuruma shrimp, *Marsupenaeus japonicus* Bate. *Aquaculture, 334–337*, 73–81. https://doi.org/10.1016/j.aquaculture.2011.11.044

Nguyen, V. N., Tran, V. K., & Duy, H. P. (2014). Study on development of formulated feed for improving growth and pigmentation of koi carp (*Cyprinus carpio* L., 1758) juveniles. *JLS, 8*(5), 433–441.

Nishimune, T., Watanabe, Y., Okazaki, H., & Akai, H. (2000). Thiamin is decomposed due to *Anaphe* spp. entomophagy in seasonal ataxia patients in Nigeria. *Journal of Nutrition, 130*(6), 1625–1628. https://doi.org/10.1093/jn/130.6.1625

Niu, J., Tian, L.-X., Liu, Y.-J., Mai, K.-S., Yang, H.-J., Ye, C.-X., & Gao, W. (2009). Nutrient values of dietary ascorbic acid (l-ascorbyl-2-polyphosphate) on growth, survival and stress tolerance of larval shrimp, *Litopenaeus vannamei*. *Aquaculture Nutrition, 15*(2), 194–201. https://doi.org/10.1111/j.1365-2095.2008.00583.x

Niu, J., Li, C. H., Liu, Y. J., Tian, L. X., Chen, X., Huang, Z., & Lin, H. Z. (2012). Dietary values of astaxanthin and canthaxanthin in *Penaeus monodon* in the presence and absence of cholesterol supplementation: Effect on growth, nutrient digestibility and tissue carotenoid composition. *British Journal of Nutrition, 108*(1), 80–91. https://doi.org/10.1017/S0007114511005423

Niu, J., Wen, H., Li, C. H., Liu, Y. J., Tian, L. X., Chen, X., Huang, Z., & Lin, H. Z. (2014). Comparison effect of dietary astaxanthin and β-carotene in the presence and absence of cholesterol supplementation on growth performance, antioxidant capacity and gene expression of *Penaeus monodon* under normoxia and hypoxia condition. *Aquaculture, 422–423*, 8–17. https://doi.org/10.1016/j.aquaculture.2013.11.013

Nockels, C. F. (1990). Mineral alterations associated with stress, trauma and infection and the effect on immunity. Cont. Educ. *Practicing Veterinarian, 12*(8), 1133–1139.

Noel, K., & Brinkhaus, F. (1998). Vitamin A retention of a high pigment broiler growing feed treated with endox or ethoxyquin. *Poultry Science, 77*(1), 144.

Noori, A., & Razi, A. (2018). Effects of dietary astaxanthin on the growth and skin and muscle pigmentation of sexually immature rainbow trout *Oncorhynchus mykiss* (Walbaum, 1792) (Teleostei: Salmonidae). *Iranian Journal of Ichthyology*, 361–374%V 364. https://doi.org/10.22034/iji.v4i4.234015

NRC. (1987). *Vitamin tolerance of animals*. National Academy of Sciences. National Research Council. https://doi.org/10.17226/949

NRC. (2011). *Nutrient requirements of fish and shrimp*. National Academy Press. https://doi.org/10.1007/s10499-011-9480-6

Núñez-Acuña, G., Gallardo-Escárate, C., Fields, D. M., Shema, S., Skiftesvik, A. B., Ormazábal, I., & Browman, H. I. (2018). The Atlantic salmon (*Salmo salar*) antimicrobial peptide cathelicidin-2 is a molecular host-associated cue for the salmon louse (*Lepeophtheirus salmonis*). *Scientific Reports, 8*(1), 13738. https://doi.org/10.1038/s41598-018-31885-6

Oduho, G. W., Chung, T. K., & Baker, D. H. (1993). Menadione nicotinamide bisulfite is a bioactive source of vitamin K and niacin activity for chicks. *Journal of Nutrition, 123*(4), 737–743. https://doi.org/10.1093/jn/123.4.737

Oginni, O., Gbore, F.A., Adewole, A.M., Eniade, A., Adebusoye, A.J., Abimbola, A.T. and Ajumobi, O.O. (2020) Influence of vitamins on flesh yields and proximate compositions of Clarias gariepinus fed diets contaminated with increasing doses of fumonisin B1. Journal of Agriculture and Food Research, 2, https://doi.org/10.1016/j.jafr.2020.100079

O'Keefe, T. M., & Noble, R. L. (1978). Storage stability of channel catfish (*Ictalurus punctatus*) in relation to dietary level of α-tocopherol. *Journal of the Fisheries Research Board of Canada, 35*(4), 457–460. https://doi.org/10.1139/f78-079

Oldfield, J. E. (1987). History of nutrition: Development of the concept of antimetabolites. Introduction. *Journal of Nutrition, 117*(7), 1322–1323. https://doi.org/10.1093/jn/117.7.1322

Oloyo, R. A. (1991). Responses of broilers fed Guinea corn/palm kernel meal based ration to supplemental biotin. *Journal of the Science of Food and Agriculture, 55*(4), 539–550. https://doi.org/10.1002/jsfa.2740550406

Olsen, A. I., MÆland, A., WaagbØ, R., & Olsen, Y. (2000). Effect of algal addition on stability of fatty acids and some water-soluble vitamins in juvenile *Artemia franciscana. Aquaculture Nutrition, 6*(4), 263–273. https://doi.org/10.1046/j.1365-2095.2000.00157.x

Olsen, R. E., Myklebust, R., Kaino, T., & Ringø, E. (1999). Lipid digestibility and ultrastructural changes in the enterocytes of Arctic char (*Salvelinus alpinus* L.) fed linseed oil and soybean lecithin. *Fish Physiology and Biochemistry, 21*(1), 35–44. https://doi.org/10.1023/A:1007726615889

Olson, J. A. (1984). Vitamin A. In L. J. Machlin (Ed.), *Handbook of vitamins*. Marcel Dekker, Inc.

Olson, J. A. (1989). Provitamin function of carotenoids: The conversion of β-carotene into vitamin A. *Journal of Nutrition, 119*(1), 105–108. https://doi.org/10.1093/jn/119.1.105

Olson, J. A., & Hayaishi, O. (1965). The enzymatic cleavage of β-carotene into vitamin A by soluble enzymes of rat liver and intestine. *Proceedings of the National Academy of Sciences of the United States of America, 54*(5), 1364–1370. https://doi.org/10.1073/pnas.54.5.1364

O'Mahony, L., Stepien, M., Gibney, M. J., Nugent, A. P., & Brennan, L. (2011). The potential role of vitamin D enhanced foods in improving vitamin D status. *Nutrients, 3*(12), 1023–1041. https://doi.org/10.3390/nu3121023

Oonincx, D. G. A. B., & Finke, M. D. (2021). Nutritional value of insects and ways to manipulate their composition. *Journal of Insects as Food and Feed, 7*(5), 639–659. https://doi.org/10.3920/JIFF2020.0050

Ørnsrud, R., Wargelius, A., Sæle, Ø., Pittman, K., & Waagbø, R. (2004a). Influence of egg vitamin A status and egg incubation temperature on subsequent development of the early vertebral column of Atlantic salmon fry. *Journal of Fish Biology, 64*(2), 399–417. https://doi.org/10.1111/j.0022-1112.2004.00304.x

Ørnsrud, R., Gil, L., & Waagbø, R. (2004b). Teratogenicity of elevated egg incubation temperature and egg vitamin A status in Atlantic salmon, *Salmo salar* L. *Journal of Fish Diseases, 27*(4), 213–223. https://doi.org/10.1111/j.1365-2761.2004.00536.x

Ørnsrud, R., Lock, E. J., Glover, C. N., & Flik, G. (2009). Retinoic acid cross-talk with calcitriol activity in Atlantic salmon (*Salmo salar*). *Journal of Endocrinology, 202*(3), 473–482. https://doi.org/10.1677/JOE-09-0199

Ørnsrud, R., Lock, E.-J., Waagbø, R., Krossøy, C., & Fjelldal, P.-G. (2013). Establishing an upper level of intake for vitamin A in Atlantic salmon (*Salmo salar* L.) post smolts. *Aquaculture Nutrition, 19*(5), 651–664. https://doi.org/10.1111/anu.12013

Ostermeyer, U., & Schmidt, T. (2001). Determination of vitamin K in the edible part of fish by high-performance liquid chromatography. *European Food Research and Technology, 212*(4), 518–528. https://doi.org/10.1007/s002170000262

Oviedo-Rondón, E. O., Ferket, P. R., & Havestein, G. B. (2006). Nutritional factors that affect leg problems in broilers and turkeys. *Avian and Poultry Biology Reviews, 17*(3), 89–103. http://doi.org/10.3184/147020606783437921

Oviedo-Rondón, E. O., Barroeta, A. C., Briz, C., Litta, R., G., & Hernandez, J. M. (2023a). Optimum vitamin nutrition for more sustainable poultry farming. ISBN 9781789182248, *5m* Books Ltd., UK.

Oviedo-Rondón, E. O., López-Bote, C., Litta, G., & Hernandez, J. M. (2023b). Optimum vitamin nutrition for more sustainable swine farming. ISBN 9781789182477, *5m* Books Ltd., UK.

Ovung, A., & Bhattacharyya, J. (2021). Sulfonamide drugs: Structure, antibacterial property, toxicity, and biophysical interactions. *Biophysical Reviews, 13*(2), 259–272. https://doi.org/10.1007/s12551-021-00795-9

Paibulkichakul, C., Piyatiratitivorakul, S., Sorgeloos, P., & Menasveta, P. (2008). Improved maturation of pond-reared, black tiger shrimp (*Penaeus monodon*) using fish oil and astaxanthin feed supplements. *Aquaculture, 282*(1–4), 83–89. https://doi.org/10.1016/j.aquaculture.2008.06.006

Pal, H., & Chakrabarty, D. (2012). Evaluations of body composition and growth performance by applying different dietary vitamin C levels in Asian catfish, *Clarias batrachus* (Linnaeus, 1758). *International Journal of Pharmacy and Biological Sciences, 3*(4). http://www.ijpbs.com.

Pan, J. H., Feng, L., Jiang, W. D., Wu, P., Kuang, S. Y., Tang, L., Zhang, Y. A., Zhou, X. Q., & Liu, Y. (2017). Vitamin E deficiency depressed fish growth, disease resistance, and the immunity and structural integrity of immune organs in grass carp (*Ctenopharyngodon idella*): Referring to NF-κB, TOR and Nrf2 signaling. *Fish and Shellfish Immunology, 60*(January), 219–236. https://doi.org/10.1016/j.fsi.2016.11.044

Pan, J. H., Feng, L., Jiang, W. D., Wu, P., Kuang, S. Y., Tang, L., Tang, W. N., Zhang, Y. A., Zhou, X. Q., & Liu, Y. (2018). Vitamin E deficiency depressed gill immune response and physical barrier referring to NF-kB, TOR, Nrf2 and MLCK signalling in grass carp (*Ctenopharyngodon idella*) under infection of *Flavobacterium columnare*. *Aquaculture, 484*, 13–27. https://doi.org/10.1016/j.aquaculture.2017.10.028

Pangantihon-Kühlmann, M. P., Millamena, O., & Chern, Y. (1998). Effect of dietary astaxanthin and vitamin A on the reproductive performance of *Penaeus monodon* broodstock. *Aquatic Living Resources, 11*, 403–409. https://doi.org/10.1016/S0990-7440(99)80006-0

Pankhurst, N. W. (2011). The endocrinology of stress in fish: An environmental perspective. *General and Comparative Endocrinology, 170*(2), 265–275. https://doi.org/10.1016/j.ygcen.2010.07.017

Parsons, J. L., & Klostermann, H. J. (1967). Dakota scientists report new antibiotic found in flaxseed. *Feedstuffs, 39*(45), 74.

Peng, S. M., Shi, Z. H., Fei, Y., Gao, Q. X., Sun, P., & Wang, J. G. (2013). Effect of high-dose vitamin C supplementation on growth, tissue ascorbic acid concentrations and physiological response to transportation stress in juvenile silver pomfret, *Pampus argenteus. Journal of Applied Ichthyology, 29*(6), 1337–1341. https://doi.org/10.1111/jai.12250

Peng, X., Shang, G., Wang, W., Chen, X., Lou, Q., Zhai, G., Li, D., Du, Z., Ye, Y., Jin, X., He, J., Zhang, Y., & Yin, Z. (2017). Fatty acid oxidation in zebrafish adipose tissue is promoted by 1α, 25(OH)2D3. *Cell Reports, 19*(7), 1444–1455. https://doi.org/10.1016/j.celrep.2017.04.066

Penn, M. H. (2011). Lipid malabsorption in "Atlantic salmon—the reoccurring problem of floating feces". *Fiskehelse. Tekna Fiskehelseforeningen*, Oslo, 6–11.

Perry, S. C. (1978). Vitamin allowances for animal feeds. In *Vitamin nutrition update – Seminar series 2, RCD 5483/1078*. Hoffmann-La Roche.

Petit, H., Nègre-Sadargues, G., Castillo, R., & Trilles, J. P. (1997). The effects of dietary astaxanthin on growth and moulting cycle of post larval stages of the prawn, *Penaeus japonicus* (Crustacea, Decapoda). *Comparative Biochemistry and Physiology Part A, 117*(4), 539–544. https://doi.org/10.1016/S0300-9629(96)00431-8

Phillips, A., Brockway, D., Rodgers, E., Sullivan, M., Cook, B., & Chipman, J. (1946). The nutrition of trout. *Fisheries Research Bulletin, 9*, 3–21. http://www.nativefishlab.net/library/textpdf/11167.pdf

Phillips, N. W. (1984). Role of different microbes and substrates as potential suppliers of specific essential nutrients to marine detritivores. *Bulletin of Marine Science, 35*, 283–298. https://www.ingentaconnect.com/contentone/umrsmas/bullmar/1984/00000035/00000003/art00004?crawler=true

Pierens, S. L., & Fraser, D. R. (2015). The origin and metabolism of vitamin D in rainbow trout. *Journal of Steroid Biochemistry and Molecular Biology, 145*, 58–64. https://doi.org/10.1016/j.jsbmb.2014.10.005

Pike, C., Crook, V., & Gollock, M. (2020). *Anguilla anguilla. IUCN Red List of Threatened Species*. https://doi.org/10.2305/IUCN.UK.2020-2.RLTS.T60344A152845178.en

Pinto, J. T., & Rivlin, R. S. (2013). Riboflavin (vitamin B2). In J. Zempleni, J. Suttie, J. Gregory & P. J. Stover (Eds.), *Handbook of vitamins* (5th ed) (pp. 191–266). CRC Press, Taylor & Francis Group. https://doi.org/10.1201/b15413

Pitaksong, T., Kupittayanant, P., & Boonanuntanasarn, S. (2013). The effects of vitamins C and E on the growth, tissue accumulation and prophylactic response to thermal and acidic stress of hybrid catfish. *Aquaculture Nutrition, 19*(2), 148–162. https://doi.org/10.1111/j.1365-2095.2012.00950.x

Polak, D. M., Elliot, J. M., & Haluska, M. (1979). Vitamin B12 binding proteins in bovine serum. *Journal of Dairy Science, 62*(5), 697–701. https://doi.org/10.3168/jds.S0022-0302(79)83312-3

Poston, H. A. (1968). The conversion of beta-carotene to vitamin A by fingerling brook trout. Cortland Hatchery Report 32, State of New York Conserv. Dept., Albany, Fish. *Research Bulletin, 32*, 42–43. https://waves-vagues.dfo-mpo.gc.ca/library-bibliotheque/115707.pdf

Poston, H. A. (1969a). The effect of excess levels of niacin on the lipid metabolism of fingerling brook trout. Cortland Hatchery Report 32, State of New York Conserv. Dept., Albany, Fish. *Research Bulletin, 32*, 9–12.

Poston, H. A. (1969b). Effects of massive doses of vitamin D3 on fingerling brook trout. Fisheries Research Bulletin, 32, 48–50.

Poston, H. A. (1971). Effect of excess vitamin K on the growth, coagulation time, and hematocrit values of brook trout fingerlings. Cortland Hatchery Report 34, State of New York Conserv. Dept., Albany, Fish. *Research Bulletin, 34*, 41–42.

Poston, H. A. (1976a). Relative effect of two dietary water-soluble analogues of menaquinone on coagulation and packed cell volume of blood of lake trout (*Salvelinus namaycush*). *Journal of the Fisheries Research Board of Canada, 33*(8), 1791–1793. https://doi.org/10.1139/f76-227

Poston, H. A. (1976b). Optimum level of dietary biotin for growth, feed utilization, and swimming stamina of fingerling lake trout (*Salvelinus namaycush*). *Journal of the Fisheries Research Board of Canada, 33*(8), 1803–1806. https://doi.org/10.1139/f76-230

Poston, H. A. (1990). Effect of body size on growth, survival, and chemical composition of Atlantic salmon fed soy lecithin and choline. *Progressive Fish-Culturist, 52*(4), 226–230. https://doi.org/10.1577/1548-8640(1990)052<0226:EOBSOG>2.3.CO;2

Poston, H. A. (1991). Choline requirement of swim-up rainbow trout fry. *Progressive Fish-Culturist, 53*(4), 220–223. https://doi.org/10.1577/1548-8640(1991)053<0220:CROSUR>2.3.CO;2

Poston, H. A., & Page, J. W. (1982). Gross and histological signs of dietary deficiencies of biotin and pantothenic acid in lake trout, *Salvelinus namaycush*. *Cornell Veterinarian, 72*(3), 242–261. PubMed: 7105759

Poston, H. A., & Wolfe, M. J. (1985). Niacin requirement for optimum growth, feed conversion and protection of rainbow trout, Salmo gairdneri Richardson, from ultraviolet-B irradiation. *Journal of Fish Diseases, 8*(5), 451–460. https://doi.org/10.1111/j.1365-2761.1985.tb01278.x

Poston, H. A., Combs, G. F., & Leibovitz, L. (1976) Vitamin E and selenium interrelations in the diet of Atlantic salmon (*Salmo salar*): Gross, histological and biochemical deficiency signs. *Journal of Nutrition, 106*(7), 892–904. https://doi.org/10.1093/jn/106.7.892

Poston, H. A., Riis, R. C., Rumsey, G. L., & Ketola, H. G. (1977). The effect of supplemental dietary amino acids, minerals and vitamins on salmonids fed cataractogenic diets. *Cornell Veterinarian, 67*(4), 472–509. PubMed: 25087300

Prabhu, P. A. J., Lock, E. J., Hemre, G. I., Hamre, K., Espe, M., Olsvik, P. A., Silva, J., Hansen, A. C., Johansen, J., Sissener, N. H., & Waagbø, R. (2019). Recommendations for dietary level of micro-minerals and vitamin D3 to Atlantic salmon (*Salmo salar*) parr and post-smolt when fed low fish meal diets. *Peer J., 7*, e6996. https://doi.org/10.7717/peerj.6996

Premkumar, V. G., Yuvaraj, S., Shanthi, P., & Sachdanandam, P. (2008). Co-enzyme Q10, riboflavin and niacin supplementation on alteration of DNA repair enzyme and DNA methylation in breast cancer patients undergoing tamoxifen therapy. *British Journal of Nutrition, 100*(6), 1179–1182. https://doi.org/10.1017/S0007114508968276

Quackenbush, F. W. (1963). Corn carotenoids: Effect of temperature and moisture on losses during storage. *Cereal Chemistry, 40*, 266.

Qian, Y., Li, X. F., Zhang, D. D., Cai, D. S., Tian, H. Y., & Liu, W. B. (2015). Effects of dietary pantothenic acid on growth, intestinal function, anti-oxidative status and fatty acids synthesis of juvenile blunt snout bream *Megalobrama amblycephala*. *PLOS ONE, 10*(3), e0119518. https://doi.org/10.1371/journal.pone.0119518

Qiang, J., Wasipe, A., He, J., Tao, Y. F., Xu, P., Bao, J. W., Chen, D. J., & Zhu, J. H. (2019). Dietary vitamin E deficiency inhibits fat metabolism, antioxidant capacity, and immune regulation of inflammatory response in genetically improved farmed tilapia (GIFT, *Oreochromis niloticus*) fingerlings following *Streptococcus iniae* infection. *Fish and Shellfish Immunology, 92*, 395–404. https://doi.org/10.1016/j.fsi.2019.06.026

Qin, D. G., Dong, X. H., Tan, B. P., Yang, Q. H., Chi, S. Y., Liu, H. Y., & Zhang, S. (2017). Effects of dietary choline on growth performance, lipid deposition and hepatic lipid transport of grouper (*Epinephelus coioides*). *Aquaculture Nutrition, 23*(3), 453–459. https://doi.org/10.1111/anu.12413

Qin, Z. H., Li, J., Wang, F., Liu, Q., & Wang, Q. (2007). Effect of L-ascorbyl-2-polyphoshate on growth and immunity of shrimp (*Penaeus chinensis*) larvae. *Fisheries Science, 26*, 21.

Qureshi, A. A., Salser, W. A., Parmar, R., & Emeson, E. E. (2001). Novel tocotrienols of rice bran inhibit atherosclerotic lesions in C57BL/6 apo E-deficient mice. *Journal of Nutrition, 131*(10), 2606–2618. https://doi.org/10.1093/jn/131.10.2606

Raederstorff, D., Wyss, A., Calder, P. C., Weber, P., & Eggersdorfer, M. (2015). Vitamin E function and requirements in relation to PUFA. *British Journal of Nutrition, 114*(8), 1113–1122. https://doi.org/10.1017/S000711451500272X

Raggi, T., Buentello, A., & Gatlin, D. M. (2016). Characterization of pantothenic acid deficiency and the dietary requirement of juvenile hybrid striped bass, *Morone chrysops×M. saxatilis*. *Aquaculture, 451*, 326–329. https://doi.org/10.1016/j.aquaculture.2015.09.028

Rahimi, M., Ahmadivand, S., Eagderi, S., & Shamohammadi, S. (2015). Effects of vitamin C and E administration on leukocyte counts in rainbow trout (*Oncorhynchus mykiss*). *Journal of Nutrition & Health, 1*, 5. https://www.avensonline.org/fulltextarticles/jnh-2469-4185-01-0009.html

Rahman, M. M. (2015). Role of common carp (*Cyprinus carpio*) in aquaculture production systems. *Frontiers in Life Science, 8*(4), 399–410. https://doi.org/10.1080/21553769.2015.1045629

Rahman, M. M., Khosravi, S., Chang, K. H., & Lee, S. M. (2016). Effects of dietary Inclusion of astaxanthin on growth, muscle pigmentation and antioxidant capacity of juvenile rainbow trout (*Oncorhynchus mykiss*). *Preventive Nutrition and Food Science, 21*(3), 281–288. https://doi.org/10.3746/pnf.2016.21.3.281

Rahman, M. M., Rahman, R., & Mamun, M. (2018). Evaluation of the effects of dietary vitamin C, E and zinc supplementation on growth performances and survival rate of rohu, *Labeo rohita* (Hamilton, 1822). *Journal of Agriculture and Veterinary Science, 11*(2), 68–74. https://doi.org/10.9790/2380-1102026874

Rama, S., & Manjabhat, S. N. (2014). Protective effect of shrimp carotenoids against ammonia stress in common carp, *Cyprinus carpio*. *Ecotoxicology and Environmental Safety, 107*, 207–213. https://doi.org/10.1016/j.ecoenv.2014.06.016

Rao, D. S., & Raghuramulu, N. (1997). Vitamin D3 in tilapia mossambica: Relevance of photochemical synthesis. *Journal of Nutritional Science and Vitaminology, 43*(4), 425–433. https://doi.org/10.3177/jnsv.43.425

Rattray, J. B. M., Schibeci, A., & Kidby, D. K. (1975). Lipids of yeasts. *Bacteriological Reviews, 39*(3), 197–231. https://doi.org/10.1128/br.39.3.197-231.1975

Reboul, E. (2017). Vitamin E bioavailability: Mechanisms of intestinal absorption in the spotlight. *Antioxidants, 6*(4), 95. https://doi.org/10.3390/antiox6040095

Reboul, E., Goncalves, A., Comera, C., Bott, R., Nowicki, M., Landrier, J. F., Jourdheuil-Rahmani, D., Dufour, C., Collet, X., & Borel, P. (2011). Vitamin D intestinal absorption is not a simple passive diffusion: Evidences for the involvement of cholesterol transporters. *Molecular Nutrition and Food Research, 55*(5), 691–702. https://doi.org/10.1002/mnfr.201000553

Reddy, M. U., & Pushpamma, P. (1986). Effect of storage and insect infestation on thiamine and niacin content in different varieties of rice, sorghum, and legumes. *Nutrition Reports International, 34*, 393–401.

Reddy, H. R. V., Naik, M. G., & Annappaswamy, T. S. (1999). Evaluation of the dietary essentiality of vitamins for *Penaeus monodon*. *Aquaculture Nutrition, 5*(4), 267–275. https://doi.org/10.1046/j.1365-2095.1999.00116.x

Reed, A. N., Rowland, F. E., Krajcik, J. A., & Tillitt, D. E. (2023). Thiamine supplementation improves survival and body condition of hatchery-reared steelhead (*Oncorhynchus mykiss*) in Oregon. *Veterinary Sciences, 10*(2), 156. https://doi.org/10.3390/vetsci10020156

Rektsen, A. M., Ho, Q. T., Nøstbakken, O. J., Markhus, M. W., Kjellevold, M., Bøkevoll, A., Hannisdal, R., Frøyland, L., Madsen, L., & Dahl, L. (2022). Temporal variations in the nutrient content of Norwegian farmed

Atlantic salmon (*Salmo salar*), 2005–2020. *Food Chemistry, 373b*, 131445. https://doi.org/10.1016/j.food chem.2021.131445

Ren, T., Koshio, S., Ishikawa, M., Yokoyama, S., Micheal, F. R., Uyan, O., & Tung, H. T. (2007). Influence of dietary vitamin C and bovine lactoferrin on blood chemistry and non-specific immune responses of Japanese eel, *Anguilla japonica*. *Aquaculture, 267*(1–4), 31–37. https://doi.org/10.1016/j.aquaculture.2007.03.033

Riaz, M. N., Asif, M., & Ali, R. (2009). Stability of vitamins through extrusion. *Critical Reviews in Food Science and Nutrition, 49*(4), 361–368. https://doi.org/10.1080/10408390802067290

Richard, N., Fernández, I., Wulff, T., Hamre, K., Cancela, L., Conceição, L. E. C., & Gavaia, P. J. (2014). Dietary supplementation with vitamin K affects transcriptome and proteome of Senegalese sole, improving larval performance and quality. *Marine Biotechnology, 16*(5), 522–537. https://doi.org/10.1007/s10126-014-9571-2

Rider, S., Verlhac-Trichet, V., Constant, D., Chenal, E., Etheve, S., Riond, B., Schmidt-Posthaus, H., & Schoop, R. (2023). Calcifediol is a safe and effective metabolite for raising vitamin D status and improving growth and feed conversion in rainbow trout. *Aquaculture, 568*, 739285. https://doi.org/10.1016/j.aquaculture.2023.739285

Ridgway, N. D. (2016). Phospholipid synthesis in mammalian cells. In N. D. Ridgway & R. S. McLeod (Eds.), *Biochemistry of lipids, lipoproteins and membranes* (6th ed) (pp. 209–236). Elsevier.

Rigotti, A. (2007). Absorption, transport, and tissue delivery of vitamin E. *Molecular Aspects of Medicine, 28*(5–6), 423–436. https://doi.org/10.1016/j.mam.2007.01.002

Rivlin, R. S. (2006). Riboflavin. In B. A. Bowman & R. M. Russell (Eds.), *Present knowledge in nutrition* (9th ed) (pp. 250–259). International Life Sciences Institute. https://doi.org/10.1093/ajcn/85.5.1439a

Rodrigues, R. A., da Silva Nunes, C., Fantini, L. E., Kasai, R. Y. D., Oliveira, C. A. L., Hisano, H., & de Campos, C. M. (2018). Dietary ascorbic acid influences the intestinal morphology and hematology of hybrid Sorubim catfish (*Pseudoplatystoma reticulatum × P. corruscans*). *Aquaculture International, 26*(1), 1–11. https://doi.org/10.1007/s10499-017-0188-0

Rodríguez, A., Latorre, M., Gajardo, M., Bunger, A., Munizaga, A., López, L., & Aubourg, S. P. (2015). Effect of the antioxidants composition in diet on the sensory and physical properties of frozen farmed Coho salmon (*Oncorhynchus kisutch*). *Journal of the Science of Food and Agriculture, 95*(6), 1199–1206. https://doi.org/10.1002/jsfa.6808

Rodríguez Lozano, A. R., Borges, P., Robaina, L., Betancor, M., Hernández-Cruz, C. M., García, J. R., Caballero, M. J., Vergara, J. M., & Izquierdo, M. (2017). Effect of different dietary vitamin E levels on growth, fish composition fillet quality and liver histology of meagre (*Argyrosomus regius*). *Aquaculture, 468*(1), 175–183. https://doi.org/10.1016/j.aquaculture.2016.10.006

Rodríguez-Meléndez, R., Cano, S., Méndez, S. T., & Velázquez, A. (2001). Biotin regulates the genetic expression of holocarboxylase synthetase and mitochondrial carboxylases in rats. *Journal of Nutrition, 131*(7), 1909–1913. https://doi.org/10.1093/jn/131.7.1909

Rohani, M. F., Bristy, A. A., Hasan, J., Hossain, M. K., & Shahjahan, M. (2022). Dietary zinc in association with vitamin E promotes growth performance of Nile tilapia. *Biological Trace Element Research, 200*(9), 4150–4159. https://doi.org/10.1007/s12011-021-03001-9

Rokey, G. J. (2007). Formulation and dietary ingredients. In M. N. Riaz (Ed.), *Extruders and expanders in pet food, aquatic and livestock feeds* (pp. 55–76). Agrimedia GmbH.

Romano, N., & Zeng, C. (2017). Cannibalism of Decapod Crustaceans and Implications for Their Aquaculture: A Review of its Prevalence, Influencing Factors, and Mitigating Methods. *Reviews in Fisheries Science and Aquaculture, 25*(1), 42–69. http://doi.org/10.1080/23308249.2016.1221379

Rønnestad, I., Hemre, G.-I., Finn, R. N., & Lie, Ø. (1998). Alternate sources and dynamic of vitamin A and its incorporation into the eyes during the early endotrophic and exotrophic larval stages of Atlantic halibut (*Hippoglossus hippoglossus* L.). *Comparative Biochemistry and Physiology Part A, 119*(3), 787–793. https://doi.org/10.1016/S1095-6433(98)01017-4

Rose, R. (1990). Vitamin absorption. In *Developments in vitamin nutrition and health applications. Proceedings of the National Feed Ingr. Ass.*, Nutr. Inst. Kansas City. National Feed Ingredients Association.

Rose, R. C., McCorrmick, D. B., Li, T. K., Lumeng, L., Haddad, J. G., & Spector, R. (1986). Transport and metabolism of vitamins. *Federation Proceedings, 45*(1), 30–39. PubMed: 3000833

Rosenberg, I. H., & Neumann, H. (1974). Multi-step mechanism in the hydrolysis of pteroyl polyglutamates by chicken intestine. *Journal of Biological Chemistry, 249*(16), 5126–5130. https://doi.org/10.1016/S0021-9258(19)42336-3

Ross, A. C. (1993). Overview of retinoid metabolism. *Journal of Nutrition, 123*(2)(Suppl.), 346–350. https://doi.org/10.1093/jn/123.suppl_2.346

Ross, A. C., & Harrison, E. H. (2013). Vitamin A: Nutritional aspects of retinoids and carotenoids. In J. Zempleni, J. Suttie, J. Gregory & P. J. Stover (Eds.), *Handbook of vitamins* (5th ed) (pp. 1–50). CRC Press, Taylor & Francis Group, LLC. https://doi.org/10.1201/b15413

Rothe, S., Gropp, J., Weiser, H., & Rambeck, W. A. (1994). The effect of vitamin C and zinc on the copper-induced increase of cadmium residues in swine. *Zeitschrift Fur Ernahrungswissenschaft, 33*(1), 61–67. https://doi.org/10.1007/BF01610579

Roy, P. K., & Lall, S. P. (2007). Vitamin K deficiency inhibits mineralization and enhances deformity in vertebrae of haddock (*Melanogrammus aeglefinus* L.). *Comparative Biochemistry and Physiology. Part B, Biochemistry and Molecular Biology, 148*(2), 174–183. https://doi.org/10.1016/j.cbpb.2007.05.006

Rucker, R. B., Morris, J., & Fascetti, A. J. (2008). Vitamins. In J. J. Kaneko, J. W. Harvey & M. L. Bruss (Eds.). https://umkcarnivores3.files.wordpress.com/2012/02/clinical-biochemistry-of-domestic-animals-sixth-edition.pdf. *Clinical biochemistry of domestic animals* (6th ed) (pp. 695–730). Academic Press, Elsevier, Inc.

Rucker, R. B., & Bauerly, K. (2013). Pantothenic acid. In J. Zempleni, J. Suttie, J. Gregory & P. J. Stover (Eds.), *Handbook of vitamins* (5th ed) (pp. 289–313). CRC Press, Taylor & Francis Group, LLC ISBN 9781466515567. https://doi.org/10.1201/b15413

Ruff, N., Fitzgerald, R. D., Cross, T. F., Lynch, A., & Kerry, J. P. (2004). Distribution of α-tocopherol in fillets of turbot (*Scophthalmus maximus*) and Atlantic halibut (*Hippoglossus hippoglossus*), following dietary α-tocopheryl acetate supplementation. *Aquaculture Nutrition, 10*(2), 75–81. https://doi.org/10.1046/j.1365-2095.2003.00280.x

Ruiz, M. A., Betancor, M. B., Robaina, L., Montero, D., Hernández-Cruz, C. M., Izquierdo, M. S., Rosenlund, G., Fontanillas, R., & Caballero, M. J. (2019). Dietary combination of vitamin E, C and K affects growth, antioxidant activity, and the incidence of systemic granulomatosis in meagre (*Argyrosomus regius*). *Aquaculture, 498*, 606–620. https://doi.org/10.1016/j.aquaculture.2018.08.078

Rumsey, G. L. (1991). Choline-betaine requirements of rainbow trout (*Oncorhynchus mykiss*). *Aquaculture, 95*(1–2), 107–116. https://doi.org/10.1016/0044-8486(91)90077-K

Russo, J. R., and Yanong R.P.E. (2010) "Molds in fish feeds and aflatoxicosis." Institute of Food and Agriculture Sciences, University of Florida

Saensukjaroenphon, M., Evans, C. E., Paulk, C. B., Gebhardt, J. T., Woodworth, J. C., Stark, C. R., Bergstrom, J. R., & Jones, C. K. (2020). Impact of storage condition and premix type on fat-soluble vitamin stability. *Translational Animal Science, 4*(3), txaa143. https://doi.org/10.1093/tas/txaa143

Saheli, M., Rajabi Islami, H., Mohseni, M., & Soltani, M. (2021). Effects of dietary vitamin E on growth performance, body composition, antioxidant capacity, and some immune responses in Caspian trout (*Salmo caspius*). *Aquaculture Reports, 21*, 100857. https://doi.org/10.1016/j.aqrep.2021.100857

Said, H. M. (2011). Intestinal absorption of water-soluble vitamins in health and disease. *Biochemical Journal, 437*(3), 357–372. https://doi.org/10.1042/BJ20110326

Said, H. M., & Derweesh, I. (1991). Carrier-mediated mechanism for biotin transport in rabbit intestine: Studies with brush–border membrane vesicles. *American Journal of Physiology, 261*(1 Pt. 2), R94–R97. https://doi.org/10.1152/ajpregu.1991.261.1.R94

Said, H. M., Redha, R., & Nylander, W. (1988). Biotin transport in the human intestine: Site of maximum transport and effect of pH. *Gastroenterology, 95*(5), 1312–1317. https://doi.org/10.1016/0016-5085(88)90366-6

Said, H. M., Hoefs, J., Mohammadkhani, R., & Horne, D. W. (1992). Biotin transport in human liver basolateral membrane vesicles: A carrier-mediated, Na+ gradient-dependent process. *Gastroenterology, 102*(6), 2120–2125. https://doi.org/10.1016/0016-5085(92)90341-u

Saito, T., Whatmore, P., Taylor, J. F., Fernandes, J. M. O., Adam, A. C., Tocher, D. R., Espe, M., & Skjærven, K. H. (2021). Micronutrient supplementation affects transcriptional and epigenetic regulation of lipid metabolism in a dose-dependent manner. *Epigenetics, 16*(11), 1217–1234. https://doi.org/10.1080/15592294.2020.1859867

Sakurai, T., Asakura, T., Mizuno, A., & Matsuda, M. (1992). Absorption and metabolism of pyridoxamine in mice. II. Transformation of pyridoxamine to pyridoxal in intestinal tissues. *Journal of Nutritional Science and Vitaminology, 38*(3), 227–233. https://doi.org/10.3177/jnsv.38.227

Saleh, N. E., Wassef, E. A., & Shalaby, S. M. (2018). The role of dietary astaxanthin in European sea bass (*Dicentrarchus labrax*) growth, immunity, antioxidant competence and stress tolerance. *Egyptian Journal of Aquatic Biology and Fisheries, 22*(5) ((Special Issue)), 189–200. https://doi.org/10.21608/ejabf.2018.21044

Salte, R., & Norberg, K. (1991). Effects of warfarin on vitamin K-dependent coagulation factors in Atlantic salmon and rainbow trout with special reference to factor X. *Thrombosis Research, 63*(1), 39–45. https://doi.org/10.1016/0049-3848(91)90268-2

Sandnes, K. (1994). Vannløselige vitaminer (Water soluble vitamins). Project Report. Research Council of Norway (Grant. *104914/110 Emæringsinstitutt*).

Sandnes, K., Ulgenes, Y., Braekkan, O. R., & Utne, F. (1984). The effect of ascorbic acid supplementation in broodstock feed on reproduction of rainbow trout (*Salmo gairdneri*). *Aquaculture, 43*(1–3), 167–177. https://doi.org/10.1016/0044-8486(84)90019-X

Sandnes, K., Torrissen, O., & Waagbø, R. (1992). The minimum dietary requirement of vitamin C in Atlantic salmon (*Salmo salar*) fry using Ca ascorbate-2-monophosphate as dietary source. *Fish Physiology and Biochemistry, 10*(4), 315–319. https://doi.org/10.1007/BF00004480

Sanjeewani, K., & Lee, K. J. (2023). Dietary riboflavin requirement of Pacific white shrimp (*Litopenaeus vannamie*). *Aquaculture Nutrition, 2023*, 6685592. https://doi.org/10.1155/2023/6685592

Sarmento, N., Martins, E. F. F., Costa, D. C., Silva, W. S., Mattioli, C. C., Luz, M. R., & Luz, R. K. (2017). Effects of supplemental dietary vitamin C on quality of semen from Nile tilapia (*Oreochromis niloticus*) breeders. *Reproduction in Domestic Animals, 52*(1), 144–152. https://doi.org/10.1111/rda.12870

Sarmento, N. L. A. F., Martins, E. F. F., Costa, D. C., Mattioli, C. C., da Costa Julio, G. S., Figueiredo, L. G., Luz, M. R., & Luz, R. K. (2018). Reproductive efficiency and egg and larvae quality of Nile tilapia fed different levels of vitamin C. *Aquaculture, 482*, 96–102. https://doi.org/10.1016/j.aquaculture.2017.08.035

Sauberlich, H. E. (1985). Bioavailability of vitamins. *Progress in Food and Nutrition Science, 9*(1–2), 1–33. PubMed: 3911266

Sauberlich, H. E. (1999). *Laboratory tests for the assessment of nutritional status* (2nd ed). Routledge. https://doi.org/10.1201/9780203749647

Savage, D. G., & Lindenbaum, J. (1995). Folate-cobalamin interactions. In L. B. Bailey (Ed.), *Folate in health and disease* (p. 237). CRC Press, Taylor & Francis Group.

Schaffer, S., Müller, W. E., & Eckert, G. P. (2005). Tocotrienols: Constitutional effects in aging and disease. *Journal of Nutrition, 135*(2), 151–154. https://doi.org/10.1093/jn/135.2.151

Schiedt, K., Leuenberger, F. J., Vecchi, M., & Glinz, E. (1985). Absorption, retention and metabolic transformations of carotenoids in rainbow trout, salmon and chicken. *Pure and Applied Chemistry, 57*(5), 685–692. https://doi.org/10.1351/pac198557050685

Schiedt, K., Vecchi, M., & Glinz, E. (1986). Astaxanthin and its metabolites in wild rainbow trout (*Salmo gairdneri R.*). *Comparative Biochemistry and Physiology Part B, 83*(1), 9–12. https://doi.org/10.1016/0305-0491(86)90324-X

Schiedt, K., Bischof, S., & Glinz, E. (1991). Recent progress on carotenoid metabolism in animals. *Pure and Applied Chemistry, 63*(1), 89–100. https://doi.org/10.1351/pac199163010089

Schiedt, K., Bischof, S., & Glinz, E. (1993). Metabolism of carotenoids and *in vivo* racemisation of (3S,3'S)-astaxanthin in the crustacean *Penaeus*. *Methods in Enzymology, 214*, 148–168. https://doi.org/10.1016/0076-6879(93)14062-N

Scott, C. G., Cohen, N., Riggio, P. P., & Weber, G. (1982). Gas chromatographic assay of the diastereomeric composition of all-rac-α-tocopheryl acetate. *Lipids, 17*(2), 97–101. https://doi.org/10.1007/BF02535182

Seale, A. P., Fiess, J. C., Hirano, T., Cooke, I. M., & Grau, E. G. (2006). Disparate release of prolactin and growth hormone from the tilapia pituitary in response to osmotic stimulation. *General and Comparative Endocrinology, 145*(3), 222–231. https://doi.org/10.1016/j.ygcen.2005.09.006

Seetharam, B., & Alpers, D. H. (1982). Absorption and transport of cobalamin (vitamin B12). *Annual Review of Nutrition, 2*, 343–369. https://doi.org/10.1146/annurev.nu.02.070182.002015

Senadheera, S. D., Turchini, G. M., Thanuthong, T., & Francis, D. S. (2012). Effects of dietary vitamin B6 supplementation on fillet fatty acid composition and fatty acid metabolism of rainbow trout fed vegetable oil based diets. *Journal of Agricultural and Food Chemistry*, *60*(9), 2343–2353. https://doi.org/10.1021/jf204963w

Sesay, D. F., Habte-Tsion, H. M., Zhou, Q., Ren, M., Xie, J., Liu, B., Chen, R., & Pan, L. (2016). Effects of dietary folic acid on the growth, digestive enzyme activity, immune response and antioxidant enzyme activity of blunt snout bream (*Megalobrama amblycephala*) fingerling. *Aquaculture*, *452*, 142–150. https://doi.org/10.1016/j.aquaculture.2015.10.026

Sesay, D. F., Habte-Tsion, H. M., Zhou, Q., Ren, M., Xie, J., Liu, B., Chen, R., & Pan, L. (2017). The effect of dietary folic acid on biochemical parameters and gene expression of three heat shock proteins (HSPs) of blunt snout bream (*Megalobrama amblycephala*) fingerling under acute high temperature stress. *Fish Physiology and Biochemistry*, *43*(4), 923–940. https://doi.org/10.1007/s10695-016-0311-6

Shahkar, E., Yun, H., Kim, D.-J., Kim, S.-K., Lee, B. I., & Bai, S. C. (2015). Effects of dietary vitamin C levels on tissue ascorbic acid concentration, hematology, non-specific immune response and gonad histology in broodstock Japanese eel, *Anguilla japonica*. *Aquaculture*, *438*, 115–121. https://doi.org/10.1016/j.aquaculture.2015.01.001

Shahkar, E., Hamidoghli, A., Yun, H., Kim, D. J., & Bai, S. C. (2018). Effects of dietary vitamin E on hematology, tissue α-tocopherol concentration and non-specific immune responses of Japanese eel, *Anguilla japonica*. *Aquaculture*, *484*, 51–57. https://doi.org/10.1016/j.aquaculture.2017.10.036

Shaik Mohamed, J., Sivaram, V., Christopher Roy, T. S., Peter Marian, M., Murugadass, S., & Raffiq Hussain, M. (2003). Dietary vitamin A requirement of juvenile greasy grouper (*Epinephelus tauvina*). *Aquaculture*, *219*(1–4), 693–701. https://doi.org/10.1016/S0044-8486(02)00665-8

Shalaby, A.M.E. (2004) The opposing effect of ascorbic acid (vitamin C) on ochratoxin toxicity in Nile tilapia (*Oreochromis niloticus*). https://api.semanticscholar.org/CorpusID:37687563

Shao, L., Zhu, X., Yang, Y., Jin, J., Liu, H., Han, D., & Xie, S. (2016). Effects of dietary vitamin A on growth, hematology, digestion and lipometabolism of on-growing gibel carp (*Carassius auratus gibelio* var. CAS III). *Aquaculture*, *460*, 83–89. https://doi.org/10.1016/j.aquaculture.2016.03.054

Shao, R., Liao, X., Lan, Y., Zhang, H., Jiao, L., Du, Q., Han, D., Ai, Q., Mai, K., & Wan, M. (2022). Vitamin D regulates insulin pathway and glucose metabolism in zebrafish (*Danio rerio*). *FASEB Journal*, *36*(5), e22330. https://doi.org/10.1096/fj.202200334RR

Shearer, K. (2000). Experimental design, statistical analysis and modelling of dietary nutrient requirement studies for fish: A critical review. *Aquaculture Nutrition*, *6*(2), 91–102. https://doi.org/10.1046/j.1365-2095.2000.00134.x

Shearer, M. J., Barkhan, P., & Webster, G. R. (1970). Absorption and excretion of an oral dose of tritiated vitamin K1 in man. *British Journal of Haematology*, *18*(3), 297–308. https://doi.org/10.1111/j.1365-2141.1970.tb01444.x

Shearer, M. J., Bach, A., & Kohlmeier, M. (1996). Chemistry, nutritional sources, tissue distribution and metabolism of vitamin K with special reference to bone health. *Journal of Nutrition*, *126*(4) (Suppl.), 1181S–1186S. https://doi.org/10.1093/jn/126.suppl_4.1181S

Shearer, M. J., & Newman, P. (2014). Recent trends in the metabolism and cell biology of vitamin K with special reference to vitamin K cycling and MK-4 biosynthesis. *Journal of Lipid Research*, *55*(3), 345–362. https://doi.org/10.1194/jlr.R045559

Shepherd, C. J., Monroig, O., & Tocher, D. R. (2017). Future availability of raw materials for salmon feeds and supply chain implications: The case of Scottish farmed salmon. *Aquaculture*, *467*, 49–62. https://doi.org/10.1016/j.aquaculture.2016.08.021

Sherriff, J. L., O 'Sullivan, T. A., Properzi, C., Oddo, J. L., & Adams, L. A. (2016). Choline, its potential role in nonalcoholic fatty liver disease, and the case for human and bacterial genes. *Advances in Nutrition*, *7*(1), 5–13. https://doi.org/10.3945/an.114.007955

Shi, B., Hu, X., Jin, M., Xia, M., Zhao, M., Jiao, L., Sun, P., & Zhou, Q. (2021). Dietary choline improves growth performance, antioxidant ability and reduces lipid metabolites in practical diet for juvenile Pacific white shrimp, *Litopenaeus vannamei*. *Aquaculture Nutrition*, *27*(1), 39–48. https://doi.org/10.1111/anu.13163

Shi, L., Feng, L., Jiang, W. D., Liu, Y., Jiang, J., Wu, P., Zhao, J., Kuang, S. Y., Tang, L., Tang, W. N., Zhang, Y. A., & Zhou, X. Q. (2015). Folic acid deficiency impairs the gill health status associated with the NF-κB,

MLCK and Nrf2 signaling pathways in the gills of young grass carp (*Ctenopharyngodon idella*). *Fish and Shellfish Immunology, 47*(1), 289–301. https://doi.org/10.1016/j.fsi.2015.09.023

Shi, L., Feng, L., Jiang, W. D., Liu, Y., Jiang, J., Wu, P., Kuang, S. Y., Tang, L., Tang, W. N., Zhang, Y. A., & Zhou, X. Q. (2016). Immunity decreases, antioxidant system damages and tight junction changes in the intestine of grass carp (*Ctenopharyngodon idella*) during folic acid deficiency: Regulation of NF-κB, Nrf2 and MLCK mRNA levels. *Fish and Shellfish Immunology, 51*, 405–419. https://doi.org/10.1016/j.fsi.2016.02.029

Shiau, S. Y., & Chen, Y. (2000). Estimation of the dietary vitamin A requirement of juvenile grass shrimp, *Penaeus monodon. Journal of Nutrition, 130*(1), 90–94. https://doi.org/10.1093/jn/130.1.90

Shiau, S. Y., & Chin, Y. H. (1998). Dietary biotin requirement for maximum growth of juvenile grass shrimp, *Penaeus monodon. Journal of Nutrition, 128*(12), 2494–2497. https://doi.org/10.1093/jn/128.12.2494

Shiau, S. Y., & Chin, Y. H. (1999). Estimation of the dietary biotin requirement of juvenile hybrid tilapia, *Oreochromis niloticus* × O. aureus. *Aquaculture, 170*(1), 71–78. https://doi.org/10.1016/S0044-8486(98)00391-3

Shiau, S. Y., & Cho, W. H. (2002). Choline requirements of grass shrimp (*Penaeus monodon*) as affected by dietary lipid level. *Animal Science, 75*(1), 97–102. https://doi.org/10.1017/S1357729800052875

Shiau, S. Y., & Hsu, T. S. (1994). Vitamin C requirement of grass shrimp, *Penaeus monodon*, as determined with L-ascorbyl-2-monophosphate. *Aquaculture, 122*(4), 347–357. https://doi.org/10.1016/0044-8486(94)90343-3

Shiau, S. Y., & Hsu, C. W. (1999). Dietary pantothenic acid requirement of juvenile grass shrimp, *Penaeus monodon. Journal of Nutrition, 129*(3), 718–721. https://doi.org/10.1093/jn/129.3.718

Shiau, S. Y., & Huang, S. Y. (2001). Dietary folic acid requirement determined for grass shrimp, *Penaeus monodon. Aquaculture, 200*(3–4), 339–347. https://doi.org/10.1016/S0044-8486(00)00598-6

Shiau, S. Y., & Hwang, J. Y. (1993). Vitamin D requirements of juvenile hybrid tilapia (*Oreochromis niloticus* × O. aureus). *Nippon Suisan Gakkaishi, 59*(3), 553–558. https://doi.org/10.2331/suisan.59.553

Shiau, S. Y., & Hwang, J. Y. (1994). The dietary requirement of juvenile grass shrimp (*Penaeus monodon*) for vitamin D. *Journal of Nutrition, 124*(12), 2445–2450. https://doi.org/10.1093/jn/124.12.445

Shiau, S. Y., & Jan, F. L. (1992). Ascorbic acid requirement of grass shrimp, *Penaeus monodon. Nippon Suisan Gakkaishi, 58*(2), 363. https://doi.org/10.2331/suisan.58.363

Shiau, S. Y., & Liu, J. S. (1994a). Quantifying the vitamin K requirement of juvenile marine shrimp (*Penaeus monodon*) with menadione. *Journal of Nutrition, 124*(2), 277–282. https://doi.org/10.1093/jn/124.2.277

Shiau, S. Y., & Liu, J. S. (1994b). Estimation of the dietary vitamin K requirement of juvenile *Penaeus chinensis* using menadione. *Aquaculture, 126*(1–2), 129–135. https://doi.org/10.1016/0044-8486(94)90254-2

Shiau, S. Y., & Lo, P. S. (2000). Dietary choline requirements of juvenile hybrid tilapia, *Oreochromis niloticus* × O. aureus. *Journal of Nutrition, 130*(1), 100–103. https://doi.org/10.1093/jn/130.1.100

Shiau, S. Y., & Lo, P. S. (2001). Dietary choline requirement of juvenile grass shrimp (*Penaeus monodon*). *Animal Science, 72*(3), 477–482. https://doi.org/10.1017/S1357729800051997

Shiau, S. Y., & Lung, C. Q. (1993). Estimation of the vitamin B12 requirement of the grass shrimp, *Penaeus monodon. Aquaculture, 117*(1–2), 157–163. https://doi.org/10.1016/0044-8486(93)90132-I

Shiau, S. Y., & Suen, G. S. (1992). Estimation of the niacin requirements for tilapia fed diets containing glucose or dextrin. *Journal of Nutrition, 122*(10), 2030–2036. https://doi.org/10.1093/jn/122.10.2030

Shiau, S. Y., & Suen, G. S. (1994). The dietary requirement of juvenile grass shrimp (*Penaeus monodon*) for niacin. *Aquaculture, 125*(1–2), 139–145. https://doi.org/10.1016/0044-8486(94)90290-9

Shiau, S. Y., & Su, S. L. (2004). Dietary inositol requirement for juvenile grass shrimp, *Penaeus monodon. Aquaculture, 241*(1–4), 1–8. https://doi.org/10.1016/j.aquaculture.2004.01.013

Shiau, S. Y., & Wu, M. H. (2003). Dietary vitamin B6 requirement of grass shrimp, *Penaeus monodon. Aquaculture, 225*(1–4), 397–404. https://doi.org/10.1016/S0044-8486(03)00304-1

Shibata, K., Mushiage, M., Kondo, T., Hayakawa, T., & Tsuge, H. (1995). Effects of vitamin B6 deficiency on the conversion ratio of tryptophan to niacin. *Bioscience, Biotechnology, and Biochemistry, 59*(11), 2060–2063. https://doi.org/10.1271/bbb.59.2060

Shideler, C. E. (1983). Vitamin B6: An overview. *American Journal of Medical Technology, 49*(1), 17–22. PubMed: 6342384

Shields, R. G., Campbell, D. R., Huges, D. M., & Dillingham, D. A. (1982). Researchers study vitamin A stability in feeds. *Feedstuffs, 54*(47), 22.

Shimada, M., Claudiano, G., Engracia Filho, J., Yunis Aguinaga, J., Moraes, F., Moreira, R., & Moraes, J. (2014). Hepatic steatosis in cage-reared young cobia, *Rachycentron Canadum* (Linnaeus, 1766), in Brazil. *Journal of Veterinary Science and Medical Diagnosis, 3*(2), 1–5. https://doi.org/10.4172/2325-9590.1000137

Shin, D. J., & McGrane, M. M. (1997). Vitamin A regulates genes involved in hepatic gluconeogenesis in mice: Phosphoenolpyruvate carboxykinase, fructose-1,6-bisphosphatase and 6-phosphofructo-2-kinase/fructose-2,6-bisphosphatase. *Journal of Nutrition, 127*(7), 1274–1278. https://doi.org/10.1093/jn/127.7.1274

Shirmohammad, F., Mehri, M., & Joezy-Shekalgorabi, S. (2016). A review on the role of inositol in aquaculture. *Iranian Journal of Fisheries Sciences, 15*(4), 1388–1409. http://hdl.handle.net/1834/12122.

Sigurgisladottir, S., Parrish, C. C., Ackman, R. G., & Lall, S. P. (1994). Tocopherol deposition in the muscle of Atlantic salmon (*Salmo salar*). *Journal of Food Science, 59*(2), 256–259. https://doi.org/10.1111/j.1365-2621.1994.tb06942.x

Simon, J. (1999). Choline, betaine and methionine interactions in chickens, pigs and fish (including crustaceans). *World's Poultry Science Journal, 55*(4), 353–374. https://doi.org/10.1079/WPS19990025

Simpson, K. L., & Chichester, C. O. (1981). Metabolism and nutritional significance of carotenoids. *Annual Review of Nutrition, 1*, 351–374. https://doi.org/10.1146/annurev.nu.01.070181.002031

Sissener, N. H., Julshamn, K., Espe, M., Lunestad, B. T., Hemre, G.-I., Waagbø, R., & Måge, A. (2013). Surveillance of selected nutrients, additives and undesirables in commercial Norwegian fish feeds in the years 2000–2010. *Aquaculture Nutrition, 19*(4), 555–572. https://doi.org/10.1111/anu.12007

Sitrin, M. D., Lieberman, F., Jensen, W. E., Noronha, A., Milburn, C., & Addington, W. (1987) Vitamin E deficiency and neurologic disease in adults with cystic fibrosis. *Annals of Internal Medicine, 107*(1), 51–54. https://doi.org/10.7326/0003-4819-107-1-51

Sivagurunathan, U., Dominguez, D., Tseng, Y., Eryalçin, K. M., Roo, J., Boglione, C., Prabhu, P. A. J., & Izquierdo, M. (2022). Effect of dietary vitamin D3 levels on survival, mineralization, and skeletal development of gilthead sea bream (*Sparus aurata*) larvae. *Aquaculture, 560*. https://doi.org/10.1016/j.aquaculture.2022.738505

Sivagurunathan, U., Dominguez, D., Tseng, Y., Zamorano, M. J., Philip, A. J. P., & Izquierdo, M. (2023). Interaction between dietary vitamin D3 and vitamin K3 in gilthead sea-bream larvae (*Sparus aurata*) in relation to growth and expression of bone development-related gene. *Aquaculture Nutrition, 2023*, 3061649. https://doi.org/10.1155/2023/3061649

Skjærven, K. H., Oveland, E., Mommens, M., Samori, E., Saito, T., Adam, A. C., & Espe, M. (2020). Out-of-season spawning affects the nutritional status and gene expression in both Atlantic salmon female broodstock and their offspring. *Comparative Biochemistry and Physiology. Part A, Molecular and Integrative Physiology, 247*, 110717. https://doi.org/10.1016/j.cbpa.2020.110717

Small, B. C. (2022). Nutritional physiology. In R. W. Hardy & S. J. Kaushik (Eds.), *Fish nutrition* (4th ed) (pp. 593–641). Academic Press.

Smith, A. D., Warren, M. J., & Refsum, H. (2018). Vitamin B12. In *Advances in Food and Nutrition Research* N. A. M. Eskin (Ed.). Academic Press, (215–279).

Smith, C. E. (1968). Hematological changes in coho salmon fed a folic acid deficient diet. *Journal of the Fisheries Research Board of Canada, 25*(1), 151–156. https://doi.org/10.1139/f68-009

Smith, C. E., & Halver, J. E. (1969). Folic acid anemia in coho salmon. *Journal of the Fisheries Research Board of Canada, 26*(1), 111–114. https://doi.org/10.1139/f69-009

Smith, C. E., Brin, M., & Halver, J. E. (1974). Biochemical, physiological, and pathological changes in pyridoxine-deficient rainbow trout (*Salmo gairdneri*). *Journal of the Fisheries Research Board of Canada, 31*(12), 1893–1898. https://doi.org/10.1139/f74-247

Smith, J. E., & Borchers, R. (1972). Environmental temperature and the utilization of β-carotene by the rat. *Journal of Nutrition, 102*(8), 1017–1024. https://doi.org/10.1093/jn/102.8.1017

Soheil, L., Hossein, K., Shabanali, N., Mohammad, B., & Firouz, A. (2013). The effects of folic acid treatment on biometric and blood parameters of fingerling rainbow trout fishes (*Oncorhynchus mykiss*). *Journal of Aquaculture Research and Development, 4*. https://doi.org/. http://doi.org/10.4172/2155-9546.1000175

Solyanik, A. V., Semenov, V. G., Tyurin, V. G., Kuznetsov, A. F., Sofronov, V. G., Kh Volkov, A. K., Solyanik, V. V., Solyanik, S. V., & Solyanik, V. A. (2021). The effect of fumaric acid, dipromonium and vitamin C on the productivity of sows. *IOP Conference Series: Earth and Environmental Science, 935*(1). https://doi.org/10.1088/1755-1315/935/1/012023

Sørensen, M., Kousoulaki, K., Hammerø, R., Kokkali, M., Kleinegris, D., Marti-Quijal, F. J., Barba, F. J., Palihawadana, A. M., Egeland, E. S., Johnsen, C. A., Romarheim, O. H., Bisa, S., & Kiron, V. (2023). Mechanical processing of *Phaeodactylum trocornutum* and *Tetraselmis chui* biomass affect phenolic and antioxidant compound availability, nutrient digestibility and deposition of carotenoids in Atlantic salmon. *Aquaculture, 569*, 739395. https://doi.org/10.1016/j.aquaculture.2023.739395

Soto-Dávila, M., Valderrama, K., Inkpen, S. M., Hall, J. R., Rise, M. L., & Santander, J. (2020). Effects of vitamin D2 (ergocalciferol) and D3 (cholecalciferol) on Atlantic salmon (*Salmo salar*) primary macrophage immune response to *Aeromonas salmonicida subsp. salmonicida* infection. *Frontiers in Immunology, 10*, 3011. https://doi.org/10.3389/fimmu.2019.03011

Sowmya, R., & Sachindra, N. M. (2015). Enhancement of non-specific immune responses in common carp, *Cyprinus carpio*, by dietary carotenoids obtained from shrimp exoskeleton. *Aquaculture Research, 46*(7), 1562–1572. https://doi.org/10.1111/are.12310

Spasevski, N. J., Vukmirovic, D., Levic, J., & Kokic, B. (2015). Influence of pelleting process and material particle size on the stability of retinol acetate. *Archiva Zootechnica., 18*(2), 67–72.

Spector, R. (1986). Pantothenic acid transport and metabolism in the central nervous system. *American Journal of Physiology, 250*(2 Pt 2), R292–R297. https://doi.org/10.1152/ajpregu.1986.250.2.R292

Spisni, E., Tugnoli, M., Ponticelli, A., Mordenti, T., & Tomasi, V. (1998). Hepatic steatosis in artificially fed marine teleosts. *Journal of Fish Diseases, 21*(3), 177–184. https://doi.org/10.1046/j.1365-2761.1998.00089.x

Stabler, S. P. (2006). Vitamin B12. In B. A. Bowman & R. M. Russell (Eds.), *Present knowledge in nutrition* (9th ed) (pp. 302–313). International Life Sciences Institute. https://doi.org/10.1093/ajcn/85.5.1439a

Stacchiotti, V., Rezzi, S., Eggersdorfer, M., & Galli, F. (2021). Metabolic and functional interplay between gut microbiota and fat-soluble vitamins. *Critical Reviews in Food Science and Nutrition, 61*(19), 3211–3232. https://doi.org/10.1080/10408398.2020.1793728

Stafforini, D. M., McIntyre, T. M., Zimmerman, G. A., & Prescott, S. M. (2003). Platelet-activating factor, a pleiotrophic mediator of physiological and pathological processes. *Critical Reviews in Clinical Laboratory Sciences, 40*(6), 643–672. https://doi.org/10.1080/714037693

Stahl, W., Schwarz, W., Von Laar, J., & Sies, H. (1995). All-trans beta-carotene preferentially accumulates in human chylomicrons and very low-density lipoproteins compared with the 9-cis geometrical isomer. *Journal of Nutrition, 125*(8), 2128–2133. https://doi.org/10.1093/jn/125.8.2128

Stancheva, M., & Dobreva, D. A. (2013). Bulgarian marine and freshwater fishes as a source of fat-soluble vitamins for a healthy human diet. *Foods, 2*(3), 332–337. https://doi.org/10.3390/foods2030332

Steenbock, H. (1924) The induction of growth promoting and calcifying properties in a ration by exposure to light. *Science* 60(1549):224–225. https://doi.org/10.1126/science.60.1549.224

Stein, J., Daniel, H., Whang, E., Wenzel, U., Hahn, A., & Rehner, G. (1994). Rapid postabsorptive metabolism of nicotinic acid in rat small intestine may affect transport by metabolic trapping. *Journal of Nutrition, 124*(1), 61–66. https://doi.org/10.1093/jn/124.1.61

Storebakken, T., & No, H. K. (1992). Pigmentation of rainbow trout. *Aquaculture, 100*(1–3), 209–229. https://doi.org/10.1016/0044-8486(92)90372-R

Su, Y., Sun, Y., Ju, D., Chang, S., Shi, B., & Shan, A. (2018). The detoxification effect of vitamin C on zearalenone toxicity in piglets. *Ecotoxicology and Environmental Safety, 158*, 284–292. https://doi.org/10.1016/j.ecoenv.2018.04.046

Sugita, H., Takahashi, J., & Deguchi, Y. (1992). Production and consumption of biotin by the intestinal microflora of cultured freshwater fishes. *Bioscience, Biotechnology, and Biochemistry, 56*(10), 1678–1679. https://doi.org/10.1271/bbb.56.1678

Sugita, H., Takahashi, J., Miyajima, C., & Deguchi, Y. (1991). Vitamin B12-producing ability of the intestinal microflora of rainbow trout (*Oncorhynchus mykiss*). *Agricultural and Biological Chemistry, 55*(3), 893–894. https://doi.org/10.1080/00021369.1991.10870683

Sun, M., Li, X. F., Ge, Y. P., Zhang, L., Liu, B., & Liu, W. B. (2022). Dietary thiamine requirement and its effects on glycolipid metabolism in oriental river prawn (*Macrobrachium nipponense*). *Aquaculture, 550*, 737824. https://doi.org/10.1016/j.aquaculture.2021.737824

Sun, X., Chang, Y., Ye, Y., Ma, Z., Liang, Y., Li, T., Jiang, N., Xing, W., & Luo, L. (2012). The effect of dietary pigments on the coloration of Japanese ornamental carp (koi, *Cyprinus carpio* L.). *Aquaculture, 342–343*, 62–68. https://doi.org/10.1016/j.aquaculture.2012.02.019

Surai, P. F., Speake, B. K., & Sparks, N. H. C. (2001). Carotenoids in avian nutrition and embryonic development. 1. Absorption, availability and levels in plasma and egg yolk. *Journal of Poultry Science, 38*(1), 1–27. https://doi.org/10.2141/jpsa.38.1

Surai, P. F., and Dvorska, J. E. (2005). Effects of mycotoxins on antioxidant status and immunity. In D. Diaz (Ed.), *The mycotoxin blue book*. Nottingham University Press.

Suter, C. (1990). Vitamins at the molecular level. In *Proceedings of the 'National Feed Ingr. Ass. Nutr. Inst.': "Developments in Vitamin Nutrition and Health Applications"* Kansas City, MO. National Feed Ingredients Association.

Suttie, J. W. (2013). Vitamin K. In J. Zempleni, J. Suttie, J. Gregory & P. J. Stover (Eds.), *Handbook of vitamins* (5th ed) (pp. 89–124). CRC Press, Taylor & Francis Group, LLC. https://doi.org/10.1201/b15413

Svihus, B., & Zimonja, O. (2011). Chemical alterations with nutritional consequences due to pelleting animal feeds: A review. *Animal Production Science, 51*(7), 590–596. https://doi.org/10.1071/AN11004

Tacon, A. G. J. (2020). Trends in global aquaculture and aquafeed production: 2000–2017. *Reviews in Fisheries Science and Aquaculture, 28*(1), 43–56. https://doi.org/10.1080/23308249.2019.1649634

Takahashi, H., Suzuki, N., Takagi, C., Ikegame, M., Yamamoto, T., Takahashi, A., Moriyama, S., Hattori, A., & Sakamoto, T. (2008). Prolactin inhibits osteoclastic activity in the goldfish scale: A novel direct action of prolactin in teleosts. *Zoological Science, 25*(7), 739–745. https://doi.org/10.2108/zsj.25.739

Takeuchi, A., Okano, T., Ayame, M., Yoshikawa, H., Teraoka, S., Murakami, Y., & Kobayashi, T. (1984). High-performance liquid chromatographic determination of vitamin D3 in fish liver oils and eel body oils. *Journal of Nutritional Science and Vitaminology, 30*(5), 421–430. https://doi.org/10.3177/jnsv.30.421

Takeuchi, T. (2001). A review of feed development for early life stages of marine finfish in Japan. *Aquaculture, 200*(1–2), 203–222. https://doi.org/10.1016/S0044-8486(01)00701-3

Taksdal, T., Poppe, T., Sivertsen, T., & Ferguson, H. (1995). Low levels of vitamin E in plasma from Atlantic salmon *Salmo salar* with acute infectious pancreatic necrosis (IPN). *Diseases of Aquatic Organisms, 22*, 33–37. http://doi.org/10.3354/dao022033

Tan, K., Zhang, H., & Zheng, H. (2022). Carotenoid content and composition: A special focus on commercially important fish and shellfish. *Critical Reviews in Food Science and Nutrition*, 1–18. https://doi.org/10.1080/10408398.2022.2106937

Tanaka, H. (2015). Progression in artificial seedling production of Japanese eel Anguilla japonica. *Fisheries Science, 81*(1), 11–19. https://doi.org/10.1007/s12562-014-0821-z

Tanphaichair, V. (1976). Thiamine. In D. M. Hegsted et al. (Eds.), *Nutrition reviews present knowledge in nutrition* (4th ed). Nutrition Foundation, Inc.

Tanumihardjo, S. A., Russell, R. M., Stephensen, C. B., Gannon, B. M., Craft, N. E., Haskell, M. J., Lietz, G., Schulze, K., & Raiten, D. J. (2016). Biomarkers of nutrition for development (BOND) – Vitamin A review. *Journal of Nutrition, 146*(9) (Suppl.), 1816S–1848S. https://doi.org/10.3945/jn.115.229708

Tao, Y. F., Qiang, J., Bao, J. W., Chen, D. J., Yin, G. J., Xu, P., & Zhu, H. J. (2018). Changes in physiological parameters, lipid metabolism, and expression of microRNAs in genetically improved farmed tilapia (*Oreochromis niloticus*) with fatty liver induced by a high-fat diet. *Frontiers in Physiology, 9*(1521), 1521. https://doi.org/10.3389/fphys.2018.01521

Tavares-Dias, M., & Oliveira, S. R. (2009). A review of the blood coagulation in fish. *Revista Brasileira de Biociencias, 7*(2), 205–224. https://www.researchgate.net/publication/291448920_A_review_of_the_blood_coagulation_system_of_fish

Tavčar-Kalcher, G., & Vengušt, A. (2007). Stability of vitamins in premixes. *Animal Feed Science and Technology, 132*(1–2), 148–154. https://doi.org/10.1016/j.anifeedsci.2006.03.001

Taylor, J. F., Vera, L. M., De Santis, C., Lock, E. J., Espe, M., Skjærven, K. H., Leeming, D., Del Pozo, J., Mota-Velasco, J., Migaud, H., Hamre, K., & Tocher, D. R. (2019). The effect of micronutrient supplementation on growth and hepatic metabolism in diploid and triploid Atlantic salmon (*Salmo salar*) parr fed a low marine ingredient diet. *Comparative Biochemistry and Physiology. Part B, Biochemistry and Molecular Biology, 227*(January), 106–121. https://doi.org/10.1016/j.cbpb.2018.10.004

Teixeira, C. P., Barros, M. M., Pezzato, L. E., Fernandes, A. C., Koch, J. F. A., & Padovani, C. R. (2012). Growth performance of Nile tilapia, *Oreochromis niloticus*, fed diets containing levels of pyridoxine and haematological response under heat stress. *Aquaculture Research, 43*(8), 1081–1088. https://doi.org/10.1111/j.1365-2109.2011.02911.x

Teshima, S., & Kanazawa, A. (1980). Lipid constituents of serum lipoproteins in the prawn. *NIPPON SUISAN GAKKAISHI, 46*(1), 57–62. https://doi.org/10.2331/suisan.46.57

Tesoriere, L., Bongiorno, A., Pintaudi, A. M., D'Anna, R., D'Arpa, D., & Livrea, M. A. (1996). Synergistic interactions between vitamin A and vitamin E against lipid peroxidation in phosphatidylcholine liposomes. *Archives of Biochemistry and Biophysics, 326*(1), 57–63. https://doi.org/10.1006/abbi.1996.0046

Thierry, M. J., Hermodson, M. A., & Suttie, J. W. (1970). Vitamin K and warfarin distribution and metabolism in the warfarin-resistant rat. *American Journal of Physiology, 219*(4), 854–859. https://doi.org/10.1152/ajplegacy.1970.219.4.854

Thode Jensen, P., Nielsen, H. E., Danielsen, V., & Leth, T. (1983). Effect of dietary fat quality and vitamin E on the antioxidant potential of pigs. *Acta Veterinaria Scandinavica, 24*(2), 135–147. https://doi.org/10.1186/BF03546742

Thorarensen, H., Kubiriza, G. K., & Imsland, A. K. (2015). Experimental design and statistical analyses of fish growth studies. *Aquaculture, 448*, 483–490. https://doi.org/10.1016/j.aquaculture.2015.05.018

Tillitt, D. E., Riley, S. C., Evans, A. N., Nichols, S. J., Zajicek, J. L., Rinchard, J., Richter, C. A., & Krueger, C. C. (2009). Dreissenid mussels from the Great Lakes contain elevated thiaminase activity. *Journal of Great Lakes Research, 35*(2), 309–312. https://doi.org/10.1016/j.jglr.2009.01.007

Tizkar, B., Soudagar, M., Bahmani, M., Hoesseini, S. A., & Chamani, M. (2013). The effects of dietary supplementation of astaxanthin and β-carotene on the reproductive performance and egg quality of female goldfish (*Carassius auratus*). *Caspian Journal of Environmental Sciences, 11*(2), 217–231. https://doi.org/10.1016/j.theriogenology.2015.06.011

Tizkar, B., Kazemi, R., Alipour, A., Seidavi, A., Naseralavi, G., & Ponce-Palafox, J. T. (2015). Effects of dietary supplementation with astaxanthin and β-carotene on the semen quality of goldfish (*Carassius auratus*). *Theriogenology, 84*(7), 1111–1117. https://doi.org/10.1016/j.theriogenology.2015.06.011

Tizkar, B., Soudagar, M., Bahmani, M., Hosseini, S. A., Chamani, M., Seidavi, A., Sühnel, S., & Ponce Palafox, J. T. (2016). Effects of dietary astaxanthin and β-carotene on gonadosomatic and hepatosomatic indices, gonad and liver composition in goldfish *Carassius auratus* (Linnaeus, 1758) broodstocks. *Latin American Journal of Aquatic Research, 44*(2), 363–370. http://doi.org/10.3856/vol44-issue2-fulltext-17

Tocher, D. R. (1995). Glycerophospholipid metabolism. In P. W. Hochachka, & T. P. Mommsen (Eds.), *Metabolic Biochemistry. Biochemistry and Molecular Biology of Fishes,* (pp. 119–157). Elsevier.

Tocher, D. R. (2003). Metabolism and functions of lipids and fatty acids in teleost fish. *Reviews in Fisheries Science, 11*(2), 107–184. https://doi.org/10.1080/713610925

Tocher, D. R. (2015). Omega-3 long-chain polyunsaturated fatty acids and aquaculture in perspective. *Aquaculture, 449*, 94–107. https://doi.org/10.1016/j.aquaculture.2015.01.010

Tocher, D. R., & Sargent, J. R. (1990). Incorporation into phospholipid classes and metabolism via desaturation and elongation of various 14C-labelled (n-3) and (n-6) polyunsaturated fatty acids in trout astrocytes in primary culture. *Journal of Neurochemistry, 54*(6), 2118–2124. https://doi.org/10.1111/j.1471-4159.1990.tb04918.x

Tocher, D. R., Bendiksen, E. Å., Campbell, P. J., & Bell, J. G. (2008). The role of phospholipids in nutrition and metabolism of teleost fish. *Aquaculture, 280*(1–4), 21–34. https://doi.org/10.1016/j.aquaculture.2008.04.034

Toyama, M., Hironaka, M., Yamahama, Y., Horiguchi, H., Tsukada, O., Uto, N., Ueno, Y., Tokunaga, F., Seno, K., & Hariyama, T. (2008). Presence of rhodopsin and porphyropsin in the eyes of 164 fishes, representing marine, diadromous, coastal and freshwater species—A qualitative and comparative study. *Photochemistry and Photobiology, 84*(4), 996–1002. https://doi.org/10.1111/j.1751-1097.2008.00344.x

Traber, M. G. (2006). Vitamin E. In B. A. Bowman & R. M. Russell (Eds.), *Present knowledge in nutrition* (9th ed) (pp. 211–219). International Life Sciences Institute. https://doi.org/10.1093/ajcn/85.5.1439a

Traber, M. G. (2013). Vitamin E. In J. Zempleni, J. Suttie, J. Gregory & P. J. Stover (Eds.), *Handbook of vitamins* (5th ed) (pp. 125–148). CRC Press, Taylor & Francis Group, LLC. https://doi.org/10.1201/b15413

Traber, M. G., Rader, D., Acuff, R. V., Ramakrishnan, R., Brewer, H. B., & Kayden, H. J. (1998). Vitamin E dose-response studies in humans using deuterated RRR-a-tocopherol. *American Journal of Clinical Nutrition, 68*(4), 847–853. https://doi.org/10.1093/ajcn/68.4.847

Turchini, G. M., Francis, D. S., Du, Z.-Y., Olsen, R. E., Ringø, E., & Tocher, D. R. (2022). The lipids. In R. W. Hardy & S. J. Kaushik (Eds.), *Fish nutrition* (4th ed) (pp. 303–468). Academic Press.

Turner, M. R., & Lumb, R. H. (1989). Synthesis of platelet activating factor by tissues from the rainbow trout, *Salmo gairdneri*. *Biochimica et Biophysica Acta, 1004*(1), 49–52. https://doi.org/10.1016/0005-2760(89)90211-7

Udagawa, M. (2000). Physiological role of vitamin K in fish – A review. *JARQ, 34*(4), 279–284. https://www.jircas.go.jp/sites/default/files/publication/jarq/34-4-279-284_0.pdf

Udagawa, M. (2001). The effect of dietary vitamin K (phylloquinone and menadione) levels on the vertebral formation in mummichog (*Fundulus heteroclitus*). *Fisheries Science, 67*(1), 104–109. https://doi.org/10.1046/j.1444-2906.2001.00205.x

Ullrey, D. E. (1981). Vitamin E for swine. *Journal of Animal Science, 53*(4), 1039–1056. https://doi.org/10.2527/jas1981.5341039x

United States Pharmacopeia (1980). (20th ed). ISBN 0912734310. Mack Printing Company.

USDA (United States Department of Agriculture). (2008). *USDA database for the choline content of common foods* Release 2. https://data.nal.usda.gov/dataset/usda-database-choline-content-common-foods-release-2-2008

Usman, U., Kamaruddin, K., Laining, A., & Trismawanti, I. (2019). Fatty acid profile of hepatopancreas and oocyte of tiger shrimp, *Penaeus monodon*, fed modified-commercial diet by supplementing vitamin C and E. *Aquacultura Indonesiana, 20*(1), 15–23. http://doi.org/10.21534/ai.v20i1.137

Valverde, J. C., Hernández, M. D., García-Garrido, S., Rodríguez, C., Estefanell, J., Gairín, J. I., Rodríguez, C. J., Tomás, A., & García, B. G. (2012). Lipid classes from marine species and meals intended for cephalopod feeding. *Aquaculture International, 20*(1), 71–89. https://doi.org/10.1007/s10499-011-9442-z

Vance, J. E. (2015). Phospholipid synthesis and transport in mammalian cells. *Traffic, 16*(1), 1–18. https://doi.org/10.1111/tra.12230

Van den Ouweland, J. M. W. (2016). Analysis of vitamin D metabolites by liquid chromatography-tandem mass spectrometry. *TrAC Trends in Analytical Chemistry, 84*(B), 117–130. https://doi.org/10.1016/j.trac.2016.02.005

Vani, T., Saharan, N., Mukherjee, S. C., Ranjan, R., Kumar, R., & Brahmchari, R. K. (2011). Deltamethrin induced alterations of hematological and biochemical parameters in fingerlings of *Catla catla* (Ham.) and their amelioration by dietary supplement of vitamin C. *Pesticide Biochemistry and Physiology, 101*(1), 16–20. https://doi.org/10.1016/j.pestbp.2011.05.007

Vera, L. M., Lock, E. J., Hamre, K., Migaud, H., Leeming, D., Tocher, D. R., & Taylor, J. F. (2019). Enhanced micronutrient supplementation in low marine diets reduced vertebral malformation in diploid and triploid Atlantic salmon (*Salmo salar*) parr and increased vertebral expression of bone biomarker genes in diploids. *Comparative Biochemistry and Physiology. Part B, Biochemistry and Molecular Biology, 237*, 110327. https://doi.org/10.1016/j.cbpb.2019.110327

Vera, L. M., Hamre, K., Espe, M., Hemre, G., Skjærven, K., Lock, E., Prabhu, A. J., Leeming, D., Migaud, H., Tocher, D. R., & Taylor, J. F. (2020). Higher dietary micronutrients are required to maintain optimal performance of Atlantic salmon (*Salmo salar*) fed a high plant material diet during the full production cycle. *Aquaculture, 528*, 735551. https://doi.org/10.1016/j.aquaculture.2020.735551

Verbeeck, J. (1975). Vitamin behaviour in premixes. *Feedstuffs, 47*, 45–48.

Verlhac, V., Obach, A., Gabaudan, J., Schüep, W., & Hole, R. (1998). Immunomodulation by dietary vitamin C and glucan in rainbow trout (*Oncorhynchus mykiss*). *Fish and Shellfish Immunology, 8*(6), 409–424. https://doi.org/10.1006/fsim.1998.0148

Vieira, C. A. S. C., Vieira, J. S., Bastos, M. S., Zancanela, V., Barbosa, L. T., Gasparino, E., & Del Vesco, A. P. (2018). Expression of genes related to antioxidant activity in Nile tilapia kept under salinity stress and fed diets containing different levels of vitamin C. *Journal of Toxicology and Environmental Health. Part A, 81*(1–3), 20–30. https://doi.org/10.1080/15287394.2017.1401968

Villeneuve, L., Gisbert, E., Le Delliou, H., Cahu, C. L., & Zambonino-Infante, J. L. (2005). Dietary levels of all-trans retinol affect retinoid nuclear receptor expression and skeletal development in European sea bass larvae. *British Journal of Nutrition, 93*(6), 791–801. https://doi.org/10.1079/BJN20051421

Waagbø, R. (2010). Water-soluble vitamins in fish ontogeny. *Aquaculture Research, 41*(5), 733–744. https://doi.org/10.1111/j.1365-2109.2009.02223.x

Waagbø, R., Sandnes, K., Glette, J., Nilsen, E. R., & Albrektsen, S. (1992). Dietary vitamin B6 and vitamin C. Influence on immune response and disease resistance in Atlantic salmon (*Salmo salar*). *Annals of the New York Academy of Sciences, 669*(1), 379–382. http://doi.org/10.1111/j.1749-6632.1992.tb17129.x

Waagbø, R., Sandnes, K., Lie, Ø., & Roem, A. (1998). Effects of inositol supplementation on growth, chemical composition and blood chemistry in Atlantic salmon, *Salmo salar* L., fry. *Aquaculture Nutrition, 4*(1), 53–59. https://doi.org/10.1046/j.1365-2095.1998.00043.x

Waagbø, R., Graff, I., & Hamre, K. (2001). Vitaminer. In R. Waagbø, K. Hamre, M. Espe & Ø. Lie (Eds.), *Fiskeernæring* (pp. 94–123). Kystnæringen forlag & bokklubb AS.

Wade, N. M., Budd, A., Irvin, S., & Glencross, B. D. (2015). The combined effects of diet, environment and genetics on pigmentation in the giant tiger prawn, *Penaeus monodon. Aquaculture, 449*, 78–86. https://doi.org/10.1016/j.aquaculture.2015.01.023

Wagner, C. (2001). Biochemical role of folate in cellular metabolism. *Clinical Research and Regulatory Affairs, 18*(3), 161–180. https://doi.org/10.1081/CRP-100108171

Wagner, C., Briggs, W. T., & Cook, R. J. (1984). Covalent binding of folic acid to dimethylglycine dehydrogenase. *Archives of Biochemistry and Biophysics, 233*(2), 457–461. https://doi.org/10.1016/0003-9861(84)90467-3

Wahli, T., Verlhac, V., Girling, P., Gabaudan, J., & Aebischer, C. (2003). Influence of dietary vitamin C on the wound healing process in rainbow trout (*Oncorhynchus mykiss*). *Aquaculture, 225*(1–4), 371–386. https://doi.org/10.1016/S0044-8486(03)00302-8

Wall, J. S., & Carpenter, K. J. (1988). Variation in availability of niacin in grain products. *Food Technology, 42*(10), 198–204.

Walton, M. J., Cowey, C. B., & Adron, J. W. (1984). Effects of biotin deficiency in rainbow trout (*Salmo gairdneri*) fed diets of different lipid and carbohydrate content. *Aquaculture, 37*(1), 21–38. https://doi.org/10.1016/0044-8486(84)90041-3

Wan, J., Ge, X., Liu, B., Xie, J., Cui, S., Zhou, M., Xia, S., & Chen, R. (2014). Effect of dietary vitamin C on non-specific immunity and mRNA expression of three heat shock proteins (HSPs) in juvenile *Megalobrama amblycephala* under pH stress. *Aquaculture, 434*, 325–333. https://doi.org/10.1016/j.aquaculture.2014.08.043

Wang, J. W., Che, Y. C., Sun, M., Guo, Y. Q., Liu, B., & Li, X. F. (2022). Optimal niacin requirement of oriental river prawn *Macrobrachium nipponense* as determined by growth, energy sensing, and glycolipid metabolism. *Aquaculture Nutrition*, 2022 (September 10), 8596427. https://doi.org/10.1155/2022/8596427

Wang, K., Wang, E., Qin, Z., Zhou, Z., Geng, Y., & Chen, D. (2016). Effects of dietary vitamin E deficiency on systematic pathological changes and oxidative stress in fish. *Oncotarget, 7*(51), 83869–83879. https://doi.org/10.18632/oncotarget.13729

Wang, L., Ma, B., Chen, D., Lou, B., Zhan, W., Chen, R., Tan, P., Xu, D., Liu, F., & Xie, Q. (2019). Effect of dietary level of vitamin E on growth performance, antioxidant ability, and resistance to *Vibrio alginolyticus* challenge in yellow drum *Nibea albiflora. Aquaculture, 507*, 119–125. https://doi.org/10.1016/j.aquaculture.2019.04.003

Wang, L., Xu, H., Wang, Y., Wang, C., Li, J., Zhao, Z., Luo, L., Du, X., & Xu, Q. (2017). Effects of the supplementation of vitamin D3 on the growth and vitamin D metabolites in juvenile Siberian sturgeon (*Acipenser baerii*). *Fish Physiology and Biochemistry, 43*(3), 901–909. https://doi.org/10.1007/s10695-017-0344-5

Wang, W., Ishikawa, M., Koshio, S., Yokoyama, S., Sakhawat Hossain, M., & Moss, A. S. (2018). Effects of dietary astaxanthin supplementation on juvenile kuruma shrimp, *Marsupenaeus japonicus. Aquaculture, 491*, 197–204. https://doi.org/10.1016/j.aquaculture.2018.03.025

Wang, W., Ishikawa, M., Koshio, S., Yokoyama, S., Dawood, M. A. O., Hossain, M. S., & Moss, A. S. (2019). Effects of dietary astaxanthin and vitamin E and their interactions on the growth performance, pigmentation, digestive enzyme activity of kuruma shrimp (*Marsupenaeus japonicus*). *Aquaculture Research, 50*(4), 1186–1197. https://doi.org/10.1111/are.13993

Wang, X., Li, Y., Hou, C., Gao, Y., & Wang, Y. (2015). Physiological and molecular changes in large yellow croaker (*Pseudosciaena crocea R.*) with high-fat diet-induced fatty liver disease. *Aquaculture Research, 46*(2), 272–282. https://doi.org/10.1111/are.12176

Wang, Y., Liu, L., Jin, Z., & Zhang, D. (2021). Microbial cell factories for green production of vitamins. *Frontiers in Bioengineering and Biotechnology, 9*, 661562. https://doi.org/10.3389/fbioe.2021.661562

Watkins, B. A., & Kratzer, F. H. (1987). Dietary biotin effects on polyunsaturated fatty acids in chick tissue lipids and prostaglandin E2 levels in freeze-clamped hearts. *Poultry Science, 66*(11), 1818–1828. https://doi.org/10.3382/ps.0661818

Wei, X., Hang, Y., Li, X., Hua, X., Cong, X., Yi, W., & Guo, X. (2023). Effects of vitamin K3 levels on growth, coagulation, calcium content, and antioxidant capacity in largemouth bass, *Micropterus salmoides*. *Aquaculture and Fisheries, 8*(2), 159–165. https://doi.org/10.1016/j.aaf.2021.08.004

Weiser, H., & Vecchi, M. (1982). Stereoisomers of α-tocopheryl acetate. II. Biopotencies of all eight stereoisomers, individually or in mixtures, as determined by rat resorption-gestation tests. *International Journal for Vitamin and Nutrition Research. Internationale Zeitschrift Fur Vitamin- und Ernahrungsforschung. Journal International de Vitaminologie et de Nutrition, 52*(3), 351–370. PubMed: 7174231

Wen, L. M., Jiang, W. D., Liu, Y., Wu, P., Zhao, J., Jiang, J., Kuang, S. Y., Tang, L., Tang, W. N., Zhang, Y. A., Zhou, X. Q., & Feng, L. (2015). Evaluation the effect of thiamin deficiency on intestinal immunity of young grass carp (*Ctenopharyngodon idella*). *Fish and Shellfish Immunology, 46*(2), 501–515. https://doi.org/10.1016/j.fsi.2015.07.001

Wen, L. M., Feng, L., Jiang, W. D., Liu, Y., Wu, P., Zhao, J., Jiang, J., Kuang, S. Y., Tang, L., Tang, W. N., Zhang, Y. A., & Zhou, X. Q. (2016). Thiamin deficiency induces impaired fish gill immune responses, tight junction protein expression and antioxidant capacity: Roles of the NF-κB, TOR, p38 MAPK and Nrf2 signaling molecules. *Fish and Shellfish Immunology, 51*, 373–383. https://doi.org/10.1016/j.fsi.2015.12.038

Wen, M., Liu, Y. J., Tian, L. X., & Wang, S. (2015). Vitamin D3 requirement in practical diet of white shrimp, *Litopenaeus vannamei*, at low salinity rearing conditions. *Journal of the World Aquaculture Society, 46*(5), 531–538. https://doi.org/10.1111/jwas.12209

Wen, Z.-P., Feng, L., Jiang, J., Liu, Y., & Zhou, X.-Q. (2010). Immune response, disease resistance and intestinal microflora of juvenile Jian carp (*Cyprinus carpio* var. Jian) fed graded levels of pantothenic acid. *Aquaculture Nutrition, 16*(4), 430–436. https://doi.org/10.1111/j.1365-2095.2009.00686.x

West, C. E., Sijtsma, S. R., Kouwenhoven, B., Rombout, J. H. W. M., & van der Zijpp, A. J. (1992). Epithelia-damaging virus infections affect vitamin A status in chickens. *Journal of Nutrition, 122*(2), 333–339. https://doi.org/10.1093/jn/122.2.333

Whanger, P. D. (1981). Selenium and heavy metal toxicity. In J. E. Spallholz, J. L. Martin & H. E. Ganther (Eds.), *In "selenium in biology and medicine"*. AVI Publishing, Co.

White, W. S., Peck, K. M., Ulman, E. A., & Erdman, J. W. (1993). The ferret as a model for evaluation of the bioavailabilities of all-trans-beta-carotene and its isomers. *Journal of Nutrition, 123*(6), 1129–1139. https://doi.org/10.1093/jn/123.6.1129

Wilson, R. P., & Poe, W. E. (1988). Choline nutrition of fingerling channel catfish. *Aquaculture, 68*(1), 65–71. http://doi.org/10.1016/0044-8486(88)90292-X

Wilson, R. P., Bowser, P. R., & Poe, W. E. (1983). Dietary pantothenic acid requirement of fingerling channel catfish. *Journal of Nutrition, 113*(10), 2124–2128. https://doi.org/10.1093/jn/113.10.2124

Winther, B., Hoem, N., Berge, K., & Reubsaet, L. (2011). Elucidation of phosphatidylcholine composition in krill oil extracted from *Euphausia superba*. *Lipids, 46*(1), 25–36. https://doi.org/10.1007/s11745-010-3472-6

Winton, J. R. (1991). Recent advances in detection and control of infectious hematopoietic necrosis virus in aquaculture. *Annual Review of Fish Diseases, 1*, 83–93. https://doi.org/10.1016/0959-8030(91)90024-E

Witkowska, D., Sedrowicz, L., & Oledzka, R. (1992). Effect of a diet with an increased content of vitamin B6 on the absorption of amino acids in the intestine of rats intoxicated with carbaryl propoxur and thiuram. Methionine. *Bromatologia i Chemia Toksykologiczna, 25*, 25.

Wold, H. L., Wake, K., Higashi, N., Wang, D., Kojima, N., Imai, K., Blomhoff, R., & Senoo, H. (2004). Vitamin A distribution and content in tissues of the lamprey, *Lampetra japonica*. *Anatomical Record. Part A, Discoveries in Molecular, Cellular, and Evolutionary Biology, 276*(2), 134–142. https://doi.org/10.1002/ar.a.10345

Wolf, G. (1991). The intracellular vitamin A-binding proteins: An overview of their functions. *Nutrition Reviews, 49*(1), 1–12. https://doi.org/10.1111/j.1753-4887.1991.tb07349.x

Wolf, G. (1995). The enzymatic cleavage of beta-carotene: Still controversial. *Nutrition Reviews, 53*(5), 134–137. https://doi.org/10.1111/j.1753-4887.1995.tb01537.x

Wolf, G. (2006). How an increased intake of α-tocopherol can suppress the bioavailability of gamma-tocopherol. *Nutrition Reviews, 64*(6), 295–299. https://doi.org/10.1111/j.1753-4887.2006.tb00213.x

Wolf, G. (2007). Identification of a membrane receptor for retinol-binding protein functioning in the cellular uptake of retinol. *Nutrition Reviews, 65*(8 Pt. 1), 385–388. https://doi.org/10.1301/nr.2007.aug.385-388

Wolf, G. and Carpenter, K.J. (1997) Early Research into the vitamins: the work of Wilhelm Stepp. *J. Nutr.* 127(7):1255–1259. https://doi.org/10.1093/jn/127.7.1255

Wolfe, D. A., & Cornwell, D. G. (1965). Composition and tissue distribution of carotenoids in crayfish. *Comparative Biochemistry and Physiology, 16*(2), 205–213. https://doi.org/10.1016/0010-406X(65)90060-5

Won, S., Lee, S., Hamidoghli, A., Lee, S., & Bai, S. C. (2019). Dietary choline requirement of juvenile olive flounder (*Paralichthys olivaceus*). *Aquaculture Nutrition, 25*(6), 1281–1288. https://doi.org/10.1111/anu.12948

Woodall, A. N., Ashley, L. M., Halver, J. E., Olcott, H. S., & Vanderveen, J. (1964). Nutrition of salmonoid fishes: XIII. The α-tocopherol requirement of chinook salmon. *Journal of Nutrition, 84*, 125–135. https://doi.org/10.1093/jn/84.2.125

Woodward, B. (1982). Riboflavin supplementation of diets for rainbow trout. *Journal of Nutrition, 112*(5), 908–913. https://doi.org/10.1093/jn/112.5.908

Woodward, B. (1984). Symptoms of severe riboflavin deficiency without ocular opacity in rainbow trout (*Salmo gairdneri*). *Aquaculture, 37*(3), 275–281. https://doi.org/10.1016/0044-8486(84)90160-1

Woodward, B. (1985). Riboflavin requirement for growth, tissue saturation and maximal flavin-dependent enzyme activity in young rainbow trout (*Salmo gairdneri*) at two temperatures. *Journal of Nutrition, 115*(1), 78–84. https://doi.org/10.1093/jn/115.1.78

Woodward, B. (1994). Dietary vitamin requirements of cultured young fish, with emphasis on quantitative estimates for salmonids. *Aquaculture, 124*(1–4), 133–168. https://doi.org/10.1016/0044-8486(94)90375-1

Woodward, B., & Frigg, M. (1989). Dietary biotin requirements of young rainbow trout (*Salmo gairdneri*) determined by weight gain, hepatic biotin concentration and maximal biotin-dependent enzyme activities in liver and white muscle. *Journal of Nutrition, 119*(1), 54–60. https://doi.org/10.1093/jn/119.1.54

Woollard, D. C., Indyk, H. E., Fong, B. Y., & Cook, K. K. (2002). Determination of vitamin K1 isomers in foods by liquid chromatography with C30 bonded-phase column. *Journal of AOAC International, 85*(3), 682–691. https://doi.org/10.1093/jaoac/85.3.682

Wu, C., Lu, B., Wang, Y., Jin, C., Zhang, Y., & Ye, J. (2020). Effects of dietary vitamin D3 on growth performance, antioxidant capacities and innate immune responses in juvenile black carp *Mylopharyngodon piceus*. *Fish Physiology and Biochemistry, 46*(6), 2243–2256. https://doi.org/10.1007/s10695-020-00876-8

Wu, F., Huang, F., Wen, H., Jiang, M., Liu, W., Tian, J., & Yang, C. G. (2015). Vitamin C requirement of adult genetically improved farmed tilapia, *Oreochromis niloticus*. *Aquaculture International, 23*(5), 1203–1215. https://doi.org/10.1007/s10499-014-9877-0

Wu, F., Zhu, W., Liu, M., Chen, C., Chen, J., & Tan, Q. (2016). Effects of dietary vitamin A on growth performance, blood biochemical indices and body composition of juvenile grass carp (*Ctenopharyngodon idellus*). *Turkish Journal of Fisheries and Aquatic Sciences, 16*(2), 339–345. http://doi.org/10.4194/1303-2712-v16_2_14

Wu, F., Jiang, M., Wen, H., Liu, W., Tian, J., Yang, C. G., & Huang, F. (2017). Dietary vitamin E effects on growth, fillet textural parameters, and antioxidant capacity of genetically improved farmed tilapia (GIFT),

Oreochromis niloticus. Aquaculture International, 25(2), 991–1003. https://doi.org/10.1007/s10499-016-0089-7

Wu, J.-P., Wu, F., Jiang, M., Wen, H., Wei, Q.-W., Liu, W., Tian, J., Huang, F., & Yang, C.-G. (2016). Dietary folic acid (FA) requirement of a genetically improved Nile tilapia *Oreochromis niloticus* (Linnaeus, 1758). *Journal of Applied Ichthyology, 32*(6), 1155–1160. https://doi.org/10.1111/jai.13151

Wu, P., Feng, L., Kuang, S.-Y., Liu, Y., Jiang, J., Hu, K., Jiang, W.-D., Li, S.-H., Tang, L., & Zhou, X.-Q. (2011). Effect of dietary choline on growth, intestinal enzyme activities and relative expressions of target of rapamycin and eIF4E-binding protein2 gene in muscle, hepatopancreas and intestine of juvenile Jian carp (*Cyprinus carpio* var. Jian). *Aquaculture, 317*(1–4), 107–116. https://doi.org/10.1016/j.aquaculture.2011.03.042

Wu, P., Jiang, J., Liu, Y., Hu, K., Jiang, W. D., Li, S. H., Feng, L., & Zhou, X. Q. (2013). Dietary choline modulates immune responses, and gene expressions of TOR and eIF4E-binding protein2 in immune organs of juvenile Jian carp (*Cyprinus carpio* var. Jian). *Fish and Shellfish Immunology, 35*(3), 697–706. https://doi.org/10.1016/j.fsi.2013.05.030

Wu, P., Jiang, W. D., Liu, Y., Chen, G. F., Jiang, J., Li, S. H., Feng, L., & Zhou, X. Q. (2014). Effect of choline on antioxidant defenses and gene expressions of Nrf2 signaling molecule in the spleen and head kidney of juvenile Jian carp (*Cyprinus carpio* var. Jian). *Fish and Shellfish Immunology, 38*(2), 374–382. https://doi.org/10.1016/j.fsi.2014.03.032

Wu, P., Liu, Y., Jiang, W. D., Jiang, J., Zhang, Y. A., Zhou, X. Q., & Feng, L. (2017). Intestinal immune responses of Jian carp against *Aeromonas hydrophila* depressed by choline deficiency: Varied change patterns of mRNA levels of cytokines, tight junction proteins and related signaling molecules among three intestinal segments. *Fish and Shellfish Immunology, 65*, 34–41. https://doi.org/10.1016/j.fsi.2017.03.053

Wu, S., & Xu, B. (2021). Effect of dietary astaxanthin administration on the growth performance and innate immunity of juvenile crucian carp (*Carassius auratus*). *3 Biotech, 11*(3), 151. https://doi.org/10.1007/s13205-021-02700-3

Wu, Y. S., Liau, S. Y., Huang, C. T., & Nan, F. H. (2016). Beta 1,3/1,6-glucan and vitamin C immunostimulate the non-specific immune response of white shrimp (*Litopenaeus vannamei*). *Fish and Shellfish Immunology, 57*, 269–277. https://doi.org/10.1016/j.fsi.2016.08.046

Xia, M., Huang, X., Wang, H., Mai, K., & Zhou, Q. (2014). Dietary biotin requirement of juvenile Pacific white shrimp (*Litopenaeus vannamei*). *Chinese Journal of Animal Nutrition., 26*, 1513–1520. https://scholar.google.it/scholar_url?url=https://www.cabdirect.org/cabdirect/abstract/20143243184&hl=it&sa=X&ei=6WYhZfgU_cqxAuCOjvgE&scisig=AFWwaebEqL_1fU4KUHwmuMXQFJPR&oi=scholarr

Xia, M.-H., Huang, X.-L., Wang, H.-L., Jin, M., Li, M., & Zhou, Q.-C. (2015). Dietary niacin levels in practical diets for *Litopenaeus vannamei* to support maximum growth. *Aquaculture Nutrition, 21*(6), 853–860. https://doi.org/10.1111/anu.12210

Xie, J. J., Chen, X., Niu, J., Wang, J., Wang, Y., & Liu, Q. Q. (2017). Effects of astaxanthin on antioxidant capacity of golden pompano (*Trachinotus ovatus*) in vivo and in vitro. *Fisheries and Aquatic Sciences, 20*(1), 6. https://doi.org/10.1186/s41240-017-0052-1

Xie, S., Fang, W., Wei, D., Liu, Y., Yin, P., Niu, J., & Tian, L. (2018). Dietary supplementation of *Haematococcus pluvialis* improved the immune capacity and low salinity tolerance ability of post-larval white shrimp, *Litopenaeus vannamei. Fish and Shellfish Immunology, 80*, 452–457. https://doi.org/10.1016/j.fsi.2018.06.039

Xie, S., Yin, P., Tian, L., Yu, Y., Liu, Y., & Niu, J. (2020). Dietary supplementation of astaxanthin improved the growth performance, antioxidant ability and immune response of juvenile largemouth Bass (*Micropterus salmoides*) fed high-fat diet. *Marine Drugs, 18*(12). https://doi.org/10.3390/md18120642

Xu, C., Liu, W. B., Zhang, D. D., Shi, H. J., Zhang, L., & Li, X. F. (2018). Benfotiamine, a lipid-soluble analog of vitamin B1, improves the mitochondrial biogenesis and function in blunt snout bream (*Megalobrama amblycephala*) fed high-carbohydrate diets by promoting the AMPK/PGC-1β/NRF-1aAxis. *Frontiers in Physiology, 9*, 1079. https://doi.org/10.3389/fphys.2018.01079

Xu, C. M., Yu, H. R., Li, L. Y., Li, M., Qiu, X. Y., Fan, X. Q., Fan, Y. L., & Shan, L. L. (2022a). Effects of dietary vitamin C on the growth performance, biochemical parameters, and antioxidant activity of coho salmon *Oncorhynchus kisutch* (Walbaum, 1792) post-smolts. *Aquaculture Nutrition, 2022*, 6866578. https://doi.org/10.1155/2022/6866578

Xu, C. M., Yu, H. R., Li, L. Y., Li, M., Qiu, X. Y., Zhao, S. S., Fan, X. Q., Fan, Y. L., & Shan, L. L. (2022b). Dietary vitamin A requirements of coho salmon *Oncorhynchus kisutch* (Walbaum, 1792) post-smolts. *Aquaculture, 560*, 738448. https://doi.org/10.1016/j.aquaculture.2022.738448

Xu, H. J., Jiang, W. D., Feng, L., Liu, Y., Wu, P., Jiang, J., Kuang, S. Y., Tang, L., Tang, W. N., Zhang, Y. A., & Zhou, X. Q. (2016a). Dietary vitamin C deficiency depressed the gill physical barriers and immune barriers referring to Nrf2, apoptosis, MLCK, NF-κB and TOR signaling in grass carp (*Ctenopharyngodon idella*) under infection of Flavobacterium columnare. *Fish and Shellfish Immunology, 58*, 177–192. https://doi.org/10.1016/j.fsi.2016.09.029

Xu, H. J., Jiang, W. D., Feng, L., Liu, Y., Wu, P., Jiang, J., Kuang, S. Y., Tang, L., Tang, W. N., Zhang, Y. A., & Zhou, X. Q. (2016b). Dietary vitamin C deficiency depresses the growth, head kidney and spleen immunity and structural integrity by regulating NF-κB, TOR, Nrf2, apoptosis and MLCK signaling in young grass carp (*Ctenopharyngodon idella*). *Fish and Shellfish Immunology, 52*, 111–138. https://doi.org/10.1016/j.fsi.2016.02.033

Xun, P., Lin, H., Wang, R., Huang, Z., Zhou, C., Yu, W., Huang, Q., Tan, L., Wang, Y., & Wang, J. (2019). Effects of dietary vitamin B1 on growth performance, intestinal digestion and absorption, intestinal microflora and immune response of juvenile golden pompano (*Trachinotus ovatus*). *Aquaculture, 506*, 75–83. https://doi.org/10.1016/j.aquaculture.2019.03.017

Yamada, S., Tanaka, Y., Sameshima, M., & Ito, Y. (1990). Pigmentation of prawn (*Penaeus japonicus*) with carotenoids: I. *Aquaculture, 87*(3–4), 323–330. https://doi.org/10.1016/0044-8486(90)90069-Y

Yamaguchi, K., Murakami, M., Nakano, H., Konosu, S., Kokura, T., Yamamoto, H., Kosaka, M., & Hata, K. (1986). Supercritical carbon dioxide extraction of oils from Antarctic krill. *Journal of Agricultural and Food Chemistry, 34*(5), 904–907. https://doi.org/10.1021/jf00071a034

Yamamoto, Y., Fujisawa, A., Hara, A., & Dunlap, W. C. (2001). An unusual vitamin E constituent (α-tocomonoenol) provides enhanced antioxidant protection in marine organisms adapted to cold-water environments. *Proceedings of the National Academy of Sciences of the United States of America, 98*(23), 13144–13148. https://doi.org/10.1073/pnas.241024298

Yang, M. C., Wang, J. X., & Shi, X. Z. (2021). Vitamin D3 identified from metabolomic analysis of intestinal contents promotes an antibacterial response in shrimp intestinal immunity. *Aquaculture, 530*, 735951. https://doi.org/10.1016/j.aquaculture.2020.735951

Yang, P., Wang, H., Zhu, M., & Ma, Y. (2020). Evaluation of extrusion temperatures, pelleting parameters and vitamin forms on vitamin stability in feed. *Animals: An Open Access Journal from MDPI, 10*(5), 894. https://doi.org/10.3390/ani10050894

Yang, P., Wang, H. K., Zhu, M., Li, L. X., & Ma, Y. X. (2021). Degradation kinetics of vitamins in premixes for pig: Effects of choline, high concentrations of copper and zinc, and storage time. *Animal Bioscience, 34*(4), 701–713. https://doi.org/10.5713/ajas.20.0026

Yang, Q. H., Zhou, X. Q., Jiang, J., & Liu, Y. (2008). Effect of dietary vitamin A deficiency on growth performance, feed utilisation and immune responses of juvenile Jian carp (*Cyprinus carpio* var. Jian). *Aquaculture Research, 39*(8), 902–906. https://doi.org/10.1111/j.1365-2109.2008.01945.x

Yang, Q., Ding, M., Tan, B., Dong, X., Chi, S., Zhang, S., & Liu, H. (2017). Effects of dietary vitamin A on growth, feed utilization, lipid metabolism enzyme activities, and fatty acid synthase and hepatic lipase mRNA expression levels in the liver of juvenile orange spotted grouper, *Epinephelus coioides*. *Aquaculture, 479*, 501–507. https://doi.org/10.1016/j.aquaculture.2017.06.024

Yazdani Sadati, M. A., Sayed Hassani, M. H., Pourkazemi, M., Shakourian, M., & Pourasadi, M. (2014). Influence of different levels of dietary choline on growth rate, body composition, Hematological indices and liver lipid of juvenile Siberian sturgeon *Acipenser baerii* Brandt, 1869. *Journal of Applied Ichthyology, 30*(6), 1632–1636. https://doi.org/10.1111/jai.12619

Yeh, S. P., Shiu, P. J., Guei, W. C., Lin, Y. H., & Liu, C. H. (2015). Improvement in lipid metabolism and stress tolerance of juvenile giant grouper, *Epinephelus lanceolatus* (Bloch), fed supplemental choline. *Aquaculture Research, 46*(8), 1810–1821. https://doi.org/10.1111/are.12334

Yerlikaya, P., Alp, A. C., Tokay, F. G., Aygun, T., Kaya, A., Topuz, O. K., & Yatmaz, H. A. (2022). Determination of fatty acids and vitamins A, D and E intake through fish consumption. *International Journal of Food Science and Technology, 57*(1), 653–661. https://doi.org/10.1111/ijfs.15435

Yi, X., Xu, W., Zhou, H., Zhang, Y., Luo, Y., Zhang, W., & Mai, K. (2014). Effects of dietary astaxanthin and xanthophylls on the growth and skin pigmentation of large yellow croaker *Larimichthys croceus*. *Aquaculture, 433*, 377–383. https://doi.org/10.1016/j.aquaculture.2014.06.038

Yin, P., Björnsson, B. T., Fjelldal, P. G., Saito, T., Remø, S. C., Edvardsen, R. B., Hansen, T., Sharma, S., Olsen, R. E., & Hamre, K. (2022). Impact of antioxidant feed and growth manipulation on the redox regulation of Atlantic salmon smolts. *Antioxidants, 11*(9), 1708. https://doi.org/10.3390/antiox11091708

Yoshikawa, K., Imai, K., Seki, T., Higashi-Kuwata, N., Kojima, N., Yuuda, M., Koyasu, K., Sone, H., Sato, M., Senoo, H., & Irie, T. (2006). Distribution of retinyl ester-storing stellate cells in the arrowtooth halibut, *Atheresthes evermanni*. *Comparative Biochemistry and Physiology. Part A, Molecular and Integrative Physiology, 145*(2), 280–286. https://doi.org/10.1016/j.cbpa.2006.06.043

Yossa, R., Sarker, P. K., & Vandenberg, G. W. (2011). Preliminary evidence of the contribution of the intestinal microflora to biotin supply in zebrafish *Danio rerio* (Hamilton-Buchanan). *Zebrafish, 8*(4), 221–227. https://doi.org/10.1089/zeb.2011.0706

Yossa, R., Sarker, P. K., Mock, D. M., Lall, S. P., & Vandenberg, G. W. (2015). Current knowledge on biotin nutrition in fish and research perspectives. *Reviews in Aquaculture, 7*(1), 59–73. https://doi.org/10.1111/raq.12053

Ytrestøyl, T., Coral-Hinostroza, G., Hatlen, B., Robb, D. H. F., & Bjerkeng, B. (2004). Carotenoid and lipid content in muscle of Atlantic salmon, *Salmo salar*, transferred to seawater as 0+ or 1+ smolts. *Comparative Biochemistry and Physiology. Part B, Biochemistry and Molecular Biology, 138*(1), 29–40. https://doi.org/10.1016/j.cbpc.2004.01.011

Ytrestøyl, T., Struksnaes, G., Koppe, W., & Bjerkeng, B. (2005). Effects of temperature and feed intake on astaxanthin digestibility and metabolism in Atlantic salmon, *Salmo salar*. *Comparative Biochemistry and Physiology. Part B, Biochemistry and Molecular Biology, 142*(4), 445–455. https://doi.org/10.1016/j.cbpb.2005.09.004

Ytrestøyl, T., Aas, T. S., & Åsgård, T. (2015). Utilisation of feed resources in production of Atlantic salmon (*Salmo salar*) in Norway. *Aquaculture, 448*, 365–374. https://doi.org/10.1016/j.aquaculture.2015.06.023

Ytrestøyl, T., Bou, M., Dimitriou, C., Berge, G. M., Østbye, T. K., & Ruyter, B. (2023). Dietary level of the omega-3 fatty acids EPA and DHA influence the flesh pigmentation in Atlantic salmon. *Aquaculture Nutrition, 2023*, 5528942. https://doi.org/10.1155/2023/5528942

Yu, H. R., Guo, M. J., Yu, L. Y., Li, L. Y., Wang, Q. H., Li, F. H., Zhang, Y. Z., Zhang, J. Y., & Hou, J. Y. (2022). Effects of dietary riboflavin supplementation on the growth performance, body composition and anti-oxidative capacity of coho salmon (*Oncorhynchus kisutch*) post-smolts. *Animals (Basel), 12*(22), 3218. https://doi.org/10.3390/ani12223218

Yu, W., Lin, H., Yang, Y., Zhou, Q., Chen, H., Huang, X., Zhou, C., Huang, Z., & Li, T. (2021). Effects of supplemental dietary *Haematococcus pluvialis* on growth performance, antioxidant capacity, immune responses and resistance to *Vibrio harveyi* challenge of spotted sea bass *Lateolabrax maculatus*. *Aquaculture Nutrition, 27*(2), 355–365. https://doi.org/10.1111/anu.13189

Yu, Y., Liu, Y., Yin, P., Zhou, W., Tian, L., Liu, Y., Xu, D., & Niu, J. (2020). Astaxanthin attenuates fish oil-related hepatotoxicity and oxidative insult in juvenile Pacific white shrimp (*Litopenaeus vannamei*). *Marine Drugs, 18*(4), 218. https://doi.org/10.3390/md18040218

Yuan, J., Feng, L., Jiang, W.-D., Liu, Y., Jiang, J., Li, S.-H., Kuang, S.-Y., Tang, L., & Zhou, X.-Q. (2016). Effects of dietary vitamin K levels on growth performance, enzyme activities and antioxidant status in the hepatopancreas and intestine of juvenile Jian carp (*Cyprinus carpio* var. Jian). *Aquaculture Nutrition, 22*(2), 352–366. https://doi.org/10.1111/anu.12264

Yuan, Z. H., Wu, P., Feng, L., Jiang, W., Liu, Y., Kuang, S., Tang, L., & Zhou, X. (2020a). Dietary choline inhibited the gill apoptosis in association with the p38MAPK and JAK/STAT3 signalling pathways of juvenile grass carp (*Ctenopharyngodon idella*). *Aquaculture, 529*, 735699. https://doi.org/10.1016/j.aquaculture.2020.735699

Yuan, Z. H., Feng, L., Jiang, W. D., Wu, P., Liu, Y., Jiang, J., Kuang, S. Y., Tang, L., & Zhou, X. Q. (2020b). Choline deficiency decreased the growth performances and damaged the amino acid absorption capacity in juvenile grass carp (*Ctenopharyngodon idella*). *Aquaculture, 518*, 734829. https://doi.org/10.1016/j.aquaculture.2019.734829

Yuan, Z. H., Feng, L., Jiang, W., Wu, P., Liu, Y., Kuang, S., Tang, L., & Zhou, X. (2021). Dietary choline deficiency aggravated the intestinal apoptosis in association with the MAPK signalling pathways of juvenile grass carp (*Ctenopharyngodon idella*). *Aquaculture, 532*, 736046. https://doi.org/10.1016/j.aquaculture.2020.736046

Yusuf, A., Huang, X., Chen, N., Apraku, A., Wang, W., Cornel, A., & Rahman, M. M. (2020). Impact of dietary vitamin C on plasma metabolites, antioxidant capacity and innate immunocompetence in juvenile largemouth bass, *Micropterus salmoides*. *Aquaculture Reports, 17*, 100383. https://doi.org/10.1016/j.aqrep.2020.100383

Yusuf, A., Huang, X., Chen, N., Li, S., Apraku, A., Wang, W., & David, M. A. (2021). Growth and metabolic responses of juvenile largemouth bass (*Micropterus salmoides*) to dietary vitamin C supplementation levels. *Aquaculture, 534*, 736243. https://doi.org/10.1016/j.aquaculture.2020.736243

Zee, J. A., Carmichael, L., Codère, D., Poirier, D., & Fournier, M. (1991). Effect of storage conditions on the stability of vitamin C in various fruits and vegetables produced and consumed in Quebec. *Journal of Food Composition and Analysis, 4*(1), 77–86. https://doi.org/10.1016/0889-1575(91)90050-G

Zehra, S., & Khan, M. A. (2012). Dietary vitamin C requirement of fingerling, *Cirrhinus mrigala* (Hamilton), based on growth, feed conversion, protein retention, hematological indices, and liver vitamin C concentration. *Journal of the World Aquaculture Society, 43*(5), 648–658. https://doi.org/10.1111/j.1749-7345.2012.00597.x

Zehra, S., & Khan, M. A. (2017). Dietary thiamin and pyridoxine requirements of fingerling Indian major carp, *Cirrhinus mrigala* (Hamilton). *Aquaculture Research, 48*(9), 4945–4957. https://doi.org/10.1111/are.13313

Zehra, S., & Khan, M. A. (2018a). Dietary niacin requirement of fingerling *Channa punctatus* (Bloch). *Journal of Applied Ichthyology, 34*(4), 929–936. https://doi.org/10.1111/jai.13604

Zehra, S., & Khan, M. A. (2018b). Dietary pyridoxine requirement of fingerling *Channa punctatus* (Bloch) based on growth performance, liver pyridoxine concentration, and carcass composition. *Journal of Applied Aquaculture, 30*(3), 238–255. https://doi.org/10.1080/10454438.2018.1456999

Zehra, S., & Khan, M. A. (2018c). Dietary riboflavin requirement of fingerling *Channa punctatus* (Bloch) based on growth, conversion efficiencies, protein retention, liver riboflavin storage, RNA/DNA ratio and carcass composition. *Aquaculture Nutrition, 24*(1), 269–276. https://doi.org/10.1111/anu.12555

Zehra, S., & Khan, M. A. (2018d). Dietary thiamin requirement of fingerling *Channa punctatus* (Bloch) based on growth, protein gain, liver thiamin storage, RNA/DNA ratio and biochemical composition. *Aquaculture Nutrition, 24*(3), 1015–1023. https://doi.org/10.1111/anu.12638

Zehra, S., & Khan, M. A. (2019). Dietary biotin requirement of fingerling *Channa punctatus* (Bloch). *Journal of Applied Aquaculture, 31*(3), 236–253. https://doi.org/10.1080/10454438.2018.1545722

Zeisel, S. H. (1981). Dietary choline: Biochemistry, physiology, and pharmacology. *Annual Review of Nutrition, 1*(1), 95–121. https://doi.org/10.1146/annurev.nu.01.070181.000523

Zeisel, S. H. (2013). Metabolic crosstalk between choline/1-carbon metabolism and energy homeostasis. *Clinical Chemistry and Laboratory Medicine, 51*(3), 467–475. https://doi.org/10.1515/cclm-2012-0518

Zeisel, S. H., & Blusztajn, J. K. (1994). Choline and human nutrition. *Annual Review of Nutrition, 14*(1), 269–296. https://doi.org/10.1146/annurev.nu.14.070194.001413

Zempleni, J., Green, G. M., Spannagel, A. W., & Mock, D. M. (1997). Biliary excretion of biotin and biotin metabolites is quantitatively minor in rats and pigs. *Journal of Nutrition, 127*(8), 1496–1500. https://doi.org/10.1093/jn/127.8.1496

Zhang, J., Liu, Y.-J., Tian, L.-X., Yang, H.-J., Liang, G.-Y., Yue, Y.-R., & Xu, D.-H. (2013). Effects of dietary astaxanthin on growth, antioxidant capacity and gene expression in Pacific white shrimp *Litopenaeus vannamei*. *Aquaculture Nutrition, 19*(6), 917–927. https://doi.org/10.1111/anu.12037

Zhang, J. Z., Henning, S. M., & Swendseid, M. E. (1993). Poly(ADP-ribose) polymerase activity and DNA strand breaks are affected in tissues of niacin-deficient rats. *Journal of Nutrition, 123*(8), 1349–1355. https://doi.org/10.1093/jn/123.8.1349

Zhang, L., Feng, L., Jiang, W. D., Liu, Y., Wu, P., Kuang, S. Y., Tang, L., Tang, W. N., Zhang, Y. A., & Zhou, X. Q. (2017). Vitamin A deficiency suppresses fish immune function with differences in different intestinal segments: The role of transcriptional factor NF-κB and p38 mitogen-activated protein kinase signalling pathways. *British Journal of Nutrition, 117*(1), 67–82. https://doi.org/10.1017/S0007114516003342

Zhang, X., Ma, Y., Xiao, J., Zhong, H., Guo, Z., Zhou, C., Li, M., Tang, Z., Huang, K., & Liu, T. (2021). Effects of vitamin E on the reproductive performance of female and male Nile tilapia (*Oreochromis niloticus*) at the physiological and molecular levels. *Aquaculture Research, 52*(8), 3518–3531. https://doi.org/10.1111/are.15193

Zhang, Y., Li, Y., Liang, X., & Gao, J. (2017). Effects of dietary vitamin E supplementation on growth performance, fatty acid composition, lipid peroxidation and peroxisome proliferator-activated receptors (PPAR) expressions in juvenile blunt snout bream *Megalobrama amblycephala*. *Fish Physiology and Biochemistry, 43*(4), 913–922. https://doi.org/10.1007/s10695-016-0224-4

Zhang, Z., & Wilson, R. P. (1999). Reevaluation of the choline requirement of fingerling channel catfish (*Ictalurus punctatus*) and determination of the availability of choline in common feed ingredients. *Aquaculture, 180*(1–2), 89–98. https://doi.org/10.1016/S0044-8486(99)00190-8

Zhao, H. F., Feng, L., Jiang, W. D., Liu, Y., Jiang, J., Wu, P., Zhao, J., Kuang, S. Y., Tang, L., Tang, W. N., Zhang, Y. A., & Zhou, X. Q. (2015). Flesh shear force, cooking loss, muscle antioxidant status and relative expression of signaling molecules (Nrf2, Keap1, TOR, and CK2) and their target genes in young grass carp (*Ctenopharyngodon idella*) muscle fed with graded levels of choline. *PLOS ONE, 10*(11), e0142915. https://doi.org/10.1371/journal.pone.0142915

Zhao, H. F., Jiang, W. D., Liu, Y., Jiang, J., Wu, P., Kuang, S. Y., Tang, L., Tang, W. N., Zhang, Y. A., Zhou, X. Q., & Feng, L. (2016). Dietary choline regulates antibacterial activity, inflammatory response and barrier function in the gills of grass carp (*Ctenopharyngodon idella*). *Fish and Shellfish Immunology, 52*, 139–150. https://doi.org/10.1016/j.fsi.2016.03.029

Zhao, H., Chen, B., Huang, Y., Cao, J., Wang, G., Chen, X., & Mo, W. (2020). Effects of dietary vitamin B1 on growth performance, blood metabolites, body composition, intestinal enzyme activities and morphometric parameters of juvenile yellow catfish (*Pelteobagrus fulvidraco*). *Aquaculture Nutrition, 26*(5), 1681–1690. https://doi.org/10.1111/anu.13113

Zhao, S., Feng, L., Liu, Y., Kuang, S.-Y., Tang, L., Jiang, J., Hu, K., Jiang, W.-D., Li, S.-H., & Zhou, X.-Q. (2012). Effects of dietary biotin supplement on growth, body composition, intestinal enzyme activities and microbiota of juvenile Jian carp (*Cyprinus carpio* var. Jian). *Aquaculture Nutrition, 18*(4), 400–410. https://doi.org/10.1111/j.1365-2095.2011.00905.x

Zhao, W., Wang, Z., Yu, Y., Qi, Z., Lü, L., Zhang, Y., & Lü, F. (2016). Growth and antioxidant status of oriental river prawn *Macrobrachium nipponense* fed with diets containing vitamin E. *Chinese Journal of Oceanology and Limnology, 34*(3), 477–483. https://doi.org/10.1007/s00343-015-4396-z

Zhao, W., Guo, Y. C., Huai, M. Y., Li, L., Man, C., Pelletier, W., Wei, H. L., Yao, R., & Niu, J. (2022). Comparison of the retention rates of synthetic and natural astaxanthin in feeds and their effects on pigmentation, growth, and health in rainbow trout (*Oncorhynchus mykiss*). *Antioxidants, 11*(12), 2473. https://doi.org/10.3390/antiox11122473

Zhao, W., Wei, H. L., Chen, M. D., Yao, R., Wang, Z. Q., & Niu, J. (2023). Effects of synthetic astaxanthin and *Haematococcus pluvialis* on growth, antioxidant capacity, immune response, and hepato-morphology of *Oncorhynchus mykiss* under cage culture with flowing freshwater. *Aquaculture, 562*, 738860. https://doi.org/10.1016/j.aquaculture.2022.738860

Zhou, J., Yun, X., Wang, J., Li, Q., & Wang, Y. (2022). A review on the ecotoxicological effect of sulphonamides on aquatic organisms. *Toxicology Reports, 9*, 534–540. https://doi.org/10.1016/j.toxrep.2022.03.034

Zhou, Q.-C., Wang, L.-G., Wang, H.-L., Wang, T., Elmada, C.-Z., & Xie, F.-J. (2013). Dietary vitamin E could improve growth performance, lipid peroxidation and non-specific immune responses for juvenile cobia (*Rachycentron canadum*). *Aquaculture Nutrition, 19*(3), 421–429. https://doi.org/10.1111/j.1365-2095.2012.00977.x

Zhou, Q., Wang, L., Wang, H., Xie, F., & Wang, T. (2012). Effect of dietary vitamin C on the growth performance and innate immunity of juvenile cobia (*Rachycentron canadum*). *Fish and Shellfish Immunology, 32*(6), 969–975. https://doi.org/10.1016/j.fsi.2012.01.024

Zou, W., Lin, Z., Huang, Y., Limbu, S. M., Rong, H., Yu, C., Lin, F., & Wen, X. (2020). Effect of dietary vitamin C on growth performance, body composition and biochemical parameters of juvenile Chu's croaker (*Nibea coibor*). *Aquaculture Nutrition, 26*(1), 60–73. https://doi.org/10.1111/anu.12967

Zwingelstein, G., Brichon, G., Bodennec, J., Chapelle, S., Abdul-Malak, N., & El Babili, M. (1998). Formation of phospholipid nitrogenous bases in euryhaline fish and crustaceans. II. Phosphatidylethanolamine methylation in liver and hepatopancreas. *Comparative Biochemistry and Physiology. Part B: Biochemistry and Molecular Biology, 120*(3), 475–482. https://doi.org/10.1016/S0305-0491(98)10032-9

Index